Sorption Enhanced
Reaction Processes

Sustainable Chemistry Series

ISSN: 2514-3042

Published

Vol. 1   Sorption Enhanced Reaction Processes
   by Alírio Egídio Rodrigues, Luís Miguel Madeira,
   Yi-Jiang Wu and Rui Faria

Sustainable
Chemistry
Series
Volume 1

Sorption Enhanced Reaction Processes

Alírio Egídio Rodrigues
University of Porto, Portugal

Luís Miguel Madeira
University of Porto, Portugal

Yi-Jiang Wu
East China University of Science and Technology, China

Rui Faria
University of Porto, Portugal

World Scientific

NEW JERSEY · LONDON · SINGAPORE · BEIJING · SHANGHAI · HONG KONG · TAIPEI · CHENNAI · TOKYO

Published by

World Scientific Publishing Europe Ltd.

57 Shelton Street, Covent Garden, London WC2H 9HE

Head office: 5 Toh Tuck Link, Singapore 596224

USA office: 27 Warren Street, Suite 401-402, Hackensack, NJ 07601

Library of Congress Cataloging-in-Publication Data
Names: Rodrigues, Alírio E., author.
Title: Sorption enhanced reaction processes / by Alírio Egídio Rodrigues
    (University of Porto, Portugal), Luís Miguel Madeira (University of Porto, Portugal),
    Yi-Jiang Wu (East China University of Science and Technology, China),
    Rui Faria (University of Porto, Portugal).
Description: New Jersey : World Scientific, 2017. | Series: Sustainable chemistry series ; volume 1 |
    Includes bibliographical references.
Identifiers: LCCN 2017009024 | ISBN 9781786343567 (hc : alk. paper)
Subjects: LCSH: Adsorption. | Heat of adsorption. | Chemical reactions. | Separation (Technology) |
    Membrane reactors.
Classification: LCC TP156.A35 S67 2017 | DDC 660/.284235--dc23
LC record available at https://lccn.loc.gov/2017009024

British Library Cataloguing-in-Publication Data
A catalogue record for this book is available from the British Library.

Desk Editors: Herbert Moses/Mary Simpson

Typeset by Stallion Press
Email: enquiries@stallionpress.com

Printed in Singapore

Preface

The book on *Sorption Enhanced Reaction Processes* (SERP) emerges from several plenary and keynote lectures (CAMURE 7 & ISMR 6, World Congress Chemical Engineering 8, 2009; CAMURE 8 & ISMR 7, 2011; All-Polish Symposium on Chemical Reaction Engineering and Multifunctional Reactors, 2012; International Congress on Catalytic Membrane Reactors ICCMR 11, 2013; CHEMREACTOR 21, 2014), teaching on Advanced Catalysis Engineering course at TU Delft (2012 and 2015) and Chemical Reaction Engineering at FEUP, and activities carried out at both LSRE (Laboratory of Separation and Reaction Engineering) and LEPABE (Laboratory for Process Engineering, Environment, Biotechnology and Energy) laboratories, located in the Department of Chemical Engineering at FEUP (Faculty of Engineering, University of Porto, Portugal), wherein the authors have been involved in research crossing the topic of sorption enhanced reactors (adsorptive reactors, chromatographic reactors, membrane reactors, simulated moving bed reactors, pressure swing adsorptive reactors).

The book addresses process intensification by combining adsorption and reaction, reaction and membranes or reaction/adsorption/membranes in a single unit in order to overcome thermodynamic limitations of conversion in reversible reactions. The key idea in such hybrid technologies is therefore to go beyond equilibrium conversion with the help of an adsorption process (in case of SERP) or membranes (in the case of membrane reactors) by removing one (or more) of the reaction products, thus shifting forward the

v

overall reaction. The book concentrates on gas phase and liquid phase processes involving different technologies: pressure swing adsorptive reactors (PSAR), membrane reactors (MR) and simulated moving bed reactors (SMBR). Gas phase processes are illustrated by hydrogen production using PSAR for sorption enhanced steam methane reforming (SE-SMR) and steam reforming of ethanol (SE-SRE) and membrane reactors for water-gas shift (WGS-MR) and sorption enhanced water-gas shift (SE-WGS-MR). Liquid phase processes using SMBR technology are illustrated by the synthesis of acetals (glycerol acetal) and esters (butyl acrylate) using an acid ion exchange resin as catalyst and selective adsorbent. A hybrid technology, PermSMBR, combining membrane permeation and SMBR is also introduced. The book emphasizes process development going from materials science (catalysts, adsorbents and membranes development and screening) and combining experimental determination of basic reaction/adsorption/permeation parameters with detailed modelling, simulation and operation; emphasis is also given to presenting experimental data and practical applications of SERP concepts. The readers will get a clear path for process development of SERP whatever the area of application will be.

The book is divided in five chapters. In Chapter 1, Introduction, a brief history of Process Intensification (PI) is presented and the integration of reaction and separation is discussed before addressing adsorptive reactors (PSAR and SMBR) and MR. Chapter 2, Gas-phase Adsorptive Reactor for Hydrogen Production Processes, starts with a review of hydrogen economy and feedstocks and processes for hydrogen production. A section is devoted for materials (catalysts and adsorbents) development. The reactor design section includes experimental studies in fixed bed reactors as well as fluidized bed reactors and modelling and simulation. The important issue of regeneration and cyclic operation is discussed for pressure swing and temperature swing strategies, the chapter ending with practical applications. Chapter 3 deals with MR for Water-Gas Shift (WGS). First the concept is introduced and thermodynamic aspects are discussed. Then catalysts, mechanisms and kinetic models are reviewed. Membrane types are then discussed and membranes for

H_2 removal are compared with membranes for CO_2 removal (both reaction products of the WGS reaction). Reactor configurations are discussed next, and some modelling and simulation studies are provided. Then, a parametric review on the effect of the main operating variables is shown. The chapter ends with practical applications and pilot scale cases. Chapter 4 is dedicated to Liquid Phase Simulated Moving Bed Reactor, starting with the concept of SMB by analogy with the true moving bed (TMB). Then the SMBR idea is discussed and process development methodology is described in detail and illustrated with lab scale examples for the synthesis of acetals and esters. The idea of integrated process is presented and finally is analyzed the combination of reaction, adsorption and membrane permeation: the PermSMBR. The last Chapter 5, Conclusions and Perspectives, discusses the difficulties encountered when going to full-scale and provides future perspectives.

Last but not the least, we would like to acknowledge the main contributions from PhD students and postdocs who helped shaping the research areas of PI addressed herein at LSRE and LEPABE. In the LSRE, research in SERP started with PhD of J.M. Loureiro (1986) on adsorptive reactors. The first deep contribution for SMBR was done by Diana Azevedo (2001), now at UFC, using enzymatic sucrose inversion and separation glucose/fructose. The whole area of green additives for diesel started with a CYTED project leading to the PhD thesis of Viviana Silva (2003), now at BASF, on diethyl acetal from biomass based ethanol and acetaldehyde. This work was winner of Solvay Ideas Challenge and IChemE award (2008) and patented. It was followed by Ganesh Gandi (2006), now at Godavari Biorefineries, on dimethyl acetal and Nuno Graça (2012) on dibutyl acetal. This line was continued by Rui Faria (2014) on glycerol acetal in the framework of EU project Eurobioref. In the area of sorption enhanced esterification reactions, Carla Pereira (2009) developed SMBR and PermSMBR for the synthesis of green solvent ethyl lactate, winner of PSE Model-based Innovation Prize in 2012. Dânia Constantino extended the approach to the synthesis of butyl acrylate. Jonathan Gonçalves (2015) followed the ideas of Mirjana Minceva

(2004) now at TU Munchen, and studied the isomerization of xylenes and separation of p-xylene in SMBR with dual bed columns.

On the gas-solid SERP research started on the PhD thesis of Rodrigo Davesac (2004) on propane/propylene separation and PSAR. Much effort was concentrated on hydrogen production by steam methane reforming with CO_2 adsorption in hydrotalcites. Dr. G.H. Xiu working on process simulation introduced the concept of reactive regeneration (2002) and experimental studies were done in the PhD thesis of Eduardo Oliveira (2009) and Naruewan Chanburanasiri (2013) in a joint program with Chulalongkorn University. Sorption enhanced steam reforming of ethanol was started with Dr. P. Vaidya (2006), now at ICT Mumbai, and Dr. A.F. Cunha (LSRE), and studied in detail by Yi-Jiang Wu (2014), now at 3M China.

From LEPABE, the contribution of the following researchers is deeply acknowledged: Diogo Mendes (PhD in Pd–Ag membrane reactors in the WGS reaction in 2010); Patrícia Pérez (PhD running in adsorptive-membrane reactors for high-purity H_2 production via WGS reaction); Carlos Miguel (PhD running on the integration of hybrid CO_2 capture by adsorption and its recycling by reactive regeneration) and Joel Silva (PhD running on hybrid sorption enhanced membrane reactor for hydrogen production with carbon dioxide separation via glycerol steam reforming). The work of postdoctoral researchers on membrane reactors/sorption enhanced membrane reactors is also acknowledged, namely Ju-Meng Zheng, Vânia Chibante and Miguel Angel Soria. The help given by Carlos Miguel in the elaboration of some figures in Chapter 3 is also sincerely appreciated.

Alírio E. Rodrigues, Luís M. Madeira,
Yi-Jiang Wu and Rui P. V. Faria

About the Authors

Alírio E. Rodrigues
LSRE, Laboratory of Separation and Reaction Engineering,
Department of Chemical Engineering,
Faculty of Engineering, University of Porto (FEUP),
Rua Dr Roberto Frias s/n, 4200-465 Porto-Portugal

Alírio E. Rodrigues graduated in Chemical Engineering (1968) at University of Porto (Portugal) and received his Dr. Ing degree (1973) from the Université de Nancy (France). After teaching at the University of Luanda and University of Évora, he joined the Chemical Engineering Department at the Faculty of Engineering, University of Porto (FEUP), Portugal, in 1976, where he is an Emeritus Professor (since 2013). He has been Visiting Professor at Université Technologie Compiègne, University of Virginia, Universidad de Oviedo and Universidade Federal Ceará. He was the founder of LSRE, Laboratory of Separation and Reaction Engineering. His research activities are focused on cyclic adsorption/reaction processes (pressure swing adsorption, simulated moving bed) for olefins/paraffins separation, CO_2 capture, hydrogen purification, etc., lignin valorization, perfume engineering and microencapsulation. His teaching activities were focused on chemical reaction engineering, separation processes, product engineering and system dynamics and process control. He has supervised more than 60 PhD students and published more than 600 papers in peer-reviewed journals, several books and patents.

Luís M. Madeira
LEPABE, Laboratory for Process Engineering, Environment,
Biotechnology and Energy,
Department of Chemical Engineering,
Faculty of Engineering, University of Porto (FEUP),
Rua Dr Roberto Frias s/n, 4200-465 Porto-Portugal

Luís M. Madeira graduated in Chemical Engineering (1993) and received his PhD (1998) from the Technical University of Lisbon (Instituto Superior Técnico), Portugal. He joined the Chemical Engineering Department at the Faculty of Engineering – University of Porto (FEUP), Portugal, in 1999, where he is an Associate Professor (since 2011). L.M. Madeira is currently the Director of Studies of the Integrated Master in Chemical Engineering undergraduate program at FEUP (since 2012) and the Vice-Scientific Coordinator of the research unit LEPABE (Laboratory for Process Engineering, Environment, Biotechnology and Energy). His teaching activities are mostly focused in chemical reaction engineering and chemical engineering laboratories while his main research interests include, among others, hydrogen production/purification processes, particularly making use of hybrid reactors — membrane reactors and sorption enhanced membrane reactors (water-gas shift, reforming of wastes, etc.), and CO_2 capture/valorization.

Yi-Jiang Wu
State Key Laboratory of Chemical Engineering,
College of Chemical Engineering,
East China University of Science and Technology (ECUST),
Meilong Road 130, Shanghai 20037, China

Yi-Jiang Wu received his BSc in Applied Chemistry (2009) from Nanjing University in China and his PhD in Chemical Engineering (2014) from the University of Porto in Portugal under the supervision of Professor A.E. Rodrigues. After two years of postdoctoral fellow at the East China University of Science and Technology, he is currently working in Air Quality Lab, 3M China Limited R&D Center as a Senior Product Development Engineer. His research interests lie in the design and preparation of multifunctional materials for process intensification and hazardous gas removal through adsorption processes with both experimental and numerical simulation approaches.

Rui P. V. Faria
LSRE, Laboratory of Separation and Reaction Engineering,
Department of Chemical Engineering,
Faculty of Engineering, University of Porto (FEUP),
Rua Dr Roberto Frias s/n, 4200-465 Porto-Portugal

Rui P. V. Faria finished his Master degree in Chemical Engineering at the Faculty of Engineering of Porto University in 2008 and started his professional activity as a Process Engineer in Fluidinova, Engenharia de Fluidos, SA, integrating the team responsible for the start-up and optimization of an industrial plant for the production of nanoparticle materials to be applied in medical devices. In 2010, he proceeded with his studies at the Faculty of Engineering of Porto University and received his PhD in 2014. Since then, he has been working as a Researcher at the Laboratory of Separation and Reaction Engineering focusing on chemical reaction engineering, chromatographic separation processes and process intensification, particularly, simulated moving bed reactor processes.

Contents

Chapter 1

Introduction

1.1. Process Intensification

1.1.1. History

The original thinking of "Process Intensification" (PI) was first introduced at Imperial Chemical Industries (ICI) by Colin Ramshaw and his colleagues during the 1970s (Reay *et al.*, 2013), as a concept to reduce the capital cost by reducing the size of unit operations dramatically, with a size reduction factor that is expected to range from 2 to 1000 times (Stankiewicz and Moulijn, 2000).

The definition and content of PI for chemical engineering changed over the last decades. Initially, size reduction by heat/mass-transfer/mixing enhancement with HiGee technology was the main focus of PI exploitation. However, as Ramshaw and co-workers pointed out (Reay *et al.*, 2013), the size reduction may not be the single objective. Besides, some early references related with the employment of HiGee forces in e.g. Podbielniak centrifugal contactor by rotating systems may even date back to 1930s, long before the PI expression appeared. Later, it became apparent that apart from cost reduction many other important advantages can also be achieved from PI. For instance, equipment miniaturization can increase the safety of chemical process without sacrificing product quality (Reay *et al.*, 2013). Consequently, the interest of both chemical industry and academia in PI increased to a significant extent since the mid-1980s, and the arena for PI was broadened to fulfil the needs of modern chemical engineering development, including some key objectives:

- Reduce cost (CAPEX, variable cost, etc.).
- Reduce energy consumption and CO_2 emission.
- Reduce raw material cost.
- Reduce environmental impact.
- Increase process flexibility and stability.
- Increase operating safety.
- Improve product quality.

Since PI is always associated with one or more of the aforementioned attractive benefits, Stankiewicz and Moulijn (2000) offered the following definition for PI:

Any chemical engineering development that leads to a substantially smaller, cleaner, and more energy efficient technology is process intensification.

However, the methodologies for achieving these benefits can be classified as PI, Process Systems Engineering or even classical process optimization, and such definition may bring difficulties to discriminate between them. A comparison of some basic features of these concepts, taken from one fundamental and comprehensive literature review by Van Gerven and Stankiewicz (2009) of PI, is given in Table 1.1, which is helpful to realize that PI differs essentially in character from the other methods.

Generally speaking, PI aims at drastically improving the performance of (or even realizing) a chemical engineering process by completely redesigning conventional operating units and unit operations. Therefore, the definition of PI in the European Roadmap for PI was described as: "Process Intensification provides radically innovative principles in process and equipment design which can benefit process and chain efficiency, capital and operating expenses, quality, wastes, process safety and more" (EFCE, 2007). Such vague expression still does not define the term PI in a rigorous way.

More recently, Lutzea *et al.* (2010) classified the three principles associated with PI: (1) integration of operations; (2) integration of

Table 1.1 A comparison between process systems engineering and PI. Reprinted with permission from Van Gerven and Stankiewicz (2009). Copyright (2016), American Chemical Society.

	Process systems engineering	PI
Aim	Multi-scale integration of existing and new concepts	Development of new concepts of processing methods and equipment
Focus	Model, software, numerical method	Experiment, phenomenon, interface
Interdisciplinarity	Modest (interface with applied mathematics and informatics, chemistry)	Strong (chemistry and catalysis, applied physics, mechanical engineering, materials science, electronics, etc.)

functions; and (3) integration of phenomena, which could be more useful to determine whether an enhancement method is a PI or not. Additionally, there is still an on-going discussion about the definition of PI (Luo, 2013); different terms of its definition clearly show that the PI is a developing field of research, which is far away from a mature status. The development and implementation of PI has gained a worldwide R&D interest in recent years. For example, the European Commission funded the 'F^3 Factory' project in 2009 (http://www.f3factory.com), which aims to deliver holistic process design methodology applying PI concepts, and a roadmap (10–40 years) for PI in Europe was published in 2008 by the Working Party on PI from the European Federation of Chemical Engineering (EFCE). For industrial applications, multiple European companies related with the chemical industry established the European Process Intensification Centre (EUROPIC) in 2008, which aims to accelerate the application of PI from an industry-driven platform. For academic research, there is an international scientific journal entitled *Chemical Engineering and Processing* with "Process Intensification" as subtitle specially focused on PI and related fields.

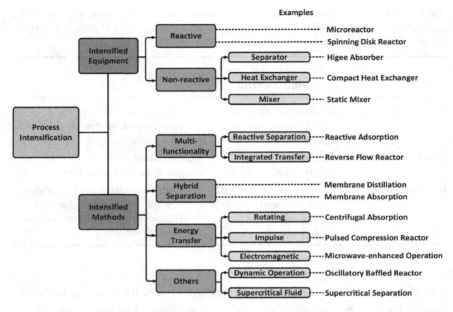

Figure 1.1 Classifications of PI equipment and methods. Adapted from Stankiewicz and Moulijn (2000).

1.1.2. Classification

As discussed in the above section, PI covers a wide range of R&D areas. As illustrated in Fig. 1.1, numerous PI technologies developed can be divided into two categories: Equipment and Methodology, which is similar as the concept of "hardware" and "software" for IT industry. The first classification, PI equipment, involves equipment for both reactive and non-reactive processes. For instance, reactive PI equipment such as structured reactors and spinning disc reactors are developing rapidly in recent years, while compact heat exchangers and static mixers are typical non-reactive PI equipment that has been employed in the chemical industry and other processes for a long time. On the other hand, one has a second category — PI Methodology, which can be classified into four different areas (Stankiewicz and Moulijn, 2002): integration of reaction and other unit operation(s) into multifunctional reactors, integration of separation and other unit operation(s) into hybrid separations, use of alternative forms of energy supply for chemical processes, and other techniques such as the use of new reaction–separation media, dynamic operating methods, etc.

As pointed out by Moulijn *et al.* (2013), the traditional classification (Fig. 1.1) focuses on the mesoscale (unit level) and macroscale (process level) PI. However, it might be even more advantageous to promote PI into smaller scales, since micro- (particle level) and nanoscale (molecule level) are the scales where the basic reaction, separation, mass and heat transfer take place and the improvement in these scales for PI can be more obvious. A good example is the development of multifunctional materials (Cunha *et al.*, 2014). Comparing with multifunctional reactors (unit level), by combining a catalyst (reaction) and an adsorbent (separation) to form a "multifunctional reactor" at the particle level one is able to intensify mass and heat transfer at the microscale. In this book, literature related with the synthesis and implementation of such multifunctional materials for PI will be discussed (cf. Section 2.3.3).

1.2. Integration of Reaction and Separation Operations

1.2.1. Multifunctionality

Integration of functions, also known as functionality intensification, is one of the essential principles for PI (Lutzea *et al.*, 2010), which can be illustrated through integration of conventional reaction with other functions into multifunctionality operation. The other functions can be mass, heat and momentum transport, or other reaction and separation unit operations (Agar, 1999). Among these integrated functions, the reactive separation process, which involves the integration of chemical reaction with other separation operation(s), is considered as the most widely applied functionality intensification method (multifunctional reactor) in the chemical industry (Sundmacher and Qi, 2010).

Most of the chemical reactions are limited by the equilibrium between reactants and products. As a result, it is necessary

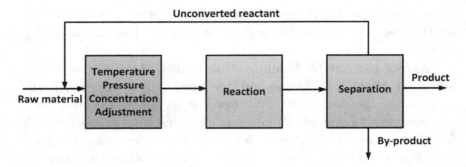

Figure 1.2 Standard process for a reversible reaction of which the conversion is limited by the thermodynamic equilibrium.

to introduce downstream separation processes to separate the equilibrium mixtures and to recycle the unconverted reactants, as shown in Fig. 1.2.

The number of separation steps is mainly dependent on the number of products produced, number of solvents used and reactants that have not been converted. The more separation steps used, the more operation procedures and energy input are required. In order to reduce the costs of separation processes, the concept of integration of reaction unit(s) and separation unit(s) into one device has been derived (Schmidt-Traub and Górak, 2006).

In this integrated unit, the product compound(s) is(are) continuously being separated from the reaction zone. In this way, a higher conversion of the reactant(s) can be achieved, and less separation steps will be required. The major characteristic of an integrated reaction and separation unit is the existence of at least two phases, a reaction phase and a transport phase (Sundmacher *et al.*, 2005). Depending on the separation properties, on the reaction phase and transport phase, a set of reactive separation methodologies developed for multifunctional reactors can be found — cf. Fig. 1.3.

When an adsorbent is used, the process is called reactive adsorption and the reactor can be classified as adsorptive reactor, chromatographic reactor or simulated moving bed reactor (SMBR), depending on the phase and operating method used. On the other hand, a permselective membrane is used for a membrane

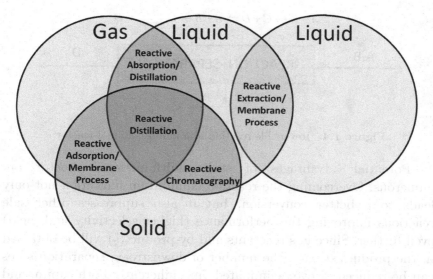

Figure 1.3 Different reactive separation methods for multifunctional reactors. Adapted from Agar (1999).

reactor (MR). Adsorptive reactors (Chapter 2), MRs (Chapter 3) and SMBRs (Chapter 4) for PI will be described and exemplified in this book.

1.2.2. Potential advantages and disadvantages

As observed in Fig. 1.4, reactive separation processes involve the integration of reaction with a separation process such as distillation, absorption, adsorption, extraction and so on, where a reversible reaction of the type $A + B \leftrightarrow C + D$ can be transformed into $A + B \rightarrow C_{(removed)} + D$. The additional degrees of freedom within the integrated unit can offer the possibility to control the concentration profiles inside the multifunctional reactor to improve the reaction performance. The most obvious benefit is the increase of yield and conversion (Paiva and Malcata, 1997), as product compound(s) is(are) separated continuously out of the reaction zone (Fig. 1.4). Nearly 100% conversion of the reactants can be achieved according to the Le Chatelier's principle (Engel and Reid, 2010), depending of course on the process and conditions used.

$$A + B \leftrightarrow C + D \Rightarrow A + B \rightarrow C_{(removed)} + D$$

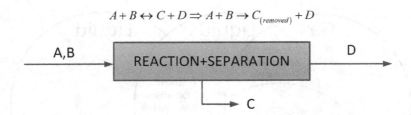

Figure 1.4 Reversible reaction in a multifunctional reactor.

Potential advantages of such multifunctional reactor are numerous. Overcoming the reaction equilibrium limitation not only leads to a better conversion, but it also suppresses other side reactions improving the performance (higher selectivity and yield) even further. Since less reactants and by-product(s) will be obtained in the product stream, the number of downstream separation steps can be reduced or even eliminated, and inherently both capital and operating costs may be reduced. Multifunctional reactors can also decrease energy demand, because the reaction might be able to be performed at lower temperature/pressure conditions and less heat exchange units will be required due to the heat integration benefits. Some further advantages are realized by improved reaction efficiency and productivity, namely the catalyst life can be extended and less amount of catalyst is required (or even no catalyst is needed) (Han and Harrison, 1994; Jansen *et al.*, 2013). Besides, Sundmacher and Qi (2010) claimed that in some cases the multifunctional reactor becomes the unique solution, such as for the separation of azeotropes by reactive distillation and for the removal of diesel soot in car exhaust by reactive filtration, which otherwise are too difficult to be solved by conventional methods. Additionally, multifunctional reactors are able to meet the requirements of green engineering and sustainable development with ecological, harmlessness and safer PI processes (Boodhoo and Harvey, 2013).

However, not all reaction and separation systems are suitable for multifunctional reactors due to the restriction of compatible reaction and separation operating conditions (i.e. temperature and pressure requirements) and robustness issues (e.g. catalyst deactivation becomes more difficult to handle in a multifunctional reactor), as well as to process efficiency and economic considerations. For instance, the MR might have higher maintenance costs due to membrane price and the special fabrication methods required (Criscuoli *et al.*, 2001), while pressure swing reactors may suffer from low gravimetric/volumetric efficiency (Wu *et al.*, 2014b).

On the other hand, development of processes based on multi-functional reactors might be even more expensive and more time consuming than that for a conventional process due to the lack of screening methods, process simulation/verification, and scale-up capability. The reaction and separation processes should be studied individually in the first place, and the complex interaction between reaction and separation must also be investigated to understand the steady state and dynamic operating behaviour within the integrated unit. Besides, the modelling and verification of multifunctional reactors are found to be much more complicated than that for conventional operating units. Consequently, careful scale-up is always needed before a commercial industrial scale multifunctional reactor can be built. As a result, there is still much to be done for academic and industrial research to make the multifunctional reactors available for technical application.

1.2.3. Examples

One of the most familiar integration strategies between reaction and separation is Reactive Distillation. In this process, an additional function such as reactor is integrated into the distillation column, which removes the reactor from the original flowsheet (Fig. 1.5).

Instead of using a conventional reactor, the reactants are fed into the reactive section of the distillation column where the reaction takes place predominately. The bottom section serves as a striping section to collect the heavier product and the unconverted reactants can return to the reaction section. The vapour leaving the reactive section consists of lighter product and the unconverted reactants which can be separated within the rectifying section. However, the most crucial advantage of such multifunctional reactor usually is not

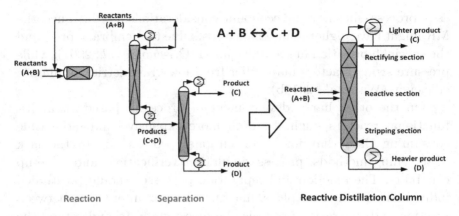

Figure 1.5 Comparison between conventional reaction and separation (distillation) with a reactive distillation column.

the reduction of the number of operating units in the plant, but allowing the overcoming of azeotrope and equilibrium limitation of the reaction.

This concept of reactive distillation was first demonstrated by the Eastman Chemical Company, who invented and employed reactive distillation for methyl acetate production from acetic acid and methanol:

$$CH_3OH + CH_3COOH \leftrightarrow CH_3COOCH_3 + H_2O \qquad (1.1)$$

The limitation of chemical equilibrium and the presence of azeotrope for the methyl acetate–water–methanol system make it very difficult to obtain high purity methyl acetate by the conventional process. Comparing with the conventional process which involves one liquid-phase reactor, one liquid–liquid extractor and eight distillation columns, the new process can produce ultra-high-purity (99.7%) methyl acetate within a single reactive distillation column (Agreda and Lilly, 1990), which is able to save 80% energy at 20% of the investment costs (Kiss, 2013). Other examples of integration of reaction and separation operations will be provided along the following sections and throughout this book.

1.3. Sorption Enhanced Reaction Processes

1.3.1. Reaction

Identifying reaction systems that would benefit from the integration of reaction and separation could be the starting point for investigation and development of multifunctional reactors. Therefore, Aida and Silveston (2008) have summarized some properties of reaction systems that would benefit from this combination to identify potential reactions that can be performed and improved in multifunctional reactors: (1) equilibrium-limited reactions, where certain temperature, pressure and other conditions are required to provide reasonable reaction rates; (2) product inhibited reactions (e.g. enzyme feedback inhibition reaction), or other reaction systems where the feed contains an inhibitor; (3) parallel reactions with a specific desired product, and the reaction producing the desired product is often an equilibrium-limited reaction, and (4) sequential reactions where the intermediate product(s) is(are) desired. Further requirements for the reaction system, especially for a catalytic system, are that the catalyst material should exhibit chemical and mechanical stability.

Accordingly, we may list numerous reaction systems that could be tested for the multifunctional reactor concept. One typical example is the hydrogen production process with methane as feedstock, for which a brief illustration can be found in Fig. 1.6. Depending on the feed composition and type of the primary reformer employed, at least three different reforming methodologies have already been used for the industrial production of hydrogen from methane (Rostrup-Nielsen, 2004): steam methane reforming (SMR), partial oxidation (PROX) and auto-thermal reforming (ATR)/oxidative steam reforming (OSR) of methane.

The effluent from the primary reformer, which mainly consists of CO, H_2, CO_2 and unconverted reactants, then undergoes a high temperature water-gas shift (WGS) reaction to convert the CO into CO_2. In traditional hydrogen production plants, the process stream goes subsequently through a low temperature WGS reactor

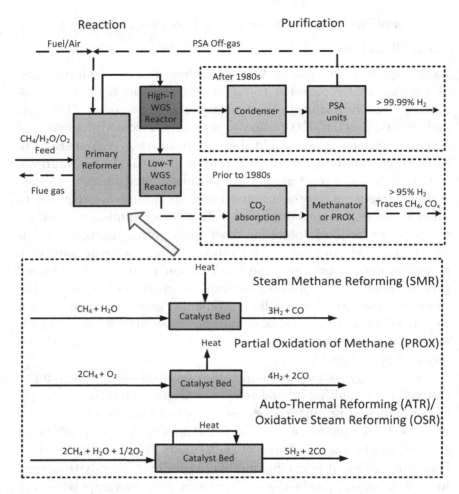

Figure 1.6 Different methodologies for hydrogen production from methane.

to further convert the remaining CO to CO_2 (see further details in Chapter 3). Afterwards, the gas-phase stream undergoes CO_2 absorption using the monoethanolamine (MEA) process (Aaron and Tsouris, 2005). The remaining CO is converted to CH_4 by the methanation reaction (Ferreira-Aparicio *et al.*, 2005) and/or to CO_2 in the PROX unit. Since the 1980s, a new CO removal process was proposed. In the new process, after the high temperature

WGS reactor, the temperature of the process stream is reduced and the water is removed in a condenser. The gas stream then goes to a pressure swing adsorption (PSA) unit (Grande *et al.*, 2008; Ribeiro *et al.*, 2008) where high purity (>99.99%) hydrogen is produced.

As a result, the overall process is quite complex. The primary reforming reaction is equilibrium-limited, and many side reactions can also be found in the reforming reactor. Therefore, a number of additional reaction, separation and purification units are always required for the production of hydrogen with high purity to be used for fuel cell applications (CO content no more than 30 ppm) (Rohland and Plzak, 1999). To reduce the capital costs and complexity of such hydrogen production technology, hybrid configurations of multifunctional reactors by coupling reaction systems with different *in situ* separation techniques have been proposed, such as adsorptive reactors (Agar, 2005; Reßler *et al.*, 2006) where a catalyst and an adsorbent have been used, also known as the sorption enhanced reaction process (SERP), described further in the next section. Use of permselective membranes has been an alternative strategy in MRs, described in Section 1.3.5.

1.3.2. Adsorption

Adsorption is an essential unit operation in the chemical industry for the separation (or purification) of gas or liquid mixtures. When a fluid mixture is contacted with a solid adsorbent, a certain component(s) within the mixture can be selectively adsorbed on the surface of the solid creating an adsorbed phase, the component being referred as adsorbate. The difference in the fluid–solid molecular forces of attraction between the components of the mixture is the dominant cause of adsorption separation, which leads to a difference in the compositions of the adsorbed and the bulk phases. Generally, it is an exothermic process. The reverse process where the adsorbed molecules are released from the solid adsorbent surface to the bulk fluid phase is referred as the desorption process, which is an endothermic process and requires different forms of energy. Both adsorption and desorption are two vital and integral steps for an effective adsorptive separation process where the adsorbent material is repeatedly used.

The concept of SERP is based on Le Chatelier's principle: in the presence of the equilibrium-limited reaction(s), selective removal of a product by adsorption may increase the yield of a desired product and bring multiple benefits at the same time. In most previous studies, a hybrid system containing a mixture with a catalyst material for reaction and an adsorbent material for separation has been extensively used, the reactor developed being commonly referred as an adsorptive reactor. One typical example of such hybrid system concept is shown in Fig. 1.7.

On the upward side of Fig. 1.7, the SERP is carried out where the catalyst and the adsorbent are arranged in a random mixing pattern system. Theoretically, only the desired product will be obtained from the outgoing stream before the adsorbent is saturated. Note that SERP requires the periodical regeneration of the adsorbent when it is saturated, as shown in the lower part of Fig. 1.7. The most common regeneration techniques are pressure swing, purge and thermal swing methods (Harrison, 2008).

Temperature swing regeneration is carried out by increasing the temperature of the sorbent material, consequently decreasing the sorption capacity of the material and releasing the adsorbed species.

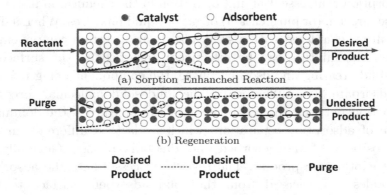

Figure 1.7   The hybrid system for a SERP.

The main drawback of this technology is an extended period of time required to cool down the sorbent. On the other hand, pressure swing desorption can be carried out by decreasing the partial pressure of the adsorbed species in the gas phase, which can be completed in a few minutes.

Identifying the suitable adsorbent and adsorption/desorption process for a SERP may require following some requisites. The required properties can be classified into the following categories: (1) adsorption capacity, namely adequate adsorption capacity for the specified product(s) in the operating temperature range of the reaction, (2) adsorption selectivity, i.e. significant differences in adsorption capacity for the different products as well as for reactants, (3) adsorption rate, such that the rate of adsorption should match the rate of reaction to shift the original reaction equilibrium, (4) regenerability, particularly aiming an adsorbent with low or moderate heat of adsorption and therefore the desorption process can be approached by reducing pressure or purging without the need for a large temperature variation and (5) stability, the adsorption performance of the adsorbent should be stable during the repeated adsorption and desorption cycles. Taking one typical SERP, the Claus reaction for instance:

$$SO_2 + 2H_2S \leftrightarrow (3/8)S_8 + 2H_2O, \text{equilibrium-limited} \qquad (1.2)$$

$$SO_2 + 2H_2S \rightarrow (3/8)S_8 + 2H_2O_{(adsorbed)}, \text{sorption enhanced} \qquad (1.3)$$

The desirable situation for compatible reaction and adsorption rates is illustrated in Fig. 1.8. One can find that the length required for the adsorption zone increases as a function of temperature due to the lower adsorption capacities at high temperature conditions. On the other hand, the length for reaction reduces along with the rise of operating temperature, since the rate of reaction increases with the temperature increase.

When the reaction rates far exceed adsorption rates, the adsorption cannot have an obvious effect on reaction, and the length for adsorption may become too long. Figure 1.8 indicates that only

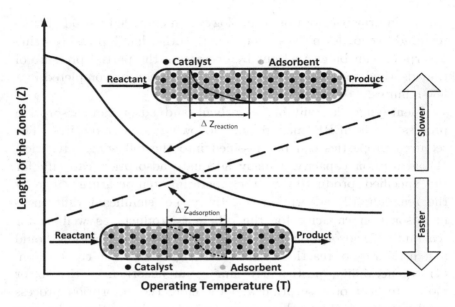

Figure 1.8 The length of the reaction (continuous line) and adsorption (dashed line) zones as a function of the operating temperature. Reprinted from Elsner *et al.* (2002). Copyright 2016, with permission from Elsevier.

under the desired ideal temperature condition (about 523 K for Claus reaction — intersection of both curves), adsorption rates and reaction rates are of comparable magnitude, the length of each section is the same, and the use of adsorption will have the largest enhancement effect on reaction rate, conversion, and thus on product yield.

Other considerations for the adsorbent are that it should not be catalytic active (unless it is a hybrid material with both adsorptive and catalytic properties) and be mechanically stable at the reaction temperature or over a range of different operating conditions employed (e.g. when thermal swing regeneration is used). Generally, and comparing with the research done so far on catalysis, discovering or developing a suitable adsorbent material is usually found as the major obstacle to develop a feasible adsorptive reactor and realize a promising SERP.

1.3.3. Adsorptive reactors

Adsorptive reactors represent a kind of multifunctional reactors in which the process is affected by the chemical reaction and the adsorption separation simultaneously. Of course, both reaction and adsorption can be performed from species present in either gas or liquid phases, but the term "chromatographic reactor" is commonly used for the liquid phase integrated reactor, which can be found in the following Section 1.3.4. As we may find in Fig. 1.9, the product component(s) can be selectively removed from the reaction mixture (gas phase) by adsorption (Carvill *et al.*, 1996), or reactant(s) can be enriched by adsorption (Salden and Eigenberger, 2001), to affect the reaction kinetics and thermodynamics equilibrium of the reaction mixture within an adsorptive reactor.

The former product removal case is most commonly employed for SERP; some typical chemical reactions in adsorptive reactors that have been extensively studied are given in Table 1.2. It can be found that compounds such as CO_2 and H_2O are preferred for adsorption since they are very stable on the adsorbent surface under the operating conditions where gas-phase catalytic reactions are performed.

Depending on the type of the reactor used, adsorptive reactors can be generally classified into fixed bed and fluidized bed adsorptive

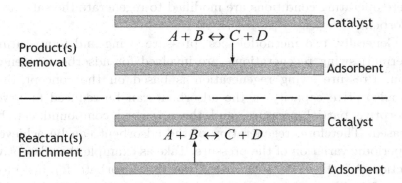

Figure 1.9 Two typical methods to enhance the reaction in an adsorptive reactor.

Table 1.2 Applications of the adsorptive reactor for SERP.

Reaction	Adsorption	References
$CO + 2H_2 \leftrightarrow CH_3OH$	CH_3OH on silica–alumina	Kruglov (1994)
$CO + 0.5O_2 \leftrightarrow CO_2$	CO_2 on 5A zeolite	Vaporciyan and Kadlec (1989)
$CO + H_2O \leftrightarrow CO_2 + H_2$	CO_2 on CaO	Han and Harrison (1994)
$CH_4 + 2H_2O \leftrightarrow CO_2 + 4H_2$	CO_2 on hydrotalcite	Ding and Alpay (2000)
$CO_2 + H_2 \leftrightarrow CO + H_2O$	H_2O on NaX–zeolite	Carvill *et al.* (1996)
$2H_2S + SO_2 \leftrightarrow 3/nS_n + 2H_2O$	H_2O on 3A zeolite	Elsner *et al.* (2002)

reactors, but fixed bed adsorptive reactors have been implemented in most cases. One important property of such fixed bed adsorptive reactor is its cyclic behaviour. Due to the limited adsorption capacity, the adsorbent will be saturated with the adsorbate after some reaction time, and periodic regeneration is required to recover the adsorption capability of the adsorptive reactor system. Therefore, each fixed bed adsorptive reactor for SERP has to swing between at least two states: (i) reaction step, where feedstock is fed to the reactor to produce the desired species; (ii) regeneration (desorption) step, where operating conditions are modified to regenerate the saturated adsorbent.

Generally, two methodologies, pressure swing and temperature (thermal) swing regenerations, are involved for adsorbent regeneration. Pressure swing regeneration is based on the concept that the adsorptive capacity of the adsorbent can be reduced at lower adsorptive partial pressure, and the adsorbed compound can be released. Therefore, regeneration of the adsorbent can be achieved by periodic variation of the pressure. Take as example the adsorptive reaction $A + B \leftrightarrow C + D$ with C as the adsorbate for instance. Figure 1.10 illustrates the whole process for carrying out the reaction in a single column reactor with four steps. This reactor is commonly known as pressure swing adsorptive reactor (PSAR). On the other

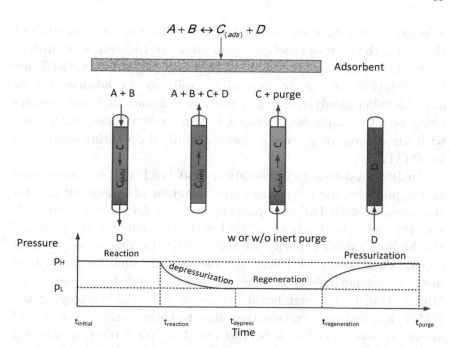

Figure 1.10 Operating a PSAR for a reversible reaction A + B ↔ C + D, where C is selectively adsorbed and D is the desired product. p_H and p_L are high pressure and low pressure, respectively.

hand, temperature swing regeneration is carried out by increasing the temperature of the adsorbent material, consequently decreasing the adsorption capacity and releasing the adsorbed material. The main drawback of this technology is an extended period of time required to cool down the reactor, while pressure swing desorption can be carried out by rapid decompression or vacuuming, which can be completed in a few minutes or even in a few seconds.

According to Carvill *et al.* (1996), the first theoretical investigation of the PSAR was carried out by Vaporciyan and Kadlec in the 1980s (Vaporciyan and Kadlec, 1987). In this study, a fixed bed reactor column has been proposed where an extremely fast equilibrium-limited reaction integrated with a rapid pressure swing process is carried out. It was found that a PSAR can be carefully tailored (packing and the flow pattern) to ensure the reactant has a longer residence time within the bed than the product(s), and thus limit the reverse reaction. Soon after the theoretical approach, Vaporciyan and Kadlec (1989) realized and performed a SERP for CO oxidation reaction by using 5A zeolite as the adsorbent with a Pt/alumina catalyst in the PSAR. It is found that the productivity of CO_2 could be increased by up to two times, comparing with the conventional reactor under optimized operating conditions for PSAR.

In 1994, a detailed mathematical model of PSAR was developed and applied for the dehydrogenation reaction of ethane by Lu and Rodrigues (1994). Different packing methods have been compared, and the well-mixed adsorbent and catalyst configuration provided the highest product purity in this SERP. In the same year, an experimental study on PSAR for the dehydrogenation reaction of methyl cyclohexane to toluene was carried out by Alpay *et al.* (1994). Due to the significant economic potential, employing the PSAR for hydrogen production has recently attracted a lot of attention since Carvill *et al.*'s (1996) work on WGS reaction, where a Pt-based catalyst was used for reaction and a hydrotalcite adsorbent for CO_2 adsorption, H_2 diluted in N_2 being obtained during the SERP step. A comprehensive PSAR for steam reforming of methane by using hydrotalcite CO_2 adsorbent and supported Ni catalyst was performed by Ding and Alpay (2000). Besides, the concept of PSAR has also been employed for the enhanced H_2 production from ethanol steam reforming (He *et al.*, 2009). Steam reforming of ethanol is highly endothermic and very strongly controlled by reaction thermodynamics. Consequently, the conventional ethanol steam reforming reactor is normally operated at a high temperature (>873 K), while in a PSAR, the reaction can be carried out at 773 K without sacrificing the reactants conversion and obtaining a high-purity H_2 product gas with CO content less than 30 ppm; a four-column unit (Fig. 1.11) can be used to run this process continuously (Wu *et al.*, 2014a).

In addition to equilibrium-limited reactions, a series reaction ($A \rightarrow B \rightarrow C$) was studied by Kodde *et al.* (2000) in a PSAR selectively adsorbing reactant A. Soon after the theoretical investigation,

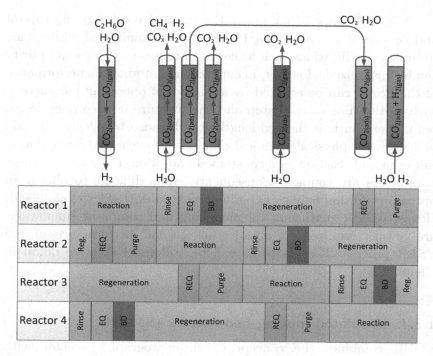

Figure 1.11 PSARs for sorption enhanced steam reforming of ethanol, EQ: pressure equalization, REQ: received pressure equalization, BD: blowdown. Reprinted from Wu *et al.* (2014a). Copyright 2016, with permission from Elsevier.

a PSAR packed with a CrO_2/Al_2O_3 catalyst and a zeolite K–Y adsorbent was designed by Sheikh *et al.* (2001) for the enhanced *i*-butane dehydrogenation series reaction.

The PSAR clearly has potential for hydrogen production from steam reforming reactions. However, the limiting factor is still the CO_2 adsorption at high temperature. Although the PSAR must be operated at a relatively high temperature ($>873\,K$) for steam reforming reactions in order to obtain a reasonable catalytic reaction rate, such high temperature is not favourable for most adsorbents. One of the challenges is to find the adsorbent that could adsorb CO_2 at a high rate under such operation condition with a high uptake capacity. At the same time, the adsorbent must be easily regenerated by pressure swing and/or by using steam purge (Sircar, 2007).

It is well known that under the optimal condition the capital and/or operating costs of a PSAR can be minimized while some technical specifications such as conversion, yield and product purity can be maximized. However, in experimental studies the performance of the PSAR can be affected by a number of operating parameters, such as the time of each step and the pressure in each one; design parameters, such as the bed length and the adsorbent/catalyst ratio and size; and physical/chemical parameters, such as the adsorption isotherm and reaction rate constant. In addition, the effects of these parameters are coupled. Consequently, it is difficult to obtain an optimal condition only by experimental studies (Reßler *et al.*, 2006). Hence, reliable mathematical modelling and computer simulations are always required to obtain the optimal performance of the PSAR process. These aspects will be addressed in more detail in Chapter 2.

1.3.4. Simulated moving bed reactor (SMBR)

SMBR combines the concept of chromatographic reactor with simulated continuous counter-current solid–liquid or solid–gas flow. It is a hybrid process, generally not energy-intensive, and is very competitive when high quality of separation is required to obtain high value-added products (e.g. fine chemicals and pharmaceuticals), while numerous devices for reaction and separation steps are required with traditional processes. To better explain the concept behind SMBR technology, we will start with the chromatographic reactor, and then will move towards the true and simulated moving bed (SMB) concepts.

Generally, higher conversions and better yields can be achieved within a chromatographic reactor, where a reversible chemical reaction is carried out in a liquid–solid system and differences between product affinities to the solid adsorbent phase are employed for their chromatographic separation. Therefore, one can overcome the limitation of the chemical equilibrium according to the Le Chatelier's principle (Engel and Reid, 2010), while at the same time high purity products can be obtained. The principle of a chromatographic reactor

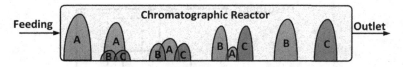

Figure 1.12 The reversible reaction A ↔ B + C in a chromatographic reactor.

can be easily explained using a simple reversible reaction A ↔ B + C, for which the reactant A has intermediate adsorption behaviour and products B and C are the more strongly and weakly adsorbed components, respectively.

A pulse of reactant A is injected into the continuous eluent stream as shown in Fig. 1.12. As soon as A enters the reactor, the reaction occurs and the products B and C are produced; then all the components interact with the surface of the adsorbent. Due to different affinities to the adsorbent, the product C is the least retained component, which travels ahead of A, while the product B is the most strongly retained component, which stays behind A. Hence the products are separated from the reaction zone, this reversible reaction can overcome the equilibrium limitation and the reactant can ideally be completely converted. As a result, two pure products can be obtained separately at the outlet at different times.

The advantages of a chromatographic reactor in comparison to conventional reaction–separation processes have been shown experimentally for various reactions (Schmidt-Traub, 2005). However, due to batch operation, only a small part of the reactor bed is used for reaction and separation processes, which leads to a low productivity. Another problem associated with batch operation is that the operation always requires large amounts of eluent/desorbent and results in highly diluted products (Schmidt-Traub and Górak, 2006). Consequently, the chromatographic batch reactor has little practical meaning from the large-scale production point of view. In order to overcome these difficulties, the chromatographic batch reactor has to be transformed into a continuous process.

It is well known that the counter-current operating mode can offer higher efficiency in separation processes such as distillation, gas absorption and liquid–liquid extraction (Ghasem and Henda, 2009).

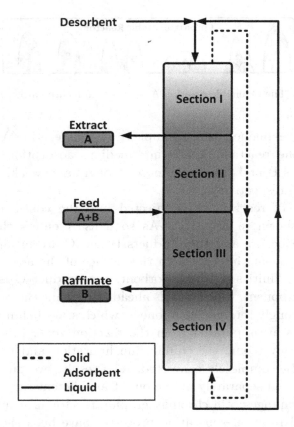

Figure 1.13 The mixture (A+B) separation in a TMB.

In counter-current processes, two contacted phases move in the opposite directions continuously. This concept can also be employed for chromatography to develop a continuous process as shown in Fig. 1.13, commonly known as the "True Moving Bed" (TMB) process, where the solid phase and the liquid phase are moving in the opposite directions continuously.

The whole process can be divided into four sections by four streams: the feed stream that contains the mixture of A and B to be separated, the desorbent stream, the extract stream that contains the more retained component A and the raffinate stream that contains the less retained component B. The separation is performed in

Sections II and III, where the more retained component A is adsorbed and carried towards the extract stream with the solid phase, while the less retained component B is carried by the liquid phase in the direction of the raffinate stream. In Section I, the solid phase is regenerated by the fresh desorbent stream, and in Section IV the liquid phase is regenerated with the fresh solid phase by adsorbing the less retained component B. The liquid and the solid phases can be recycled to Sections I and IV, respectively.

It can be seen from Fig. 1.13 that when the TMB process reaches the steady state, the feed mixture (A and B) can theoretically be completely separated. The high purity component A can be obtained from the extract stream, while the raffinate stream contains component B only.

If the feed mixture (A and B) is replaced with the reactant A, a true moving bed reactor (TMBR) can be easily developed. A TMBR for the reversible reaction $A \leftrightarrow C + D$ is illustrated in Fig. 1.14, where the reactant A has intermediate adsorption behaviour and products C and D are the more strongly and weakly adsorbed components, respectively. Sections I and IV have the same function as in the TMB process, both the solid and the liquid phase are regenerated and recycled. The reaction and separation steps are performed in Sections II and III simultaneously. The more retained product C is carried by the solid phase to the extract stream and the less retained product D is carried by the liquid phase towards the raffinate stream. When the reactant A is completely converted, high purity products C and D can be obtained from the extract and the raffinate stream separately. If we replace the desorbent stream with the reactant B, a reversible reaction $A + B \leftrightarrow C + D$ can be carried out in a TMBR. Compared with the chromatographic batch reactor, the TMBR has higher productivities and lower eluent consumption. Unfortunately, a counter-current between solid and liquid phase is too difficult to realize due to severe problems related with the transport of the solid. As a result, an alternative solution, the SMB process, was developed.

The concept of SMB was first developed by Broughton and Gerhold from UOP in 1961, and the first industrial process was

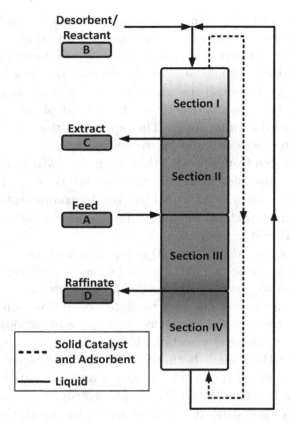

Figure 1.14 Reversible reaction A ↔ C + D (or A + B ↔ C + D) in a TMBR.

called "Sorbex" (Broughton *et al.*, 1975), where a rotating valve was employed to achieve the simulated movement of the beds with regard to the inlet/outlet streams. As demonstrated in Fig. 1.15, a number of conventional fixed bed columns are used in a SMB process, and the counter-current between the solid and the liquid phase is simulated by switching all the streams periodically in the direction of the liquid flow.

Similarly to the TMB, the whole unit can also be generally divided into four sections by four streams, but now each section has two columns. The function of each section remains the same as described for the TMB. By switching all the streams periodically in

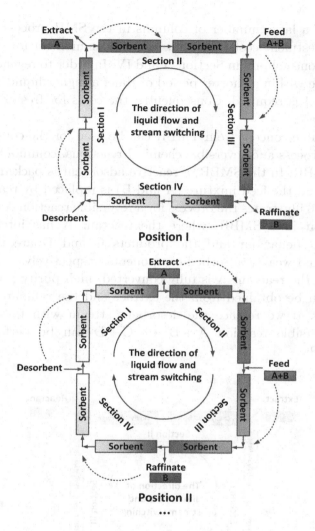

Figure 1.15 SMB process.

the direction of the liquid flow, components A and B can be obtained continuously with high purity at the extract and raffinate streams. The total number of columns is variable and not always fixed to eight with two columns per section (2-2-2-2) as shown in Fig. 1.15 (Schmidt-Traub, 2005). The SMB is theoretically equivalent to a TMB if the number of columns in each section is infinite. However, the use of a large number of columns in the SMB process leads to higher investment costs, and thus a very common set-up is 1-2-2-1 with only one column in Sections I and IV. In order to regenerate the solid phase within a shorter period of time, a higher liquid flow rate in Section I is required. Consequently, the costs for fresh desorbent may increase.

The same concept from TMBR can be used for the combination of SMB process and reversible chemical reactions, commonly known as the SMBR. In the SMBR, a reactive adsorbent is packed into the columns and the feed mixture (A and B) is replaced by reactant A. Figure 1.16 illustrates the process for a reversible reaction $A \leftrightarrow C + D$ carried out in a SMBR, where the reactant A has intermediate adsorption behaviour and the products C and D are the more strongly and weakly adsorbed components, respectively.

When the reactant A is fully converted, high purity products C and D can be obtained from the extract and the raffinate streams separately. If we replace the desorbent stream with the reactant B, a reversible reaction $A + B \leftrightarrow C + D$ can be performed in the SMBR.

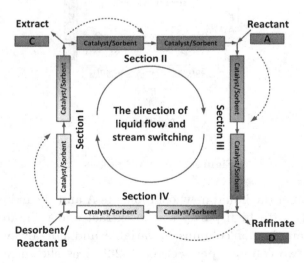

Figure 1.16   Reversible reaction $A \leftrightarrow C + D$ (or $A + B \leftrightarrow C + D$) in a SMBR.

The SMBR process is useful particularly for reaction systems with two products that are difficult to separate. Examples include the enzyme catalyzed sucrose inversion reaction (Azevedo and Rodrigues, 2001) and the production of dextran (Bechtold *et al.*, 2006), with a stoichiometry of the form $A \leftrightarrow C + D$ and that have been studied in SMBRs. For reactions with a stoichiometry of the form $A + B \leftrightarrow C + D$, examples studied with SMBRs are the synthesis of dimethylacetal (Silva *et al.*, 2011) and diethylacetal (Silva and Rodrigues, 2005). In the case of dimethylacetal synthesis, it was found that the purity of dimethylacetal can reach 91.41% in a SMBR (Pereira *et al.*, 2007). As for the ethyl lactate synthesis reaction, it was found that under appropriate conditions the lactic acid conversion can be driven to completion and ethyl lactate with the purity of 95% can be obtained (Pereira *et al.*, 2009). In addition, the isomerization of glucose to fructose with a stoichiometry of the form $A \leftrightarrow B$ has also been investigated in a SMBR (Silva *et al.*, 2005; Zhang *et al.*, 2004); it was found that a fructose-rich product (90% in purity) can be obtained with a new configuration of the SMBR (Silva *et al.*, 2005).

In experimental studies, the performance of the SMBR can be affected by a number of operating parameters, such as flow rates and feed concentration; design parameters, such as column length and diameter, column configuration and adsorbent/catalyst ratio and size, are also important. Besides, the effects of these parameters are coupled. Consequently, an optimal design of the SMBR process is a challenging task. Taking the effect of reaction kinetics into consideration, an optimal design for the SMBR will be even more challenging. As a result, the SMBR has not yet become extensively employed in commercial applications due to the complexity of design and operation (Kaspereit, 2009). However, the roadmap for PI in Europe (EFCE, 2007) has estimated the ripeness of SMBR application in few years, with both high potential to improve cost competitiveness and high potential for innovative high quality products. This technology will be addressed in Chapter 4.

1.3.5. Membrane reactors (MRs)

A multifunctional reactor with the adsorption separation technique to enhance the reaction system can be replaced by, or even further integrated with, other separation technologies to enhance the reaction performance as well. For example, a selective membrane separation process can be employed to develop another type of multifunctional reactor, the MR, where only the selected product(s) permeates through the membrane to overcome equilibrium constraints in reversible reactions by *in situ* removing the selected product(s). The well-known mechanism of sorption–diffusion (or solution–diffusion) in dense membranes acts in a similar way as an adsorbent in a multifunctional reactor where a thermodynamic equilibrium-limited reaction is carried out; in this case the species with higher permeation rate (commonly a reaction product) is removed from the reaction medium, shifting the reaction in the forward direction.

A membrane is defined as a semipermeable barrier between two phases. Membranes were originally developed for filtration. Depending on the pore size, membranes can be classified into microfiltration (MF), ultrafiltration (UF), nanofiltration (NF), and reverse osmosis (RO) membranes. Liquid or gaseous component(s) can permeate from one side of the membrane to the other side with the help of a driving force of concentration, pressure, temperature or electrical potential. Component(s) that can permeate through the membrane constitute another stream, referred as permeate, while the retained component(s) is called retentate. The performance of a membrane is generally characterized by the selectivity and the permeability. Many kinds of PI applications can be found in literature for reaction systems with MRs based on their selectivity and permeability features. The potentialities of using MRs are diverse, and a brief description is given in Fig. 1.17. They can be useful for instance in enhancing reactant conversion/product yield in reversible reactions, in increasing selectivity towards intermediate species in consecutive reactions, or in controlled addition of some reactants.

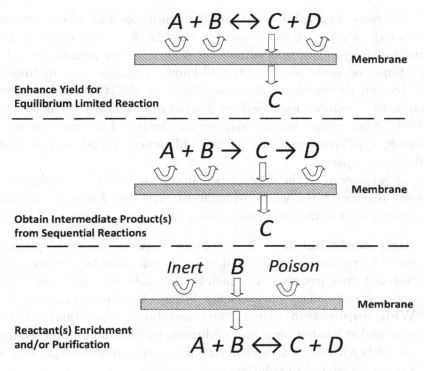

Figure 1.17 Applications of MRs for PI.

In most cases, the purpose of using selective membranes in MRs is to obtain a higher conversion in equilibrium-limited reaction systems by removing continuously and selectively one of the products, thus favouring the forward reaction according to Le Chatelier's principle (Gallucci *et al.*, 2013). Additionally, and as stated above, the differences in permeability of components can be employed to obtain the desired intermediate product(s) from a sequential reaction. Besides, when the membrane inside the MR has catalytic activity, it can be used to perform the reaction and separation functions simultaneously, which is sometimes referred as "reactive membrane". Still, if the bed is composed by an adsorbent-catalyst mixture, the reactor can be classified as a sorption enhanced membrane reactor (SEMR) (cf. Chapter 3, Section 3.6.4).

There are many different types of membranes and selectivity can be based on many different mechanisms. In dense membranes, the functional layer material transports a specific component, usually in atomic or ionic form. Two well-known examples are hydrogen conducting Pd-based membranes (Basile *et al.*, 2011) and oxygen conducting membranes based on perovskite materials (Zeng *et al.*, 1998). Other separation mechanisms can be based on the molecular sieving, capillary condensation and differences in adsorption and diffusion properties.

In additional to the general benefits related with the multifunctional reactor, MRs are considered to have the following special advantages according to Vankelecom (2007):

- **High controllability**: The contact between unstable (highly reactive) reactants can be mediated and controlled by a membrane.
- **Solvent-free process**: The membrane can be used as a contactor between two phases, and solvents will not be required anymore.
- **Wide application range**: Some membrane separations can be operated at low temperature conditions, which can be employed for products with limited thermal stability, and membrane separations are not restricted to volatile components.
- **Integrated heat supply**: The heat generated from the exothermic reaction can be used for the endothermic reaction taking place at the other side of the membrane.

One classical example of the use of a MR is the hydrogen production from SMR (Silva and de Abreu, 2016), with the following reactions (partial steam reforming of methane, WGS reaction and overall steam reforming of methane) taking place inside the MR:

$$CH_4 + H_2O \leftrightarrow CO + 3H_2 \quad \Delta H^0_{298K} = 205.8 \text{ kJ/mol} \quad (1.4)$$

$$CO + H_2O \leftrightarrow CO_2 + H_2 \quad \Delta H^0_{298K} = -41.1 \text{ kJ/mol} \quad (1.5)$$

$$CH_4 + 2H_2O \leftrightarrow CO_2 + 4H_2 \quad \Delta H^0_{298K} = 164.9 \text{ kJ/mol} \quad (1.6)$$

As shown in Fig. 1.18, the MR consists of a porous ceramic tube as the internal tube and an external tube constructed in

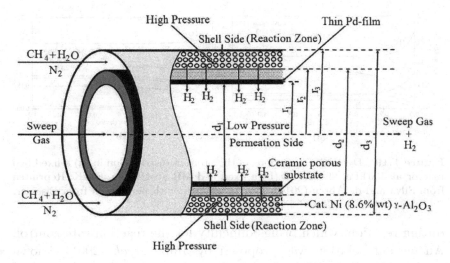

High Pressure

Thin Pd-film

Shell Side (Reaction Zone)

CH$_4$+H$_2$O
N$_2$

H$_2$ H$_2$ H$_2$ H$_2$

Low Pressure
Permeation Side

Sweep
Gas

Sweep Gas
+
H$_2$

H$_2$ H$_2$ H$_2$ Ceramic porous substrate

CH$_4$+H$_2$O
N$_2$

Cat. Ni (8.6% wt) γ-Al$_2$O$_3$

Shell Side (Reaction Zone)

High Pressure

Figure 1.18 Schematic diagram of a fixed bed MR. Reprinted from Silva and de Abreu (2016). Copyright 2016, with permission from Elsevier.

steel. The catalyst is placed in the annular space between the two tubes, the internal tube being externally covered with a thin Pd-film as the membrane showing 100% permselectivity toward hydrogen. Therefore, it is possible to shift the equilibrium towards the products side by removal of hydrogen.

It has been shown, as illustrated in Fig. 1.19, that with this reactor the conversion of methane, far beyond thermodynamic equilibrium, can be realized even under lower temperature condition (823 K), and the molar flow rates of undesired compounds become very low in the fixed bed MR, comparing with that in the conventional fixed bed reactor at 998 K.

In addition to selective product removal, another major application of MRs is gradually supplying one of the reactants through the membrane as shown in Fig. 1.17. It could be particularly useful for PROX reactions, such as the oxidation of methane into syngas:

$$0.5O_2 + CH_4 \rightarrow CO + 2H_2 \quad \Delta H^0_{298K} = -71.9 \, \text{kJ/mol} \tag{1.7}$$

In a conventional fixed bed reactor, the oxygen-to-methane molar ratio will change along the reactor length due to the depletion

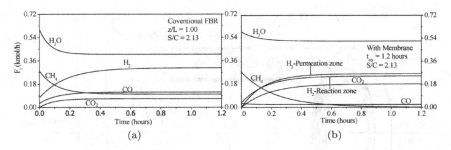

Figure 1.19 Dynamics evolutions of the product distribution in (a) a fixed bed reactor at 450 kPa, 998 K and (b) a fixed bed MR at 450 kPa, 823 K. Reprinted from Silva and de Abreu (2016). Copyright 2016, with permission from Elsevier.

during reaction, which brings difficulty for the reaction rate control. An elegant solution was proposed by Julbe *et al.* (2001), known as the "chemical valve membrane" concept, which is considered as attractive for an auto-regulation of the O_2 concentration in a MR for the partial oxidation of alkanes. Besides, air can be used as the source of oxygen instead of pure oxygen with the help of a membrane that is permselective towards oxygen in this case (Moulijn *et al.*, 2013).

In spite of many opportunities, additional challenges have to be faced in the design and utilization of MRs for large scale industrial applications (Vankelecom, 2007). One of the most significant problems related with MRs is the membrane poisoning by trace components in the feed gas, particularly sulphides. Membranes resistance and cost are other drawbacks for some applications. The low permeation rate is another major obstacle, e.g. the flow rate through the membrane and the production rate of hydrogen must be of similar magnitude for a feasible hydrogen production process with an acceptable reactor volume. Besides, an additional separation step will be required, because the sweep gas is required to remove the component(s) at the permeate side. However, strong R&D effort is being put in this technology, as described in further detail in Chapter 3, focused mostly on the WGS reaction.

1.3.6. Applications

Along the book, several applications will be illustrated for each of the multifunctional reactors addressed in each chapter. The application of adsorptive reactors for sorption enhanced WGS reaction for H_2 production in a power plant will be introduced briefly in this section. As described in Section 1.3.1, the H_2-rich gas can be obtained from the primary reforming reactor where CH_4 is converted into H_2. The overall conversion, from the thermodynamic point of view, is favoured by low pressure and high temperature operating condition. But the reforming reaction under a low pressure condition may lead to an oversized reactor. Therefore, high pressures are always used to meet the economics for equipment sizing requirements in large-scale H_2 production processes. The outlet stream obtained from the primary reforming reactor usually contains (dry molar basis): CO 10.3%, CO_2 11.4%, H_2 74.6% and 3.7% of unconverted CH_4 reactant (Ratnasamy and Wagner, 2009). As a result, subsequently converting CO via the WGS reaction would help the production of further H_2. Enhancement of the conversion with adsorption (SE-WGS) can be achieved by removing the CO_2 from the gas *in situ* with an adsorptive reactor, as predicted by the Le Chatelier's principle. Besides, the concept of SE-WGS process is particularly attractive for pre-combustion decarbonization applications in Integrated Gasification Combined Cycle (IGCC), where the H_2 product can then be fed to a gas turbine to generate power, and the CO_2 is sequestered at the same time (Manzolini *et al.*, 2015).

The appropriate CO_2 adsorbent is the crucial part of a SE-WGS process. The material must have promising CO_2 adsorption performance under cyclic pressure swing cycles (between $\sim$30 bar and atmospheric pressure) at operating temperature $\sim$673 K. Lithium zirconate and CaO-based adsorbents may fulfil the criterion of high capacity under such conditions, but the kinetics and stability of these materials for CO_2 adsorption are poor.

Allam *et al.* (2005) from Air Products along with the US Department of Energy under the program "CO_2 Capture Project" evaluated some adsorbents and developed the SE-WGS process since the 1990s; they first compared the CO_2 adsorption performances of commercial sodium oxides, K_2CO_3-promoted hydrotalcites (HTlcs), lead oxide adsorbents and double salt adsorbents for SE-WGS. The best material was found to be the K-promoted HTlcs with a TGA determined capacity of 1.6 mmol/g at 673–723 K. Afterwards, HTlcs-based materials have been considered as the most attractive adsorbent for SE-WGS. In the more recent EU FP6 project CACHET and EU FP7 project CAESAR, K_2CO_3-promoted hydrotalcite material in pellet form was prepared and employed as the CO_2 adsorbent. The long term test of the new adsorbent can be found from Fig. 1.20; the cycle-averaged CO_2 level of the top product is given, showing cyclic stability after 450 cycles, and the new adsorbent showed good stability for the working capacity for 1400 cycles. Besides, Jansen *et al.* (2013) found that such type of adsorbent material (K-promoted MG30) has sufficient catalytic WGS activity, and the need for a WGS catalyst can be eliminated in the SE-WGS process, which also brings substantial economic benefits.

The application of SE-WGS was also carried out in an IGCC power plant by Gazzani *et al.* (2013) from Politecnico di Milano (Italy), and the layout of the integrated power plant and SE-WGS

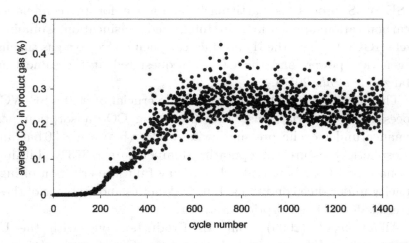

Figure 1.20 Long term stability test of the PURALOX MG70 adsorbent during SE-WGS runs. Reprinted with permission from van Selow *et al.* (2009). Copyright (2016), American Chemical Society.

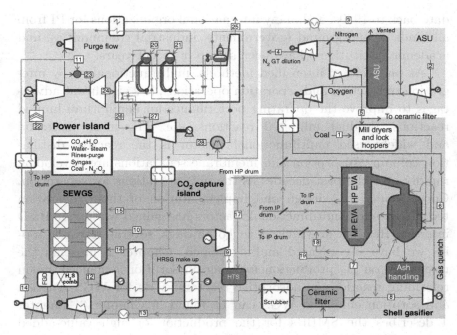

Figure 1.21 Layout of the IGCC power plant with CO_2 capture by SE-WGS. Reprinted from Gazzani *et al.* (2013). Copyright 2016, with permission from Elsevier.

can be found from Fig. 1.21. The effects of operating conditions and adsorbent types were investigated, where the specific energy consumption for CO_2 capture is able to reach as low as $2.1\,\mathrm{MJ/kgCO_2}$. Additionally, the calculated cost of CO_2 capture is about $30\text{€}/t\mathrm{CO_2}$ under the base adsorbent performance, which was already lower than IGCC with CO_2 capture by Selexol technology. Moreover, the cost can be reduced further to $23\text{€}/t\mathrm{CO_2}$ with an improved adsorbent material according to the economic assessment (Manzolini *et al.*, 2013).

1.4. Conclusions

There is no doubt that PI is and will continue to be a valuable tool for the sustainable development of the chemical industry. The use of HiGee technology for intensified mass transfer processes can even date back to 1930s, and many international organizations for PI from all kinds of background (governments, industrial communities, and academic institutions) have been set up in recent years.

As one kind of intensified methods, multifunctionality, by integration of reaction and separation operations, has significant advantages to improve the operating performance for equilibrium limited reactions, product inhibited reactions, etc. The most successful and widely used multifunctional reactor is the reactive distillation column. Adsorption and membrane separations are developing rapidly over the past decades as the unit operations for separation. Consequently, the design and implementation of these separation technologies for multifunctional reactors has gained worldwide R&D interest. Readers are invited to go through several research topics we have investigated and developed. Chapter 2 presents the researches on H_2 production focusing on sorption enhanced steam reforming reactions in adsorptive reactors. MRs for WGS reaction and sorption enhanced WGS can be found in Chapter 3. Chapter 4 describes the SMBRs for the production of high value-added compounds.

Nomenclature

ATR	Auto-thermal reforming
HTlc	Hydrotalcite
IGCC	Integrated gasification combined cycle
MR	Membrane reactor
OSR	Oxidative steam reforming
PI	Process Intensification
PROX	Partial oxidation
PSA	Pressure swing adsorption
PSAR	Pressure swing adsorptive reactor
SEMR	Sorption enhanced membrane reactor
SERP	Sorption enhanced reaction process
SE-WGS	Sorption enhanced water-gas shift
SMB	Simulated moving bed
SMBR	Simulated moving bed reactor
SMR	Steam methane reforming
TMB	True moving bed
TMBR	True moving bed reactor
WGS	Water-gas shift

References

Aaron, D., Tsouris, C., 2005. Separation of CO_2 from flue gas: A review, *Separation Science and Technology*, 40, 321–348.

Agar, D.W., 1999. Multifunctional reactors: old preconceptions and new dimensions, *Chemical Engineering Science*, 54, 1299–1305.

Agar, D.W., 2005. The Dos and Don'ts of Adsorptive Reactors, in *Integrated Chemical Processes: Synthesis, Operation, Analysis, and Control*, Sundmacher, K., Kienle, A., Seidel-Morgenstern, A. (Eds.), Wiley, Weinheim, 203–231.

Agreda, V.H., Lilly, R.D., 1990. Preparation of ultra high purity methyl acetate, US Patent No. 4,939,294.

Aida, T., Silveston, P.L., 2008. *Cyclic Separating Reactors*, John Wiley & Sons, Oxford.

Allam, R.J., Chiang, R., Hufton, J.R., Middleton, P., Weist, E.L., White, V., 2005. Development of the Sorption Enhanced Water Gas Shift Process, in *Carbon Dioxide Capture for Storage in Deep Geologic Formations*, Thomas, D.C., Benson, S.M. (Eds.), Elsevier, Amsterdam, 227–256.

Alpay, E., Chatsiriwech, D., Kershenbaum, L.S., Hull, C.P., Kirkby, N.F., 1994. Combined reaction and separation in pressure swing processes, *Chemical Engineering Science*, 49, 5845–5864.

Azevedo, D.C., Rodrigues, A.E., 2001. Design methodology and operation of a simulated moving bed reactor for the inversion of sucrose and glucose–fructose separation, *Chemical Engineering Journal*, 82, 95–107.

Basile, A., Iulianelli, A., Longo, T., Liguori, S., De Falco, M., 2011. Pd-based Selective Membrane State-of-the-Art, in *Membrane Reactors for Hydrogen Production Processes*, Falco, M.D., Marrelli, L., Iaquaniello, G. (Eds.), Springer, London, 21–55.

Bechtold, M., Makart, S., Heinemann, M., Panke, S., 2006. Integrated operation of continuous chromatography and biotransformations for the generic high yield production of fine chemicals, *Journal of Biotechnology*, 124, 146–162.

Boodhoo, K., Harvey, A., 2013. *Process Intensification Technologies for Green Chemistry: Engineering Solutions for Sustainable Chemical Processing*, John Wiley & Sons, Oxford.

Broughton, D., Bieser, H., Persak, R., 1975. Sixty years of Sorbex^{TM} operations, UOP Process Division Technology Conference. UOP Process Division, Des Plaines, IL.

Carvill, B.T., Hufton, J.R., Anand, M., Sircar, S., 1996. Sorption-enhanced reaction process, *AIChE Journal*, 42, 2765–2772.

Criscuoli, A., Basile, A., Drioli, E., Loiacono, O., 2001. An economic feasibility study for water gas shift membrane reactor, *Journal of Membrane Science*, 181, 21–27.

Cunha, A.F., Wu, Y.-J., Li, P., Yu, J., Rodrigues, A.E., 2014. Sorption-enhanced steam reforming of ethanol on a novel K-Ni-Cu-hydrotalcite hybrid material, *Industrial & Engineering Chemistry Research*, 53, 3842–3853.

Ding, Y., Alpay, E., 2000. Adsorption-enhanced steam-methane reforming, *Chemical Engineering Science*, 55, 3929–3940.

EFCE, 2007. European Roadmap for Process Intensification. Creative Energy, The Netherlands.

Elsner, M.P., Dittrich, C., Agar, D.W., 2002. Adsorptive reactors for enhancing equilibrium gas-phase reactions — two case studies, *Chemical Engineering Science*, 57, 1607–1619.

Engel, T., Reid, P., 2010. *Physical Chemistry*, Prentice Hall, New York.

Ferreira-Aparicio, P., Benito, M.J., Sanz, J.L., 2005. New trends in reforming technologies: From hydrogen industrial plants to multifuel microreformers, *Catalysis Reviews: Science and Engineering*, 47, 491–588.

Gallucci, F., Fernandez, E., Corengia, P., van Sint Annaland, M., 2013. Recent advances on membranes and membrane reactors for hydrogen production, *Chemical Engineering Science*, 92, 40–66.

Gazzani, M., Macchi, E., Manzolini, G., 2013. CO_2 capture in integrated gasification combined cycle with SEWGS — Part A: Thermodynamic performances, *Fuel*, 105, 206–219.

Ghasem, N., Henda, R., 2009. *Principles of Chemical Engineering Processes*, CRC Press, Boca Raton.

Grande, C.A., Lopes, F.V., Ribeiro, A.M., Loureiro, J.M., Rodrigues, A.E., 2008. Adsorption of off-gases from steam methane reforming (H_2, CO_2, CH_4, CO and N_2) on activated carbon, *Separation Science and Technology*, 43, 1338–1364.

Han, C., Harrison, D.P., 1994. Simultaneous shift reaction and carbon dioxide separation for the direct production of hydrogen, *Chemical Engineering Science*, 49, 5875–5883.

Harrison, D.P., 2008. Sorption-enhanced hydrogen production: A review, *Industrial & Engineering Chemistry Research*, 47, 6486–6501.

He, L., Berntsen, H., Chen, D., 2009. Approaching sustainable H_2 production: Sorption enhanced steam reforming of ethanol, *Journal of Physical Chemistry A*, 114, 3834–3844.

Jansen, D., van Selow, E., Cobden, P., Manzolini, G., Macchi, E., Gazzani, M., Blom, R., Henriksen, P.P., Beavis, R., Wright, A., 2013. SEWGS technology is now ready for scale-up!, *Energy Procedia*, 37, 2265–2273.

Julbe, A., Farrusseng, D., Guizard, C., 2001. Porous ceramic membranes for catalytic reactors — overview and new ideas, *Journal of Membrane Science*, 181, 3–20.

Kaspereit, M., 2009. Advanced Operating Concepts for Simulated Moving Bed Processes, in *Advances in Chromatography*, Grushka, E., Grinberg, N. (Eds.), CRC Press, New York, 165–192.

Kiss, A.A., 2013. Reactive Distillation Technology, in *Process Intensification for Green Chemistry*, Boodhoo, K., Harvey, A. (Eds.), John Wiley & Sons, Oxford, 251–274.

Kodde, A.J., Fokma, Y.S., Bliek, A., 2000. Selectivity effects on series reactions by reactant storage and PSA operation, *AIChE Journal*, 46, 2295–2304.

Kruglov, A.V., 1994. Methanol synthesis in a simulated countercurrent moving-bed adsorptive catalytic reactor, *Chemical Engineering Science*, 49, 4699–4716.

Lu, Z.P., Rodrigues, A.E., 1994. Pressure swing adsorption reactors: Simulation of three-step one-bed process, *AIChE Journal*, 40, 1118–1137.

Luo, L., 2013. *Heat and Mass Transfer Intensification and Shape Optimization: A Multi-scale Approach*, Springer, London.

Lutzea, P., Ganib, R., Woodleya, J.M., 2010. Application of a synthesis and design methodology to achieve process intensification, *Chemical Engineering and Processing*, 49, 547–558.

Manzolini, G., Jansen, D., Wright, A., 2015. Sorption-enhanced fuel conversion, in *Process Intensification for Sustainable Energy Conversion*, Gallucci, F., Annaland, M.V.S. (Eds.), 175–208.

Manzolini, G., Macchi, E., Gazzani, M., 2013. CO_2 capture in integrated gasification combined cycle with sewgs — part B: Economic assessment, *Fuel*, 105, 220–227.

Moulijn, J.A., Makkee, M., Van Diepen, A.E., 2013. Process intensification, in *Chemical Process Technology*, John Wiley & Sons, Oxford.

Paiva, A.L., Malcata, F.X., 1997. Integration of reaction and separation with lipases: An overview, *Journal of Molecular Catalysis B: Enzymatic*, 3, 99–109.

Pereira, C.S.M., Gomes, P.S., Gandi, G.K., Silva, V.M.T.M., Rodrigues, A.E., 2007. Multifunctional reactor for the synthesis of dimethylacetal, *Industrial & Engineering Chemistry Research*, 47, 3515–3524.

Pereira, C.S.M., Zabka, M., Silva, V.M.T.M., Rodrigues, A.E., 2009. A novel process for the ethyl lactate synthesis in a simulated moving bed reactor (SMBR), *Chemical Engineering Science*, 64, 3301–3310.

Ratnasamy, C., Wagner, J.P., 2009. Water gas shift catalysis, *Catalysis Reviews: Science and Engineering*, 51, 325–440.

Reßler, S., Elsner, M.P., Dittrich, C., Agar, D.W., Geisler, S., Hinrichsen, O., 2006. Reactive Gas Adsorption, in *Integrated Reaction and Separation Operations*, Schmidt-Traub, H., Górak, A. (Eds.), Springer, Berlin, 149–190.

Reay, D., Ramshaw, C., Harvey, A., 2013. *Process Intensification: Engineering for Efficiency, Sustainability and Flexibility*, Butterworth-Heinemann, Amsterdam.

Ribeiro, A.M., Grande, C.A., Lopes, F.V., Loureiro, J.M., Rodrigues, A.E., 2008. A parametric study of layered bed PSA for hydrogen purification, *Chemical Engineering Science*, 63, 5258–5273.

Rohland, B., Plzak, V., 1999. The PEMFC-integrated CO oxidation — a novel method of simplifying the fuel cell plant, *Journal of Power Sources*, 84, 183–186.

Rostrup-Nielsen, J.R., 2004. Fuels and energy for the future: The role of catalysis, *Catalysis Reviews: Science and Engineering*, 46, 247–270.

Salden, A., Eigenberger, G., 2001. Multifunctional adsorber/reactor concept for waste-air purification, *Chemical Engineering Science*, 56, 1605–1611.

Schmidt-Traub, H., Górak, A., 2006. *Integrated Reaction and Separation Operations*, Springer, Berlin.

Schmidt-Traub, H., 2005. *Preparative Chromatography of Fine Chemicals and Pharmaceutical Agents*, Wiley-VCH, Weinheim.

Sheikh, J., Kershenbaum, L.S., Alpay, E., 2001. 1-butene dehydrogenation in rapid pressure swing reaction processes, *Chemical Engineering Science*, 56, 1511–1516.

Silva, E.A.B.D., De Souza, A.A.U., De Souza, S.G.U., Rodrigues, A.E., 2005. Simulated moving bed technology in the reactive process of glucose isomerization, *Adsorption*, 11, 847–851.

Silva, J.D., de Abreu, C.A.M., 2016. Modelling and simulation in conventional fixed-bed and fixed-bed membrane reactors for the steam reforming of methane, *International Journal of Hydrogen Energy*, 41, 11660–11674.

Silva, V.M.T.M., Pereira, C.S.M., Rodrigues, A.E., 2011. PermSMBR — A new hybrid technology: Application on green solvent and biofuel production, *AIChE Journal*, 57, 1840–1851.

Silva, V.M.T.M., Rodrigues, A.E., 2005. Novel process for diethylacetal synthesis, *AIChE Journal*, 51, 2752–2768.

Sircar, S., 2007. Emerging technologies for bulk gas separation by adsorption, *Indian Chemical Engineer*, 49, 430–442.

Stankiewicz, A., Moulijn, J.A., 2002. Process intensification, *Industrial & Engineering Chemistry Research*, 41, 1920–1924.

Stankiewicz, A.I., Moulijn, J.A., 2000. Process intensification: transforming chemical engineering, *Chemical Engineering Progress*, 96, 22–34.

Sundmacher, K., Kienle, A., Seidel-Morgenstern, A., 2005. *Integrated Chemical Processes: Synthesis, Operation, Analysis, and Control*, Wiley-VCH, Weinheim.

Sundmacher, K., Qi, Z., 2010. Multifunctional Reactors, in *Chemical Engineering and Chemical Process Technology — Volume III: Chemical Reaction Engineering*, Encyclopedia of Life Support Systems, Developed under the auspices of the UNESCO, Eolss Publishers, Paris (http://www.eolss.net).

Van Gerven, T., Stankiewicz, A., 2009. Structure, energy, synergy, time — The fundamentals of process intensification, *Industrial & Engineering Chemistry Research*, 48, 2465–2474.

van Selow, E.R., Cobden, P.D., Verbraeken, P.A., Hufton, J.R., van den Brink, R.W., 2009. Carbon capture by sorption-enhanced water-gas shift reaction process using hydrotalcite-based material, *Industrial & Engineering Chemistry Research*, 48, 4184–4193.

Vankelecom, I.F.J., 2007. Membrane Reactors, in *Encyclopedia of Chemical Processing*, Lee, S. (Ed.), Taylor & Francis, New York, 1575–1586.

Vaporciyan, G.G., Kadlec, R.H., 1987. Equilibrium-limited periodic separating reactors, *AIChE Journal*, 33, 1334–1343.

Vaporciyan, G.G., Kadlec, R.H., 1989. Periodic separating reactors: Experiments and theory, *AIChE Journal*, 35, 831–844.

Wu, Y.-J., Li, P., Yu, J.-G., Cunha, A.F., Rodrigues, A.E., 2014a. Sorption-enhanced steam reforming of ethanol for continuous high-purity hydrogen production: 2D adsorptive reactor dynamics and process design, *Chemical Engineering Science*, 118, 83–93.

Wu, Y.-J., Li, P., Yu, J., Cunha, A.F., Rodrigues, A.E., 2014b. High-purity hydrogen production by sorption-enhanced steam reforming of ethanol: A cyclic operation simulation study, *Industrial & Engineering Chemistry Research*, 53, 8515–8527.

Zeng, Y., Lin, Y., Swartz, S., 1998. Perovskite-type ceramic membrane: synthesis, oxygen permeation and membrane reactor performance for oxidative coupling of methane, *Journal of Membrane Science*, 150, 87–98.

Zhang, Y., Hidajat, K., Ray, A.K., 2004. Optimal design and operation of SMB bioreactor: Production of high fructose syrup by isomerization of glucose, *Biochemical Engineering Journal*, 21, 111–121.

Chapter 2

Gas-phase Adsorptive Reactor
for Hydrogen Production Processes

2.1. Hydrogen Economy

2.1.1. Energy demand

Along with the development of economy, the demand for energy has increased dramatically. The International Energy Outlook (IEO) analysis in 2016 indicates a continuous world energy demand growth during the first three decades of this century, as shown in Fig. 2.1 (Doman, 2016). The world energy consumption has expanded from 495 quadrillion kJ in 2005 to 608 quadrillion kJ in 2015 and is expected to expand to 758 quadrillion kJ in 2030 (an increase by more than 50% within three decades). Additionally, the major proportion of energy consumption grows up in developing countries outside the Organisation for Economic Cooperation and Development (OECD), particularly in Asia, where the energy demands are driven by a robust long-term economic growth.

Nowadays, over 80% of the energy consumption still relies on fossil fuels such as oil, natural gas and coal, as depicted in Fig. 2.2. According to the IEO report (Doman, 2016), these non-renewable energy sources will continue to dominate the energy system in the first half of the 21st century. However, the use of fossil fuels for energy production leads to numerous environmental impacts. Billion tonnes of greenhouse gases (GHG) are annually released into the atmosphere due to various activities that use energy from fossil fuels.

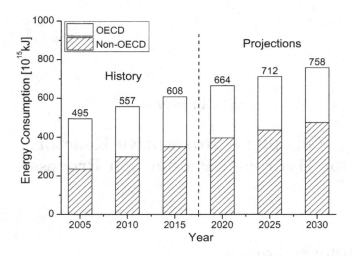

Figure 2.1 World energy consumption; data taken from Doman (2016).

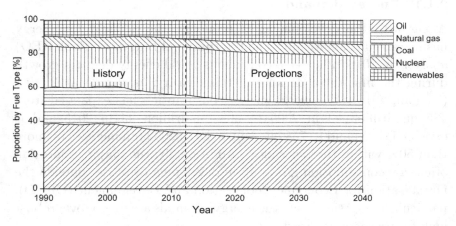

Figure 2.2 World energy consumption by type of fuel (Doman, 2016).

In 2011, the three CO_2 top emitters were China (8.1 billion tons) followed by U.S.A. (5.5 billion tons) and Europe (4.3 billion tons) (EIA and US, 2011). The release of GHG such as carbon dioxide becomes the main concern for global warming up. On the other hand, the emission of nitrogen oxides and sulphur–oxygen species from fossil fuel usage contributes to acid rain. Consequently, clean utilization of conventional energy and decarbonization have become major research areas worldwide for sustainable energy development. Particularly, finding alternative ways to meet the energy demands and solving the ecological impacts at the same time is a major challenge for researchers nowadays. Recently, the "hydrogen economy" (Barreto *et al.*, 2003), which is considered as a promising solution, has attracted worldwide interest.

2.1.2. Hydrogen as a clean fuel

The "hydrogen economy", which was proposed by Lawrence W. Jones in 1970 (Jones, 1970), is a "system" where hydrogen is employed instead of using fossil fuels for the current hydrocarbon economy. Hydrogen has many advantageous properties to be used as a fuel (Vernon and Paul, 2014), including high flame speed ($\sim2.8\,\mathrm{m\cdot s^{-1}}$), low ignition energy ($\sim0.02\,\mathrm{mJ}$), high auto-ignition temperature ($844\,\mathrm{K}$), and high octane value (>130). For instance, internal combustion engines with hydrogen as fuel can achieve thermal efficiencies up to 35% (Gandía *et al.*, 2013), which are better than the typical values for conventional gasoline engines due to the high octane number allowing high compression ratios. Besides, the hydrogen fuel cell system can be two to three times more efficient than a conventional internal combustion engine dictated by the maximum efficiency of the Carnot cycle. But most of all, hydrogen fuel is environmentally friendly. Comparing with the combustion of other chemical fuels, the combustion of hydrogen–air mixtures emits almost nothing other than water, with only very small amounts of NO_x pollutants being formed under high combustion temperatures.

2.1.3. Hydrogen as an energy carrier

Hydrogen is not readily available in nature; in fact, only small amounts of hydrogen can be found in crustal reservoirs mixed with natural gas. Significant amounts of energy and feedstock are required to produce hydrogen in the "hydrogen economy". Therefore, unlike traditional fossil fuels, hydrogen should be considered as an energy carrier similar to electricity. Generally, the whole process where hydrogen is used as an energy carrier includes production, storage, transportation and utilization (conversion) steps. In addition to the environmental concerns mentioned in the previous section, hydrogen can offer many advantages for a sustainable future development as a clean energy carrier. Hydrogen is an universal fuel (Holladay *et al.*, 2009), which can be produced from a wide range of feedstocks including natural gas (reforming), coal (gasification) as well as water (electrolysis), while the production plants and supply infrastructures can be developed according to the feedstock available in local regions (Barreto *et al.*, 2003). Besides, hydrogen can also be produced from renewable bio-fuels such as bio-ethanol or biomass. The contribution to net-zero carbon dioxide emissions and a long term sustainable production are the major advantages (Murray *et al.*, 2007). A summary of various hydrogen production methods and feedstocks can be found in the following section.

Hydrogen has many advantageous physical properties for storage and transportation, including low molecular weight ($2\,\text{g}\cdot\text{mol}^{-1}$), large diffusion coefficient ($\sim$$0.61\,\text{cm}^2\cdot\text{s}^{-1}$ in air), high thermal conductivity ($187\,\text{W}\cdot\text{K}^{-1}$ at $300\,\text{K}$) and low viscosity ($9\,\mu\text{Pa}\cdot\text{s}$ at $300\,\text{K}$). It can be stored in gaseous form, in liquid form or even in the form of metal hydrides. In addition, using hydrogen as an energy carrier is advantageous in terms of energy storage density compared to electricity. As for the utilization, hydrogen is especially applicable as fuel for fuel cells, which have higher energy conversion efficiencies than conventional combustion engines (Carrette *et al.*, 2000). The whole process is environmentally compatible since the production, storage, transportation, and end use do not have harmful effects on the environment (Sherif *et al.*, 2014).

2.2. Hydrogen Production

2.2.1. Feedstock

The total annual hydrogen production worldwide in 2013 was in excess of 50 million tons for industrial and commercial purposes (DOE, 2013). Depending on the feedstock and energy sources, hydrogen products can be generally classified into non-renewable

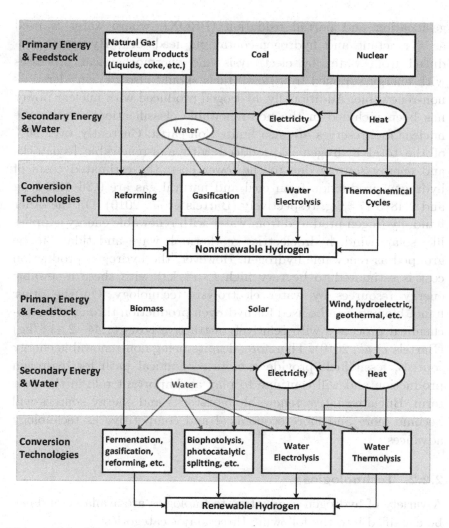

Figure 2.3 Scheme of the non-renewable and renewable hydrogen production pathways. Adapted from Gandía *et al.* (2013).

hydrogen and renewable hydrogen (Gandía *et al.*, 2013) as depicted in Fig. 2.3.

As shown in Fig. 2.3, non-renewable hydrogen can be produced from natural gas, coal and other hydrocarbons such as naphtha with different conversion technologies such as steam reforming (SR), gasification and partial oxidation (PROX), where water is used as a reactant and hydrogen-containing feedstock. Hydrogen produced from water by electrolysis and thermochemical processes with energy supplied from fossil fuels should also be considered as non-renewable. Additionally, hydrogen produced with nuclear power has been included in the non-renewable classification because the nuclear fuel reserves are also limited on earth. Currently, over 96% of the total hydrogen is produced with non-renewable feedstocks and energy sources due to the lower cost; the estimated costs of hydrogen production from coal and natural gas are 0.36–1.83 $/kg and 2.48–3.17 $/kg, respectively (Bartels *et al.*, 2010). On the other hand, hydrogen produced from water with renewable energy supplies like solar, wind, hydroelectric, geothermal, wave and tidal can be grouped as renewable hydrogen. However, the hydrogen production cost is estimated to be very high (>5 $/kg) with these renewable energy resources by water electrolysis technology. On the other hand, biomass can be used for hydrogen production through thermochemical processes with relatively attractive costs (1.44−2.83 $/kg) (Bartels *et al.*, 2010). Therefore, despite being non-renewable energy sources, fossil fuels provide a more economical path for hydrogen production and will continue to play an important role in the near term. But alternative renewable feedstocks and energy sources will become more and more economical and competitive as technology advances.

2.2.2. Technologies

A variety of hydrogen production technologies are available and can be classified into the following three major categories:

- **Thermochemical**, including hydrocarbons reforming, coal/biomass gasification, catalytic cracking/decomposition of hydrocarbons, water-gas shift (WGS), thermolysis of water, etc.;
- **Electrochemical**, including water electrolysis, water vapour electrolysis, chlor-alkali process, etc.;
- **Photochemical**, including photocatalytic processes for water cleavage and photobiological with microbial for hydrogen production from water, etc.

On one hand, hydrogen production from water by using electricity is a very typical electrochemical method, where hydrogen with very high purity can be obtained directly. Due to the relative high cost, the share of water electrolysis for hydrogen production is 4% (Gandía *et al.*, 2013). Other technologies, such as photobiological (Sakurai *et al.*, 2013) or photocatalytic (Ni *et al.*, 2007) methods to produce hydrogen from water that use the photosynthetic activity of microorganisms or catalysts are still in laboratory research stage.

On the other hand, the gasification process is a commonly used thermochemical methodology when the feedstocks are solids such as coal, coke and other biomass derivatives; 18% of hydrogen production around the globe is obtained from coal gasification (Gandía *et al.*, 2013), being of interest the following reactions.

Gasification (carbon):

$$C + H_2O \leftrightarrow CO + H_2 \quad \Delta H^0_{298K} = +131\,kJ \cdot mol^{-1} \tag{2.1}$$

The produced synthesis gas from the gasification unit is usually subjected to a WGS reaction with additional steam to produce more hydrogen and adjust the H_2:CO ratio:

$$CO + H_2O \leftrightarrow CO_2 + H_2 \quad \Delta H^0_{298K} = -41.4\ kJ \cdot mol^{-1} \tag{2.2}$$

Finally, reforming is the most widely employed thermochemical methodology for hydrogen production, which includes the catalytic SR, PROX and auto-thermal reforming (ATR)/oxidative steam reforming (OSR) of a variety of gaseous and liquid fuels such as methane, naphtha, ethanol, etc. Current commercial processes for hydrogen production largely (~50% in the world) depend on natural gas as feedstock. Most of these production processes employ the reforming technology (Rostrup-Nielsen, 2004). Depending on the feed composition and type of the reactor employed, at least three different methodologies are employed for the industrial production of hydrogen from methane: SR, PROX and ATR/OSR (Ferreira-Aparicio *et al.*, 2005).

The following equations represent the global reactions involving two representative feedstocks: natural gas (CH_4) as a typical non-renewable fossil fuel and ethanol (C_2H_5OH) as an emerging renewable biomass-derived fuel and liquid alcohol for mobile applications.

SR:

$$CH_4 + H_2O \leftrightarrow CO + 3H_2 \quad \Delta H^0_{298K} = +206 \, \text{kJ} \cdot \text{mol}^{-1} \quad (2.3)$$

$$C_2H_5OH + H_2O \leftrightarrow 2CO + 4H_2 \quad \Delta H^0_{298K} = +255 \, \text{kJ} \cdot \text{mol}^{-1}$$

$$(2.4)$$

PROX:

$$CH_4 + 1/2O_2 \leftrightarrow CO + 2H_2 \quad \Delta H^0_{298K} = -35.9 \, \text{kJ} \cdot \text{mol}^{-1} \quad (2.5)$$

$$C_2H_5OH + 1/2O_2 \leftrightarrow 2CO + 3H_2 \quad \Delta H^0_{298K} = +13.7 \, \text{kJ} \cdot \text{mol}^{-1}$$

$$(2.6)$$

ATR/OSR:

$$CH_4 + 1/4O_2 + 1/2H_2O \leftrightarrow CO + 5/2H_2$$

$$\Delta H^0_{298K} = +85.5 \, \text{kJ} \cdot \text{mol}^{-1} \quad (2.7)$$

$$C_2H_5OH + 1/4O_2 + 1/2H_2O \leftrightarrow 2CO + 7/2H_2$$

$$\Delta H^0_{298K} = +135 \, \text{kJ} \cdot \text{mol}^{-1} \quad (2.8)$$

In the SR process, a mixture of methane (ethanol) and steam reacts over the catalyst bed to produce carbon monoxide and hydrogen, known as the syngas product. Among the three different reforming processes discussed above, theoretically the SR process can generate the highest hydrogen yield, but the reaction is highly endothermic and requires external heat supply. On the other hand, methane (ethanol) and oxygen are used as the reactants in a PROX process. In this process, a part of the methane (ethanol) is first combusted over the catalyst bed to produce carbon dioxide and water, and then subsequently reacts with the remaining methane (ethanol) to produce carbon monoxide and hydrogen (Rostrup-Nielsen, 2002). PROX requires the least amount of external heat supply, but it also has the lowest theoretical hydrogen yield. The ATR process is a combination of SR and PROX. The amount of air/oxygen depends on the reaction temperature required to achieve a condition of thermal balance. This process is also known as the

OSR process. In ATR process, a mixture of methane (ethanol), steam and oxygen is fed into a combustion zone first, where a part of the methane (ethanol) is converted to generate a mixture of carbon monoxide and water. At the same time, the heat generated in the combustion zone can be used for the endothermic catalytic SR reaction to produce hydrogen. In some cases, the combustion and the catalyst bed sections are integrated into one section. However, the ATR process also has its own drawbacks; an expensive and complex oxygen separation unit is always required in order to feed pure oxygen into the reactor, or a similar equipment should be used to separate the product gas diluted in nitrogen when air is used in the feed (Holladay *et al.*, 2009).

By far SR is the most dominant process employed to produce hydrogen in the chemical industry. In practice, the SR is not really just one reaction as indicated in Eqs. (2.3) or (2.4), but involves contributions from a series of different equilibrium-limited reversible reactions as the reforming reaction usually takes place at very high temperatures ($\sim$1000 K). As a result, when SR of hydrocarbons is used for the production of hydrogen, the major product at the outlet stream from the reformer is hydrogen (70–72%), but it is contaminated with 10–15% of carbon dioxide, 8–10% of carbon monoxide, $\sim$5% of unreacted methane and is saturated with water. Other impurities (less than 1%) that might be present in the hydrogen-rich stream are N_2, O_2, Ar, etc. (Grande, 2016). Therefore, to obtain high purity hydrogen product as raw material for chemical industry (usually >98%), and ultra-pure hydrogen for fuel cell application (>99.99% with less than 10 ppm of carbon monoxide content), separation and purification steps are always required in hydrogen production processes with SR technology.

2.2.3. Hydrogen separation and purification

To suppress the carbon monoxide content and improve the hydrogen purity, a WGS process is generally employed to adjust the stoichiometric ratio of syngas, as given in Eq. (2.2). Generally, and as described in further detail in Chapter 3, two fixed bed

WGS reactors in series are used in most cases to achieve a higher hydrogen purity. A high-temperature (623 K) WGS reactor is firstly used for rapid reaction, where the conversion of carbon monoxide cannot exceed 75–80% due to the equilibrium limitation. A low-temperature (463–483 K) WGS reactor placed afterwards can shift thermodynamic equilibrium to the products side to further decrease the concentration of carbon monoxide; the carbon monoxide conversion by the combination of high- and low-temperature WGS reactors is able to reach 95% or even higher (Gnanapragasam *et al.*, 2014). As a result, the carbon monoxide content of 13–15% at the outlet stream from the primary reformer can be reduced to less than 0.5%.

In a traditional hydrogen production plant, the gas-phase stream from WGS reactors undergoes carbon dioxide absorption process using monoethanolamine, while the remaining traces of carbon monoxide can be transformed to methane by methanation reaction. The hydrogen product obtained can reach a purity ~95%. Since the 1980s, a new purification process was proposed and became widely employed — the removal of the remaining contaminants for highly purified hydrogen production can be accomplished in a pressure swing adsorption (PSA) unit that follows the WGS reactors.

The PSA process is based on the principle that adsorbents are capable of adsorbing more impurities at a higher partial pressure than at a lower partial pressure. A PSA separation process is a gas-solid system operated in cycles where fixed bed columns are packed with an appropriate solid adsorbent that preferentially adsorbs one or more components from a gas mixture. The adsorptive separation process is performed by an ordered sequence of adsorption–desorption steps; for the PSA process the pressure swings between two different points in an equilibrium isotherm curve (see Fig. 2.4(a)). The two main steps of a PSA process are: (1) adsorption step, carried out at the higher pressure, where the preferentially adsorbed compounds are separated from the feed mixture; (2) desorption (regeneration) step, accomplished at lower pressure, where the most adsorbed species are partially removed from the adsorbent preparing the column for the next cycle (see Fig. 2.4(b)).

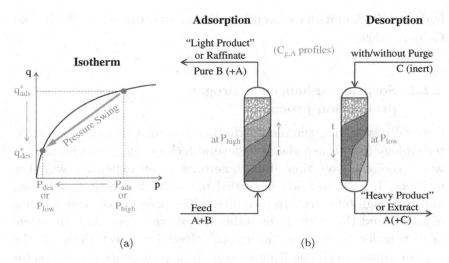

Figure 2.4 The PSA process: (a) change in equilibrium loading with pressure; (b) adsorbed phase concentration profile in a PSA process (adsorption and desorption). Adapted from Lopes (2010).

In most cases, the impurities (CO_2, CO, CH_4, etc.) are adsorbed at high pressure in a PSA unit as heavy compounds for H_2 purification, while H_2 passes through the fixed bed column as the light product. When the column is saturated with the impurities, it is disconnected from the process and the pressure is decreased, thus releasing most of the impurities (Lopes *et al.*, 2011). Additionally, a small fraction of the produced hydrogen is needed for purging and regeneration of the adsorbent, so the H_2 recovery is close to 90%. PSA technology can produce hydrogen with a purity from 98% up to 99.999+% (Grande, 2016).

The PSA technology was introduced as a commercial process for air separation in the 1960s by Skarstrom, and was developed for hydrogen purification and recovery from different sources, such as steam methane reforming off-gases, refinery fuel gases, coke oven gases and others since the 1980s, and became standard in modern hydrogen production plants for H_2 purification. Currently, more than 85% of hydrogen production units use PSA technology for hydrogen separation and purification, and hundreds of large-scale hydrogen PSA units have been installed around the world (Sircar and Golden, 2009).

2.2.4. Sorption enhanced hydrogen production processes

Generally speaking, the most extensively used hydrogen production technology (SR) is associated to several technological difficulties that were overcome over time, but sometimes in an expensive way. For instance, the SR reaction is carried out at high pressure (>20 bar) and very high temperature ($\sim$1000 K). The presence of large amount of steam and the strongly endothermic SR reaction result in severe heat transfer limitations, normally solved by direct firing of the reactor tubes. Such conditions result in a very short life of reactor tubes (3–4 years depending on the operating conditions and tube wall thickness) that are made of expensive materials (chromium and nickel). On the other hand, for the separation and purification steps, the hydrogen-rich stream from the primary reformer must be purified by some cost intensive separation units (high- and low-temperature WGS reactors, methanation reactor, PSA, etc.) before being used in fuel cells.

In order to reduce the severe reaction conditions required and the costs of separation processes, the integration of reaction unit(s) and separation unit(s) into one is proposed (Schmidt-Traub and Górak, 2006). In the case of hydrogen production via SR, a selective solid adsorbent is used to adsorb carbon dioxide from the product gas phase, this integrated process being commonly referred as sorption enhanced reaction process (SERP) (Harrison, 2008). The concept of SERP is based on Le Chatelier's principle: when one of the products is removed, the equilibrium can be shifted towards the formation of more products. Therefore, if the carbon dioxide can be selectively removed from the product gas phase by an appropriated solid adsorbent, the normal equilibrium limit can be displaced and a complete conversion can be closely approached. An illustration with H_2 production from methane reforming (global reaction) and sorption enhanced methane reforming is given in Fig. 2.5.

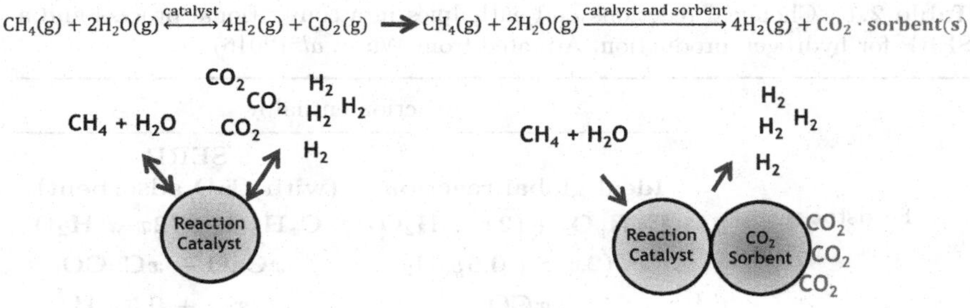

$$CH_4(g) + 2H_2O(g) \xleftrightarrow{\text{catalyst}} 4H_2(g) + CO_2(g) \longrightarrow CH_4(g) + 2H_2O(g) \xrightarrow{\text{catalyst and sorbent}} 4H_2(g) + CO_2 \cdot \text{sorbent(s)}$$

Figure 2.5 H_2 production from conventional methane reforming vs. sorption enhanced methane reforming.

A comparison of the chemical reactions and enthalpies with and without CO_2 removal for hydrogen production can be found in Table 2.1. According to Mayorga *et al.* (1997), SERP has several potential advantages including:

- Replacement of high temperatures required in the reforming reactor, so that less energy will be spent in heating and cheaper materials can be used;
- Simplification of the hydrogen purification section due to higher hydrogen purity and lower concentrations of CO_x;
- Suppression of carbon deposition in the reactor;
- Lower investment costs due to a reduced number of operating units.

To date, extensive research has been done on the SERP of hydrogen production from methane, but only few studies have been done on the sorption enhanced steam reforming of ethanol (SE-SRE). Therefore, some part of the knowledge gained and concepts used in the sorption enhanced steam methane reforming (SE-SMR) process are useful for hydrogen production from ethanol. SE-SMR process has been investigated experimentally and theoretically. SE-SMR simulations and optimization at LSRE were first reported by Xiu *et al.* (2002a, 2002b, 2003). Subsequently, the SMR kinetics of the nickel-based catalysts (Oliveira *et al.*, 2009, 2010) and the adsorption performance of modified carbon dioxide sorbents (Oliveira *et al.*, 2008) were studied, and used to investigate SE-SMR for hydrogen

Table 2.1 Chemical reactions and enthalpies in a conventional process and a SERP for hydrogen production. Adapted from Wu *et al.* (2016).

	Reaction enthalpy	
Feedstocks	**Ideal global reaction** $\mathbf{C}_x\mathbf{H}_y\mathbf{O}_z + (\mathbf{2}x\mathbf{-}z)\mathbf{H_2O}$ $\leftrightarrow (\mathbf{2}x\mathbf{-}z + \mathbf{0.5}y)\mathbf{H_2}$ $+x\mathbf{CO_2}$	**SERP (with CaO adsorbent)** $\mathbf{C}_x\mathbf{H}_y\mathbf{O}_z + (\mathbf{2}x\mathbf{-}z)\mathbf{H_2O}$ $+x\mathbf{CaO} \rightarrow x\mathbf{CaCO_3}$ $+(\mathbf{2}x\mathbf{-}z + \mathbf{0.5}y)\mathbf{H_2}$
Carbon monoxide/ syngas ($x = 1$, $y = 0$, $z = 1$)	$\Delta H^0_{298} = -41\,\mathrm{kJ \cdot mol^{-1}}$	$\Delta H^0_{298} = -219\,\mathrm{kJ \cdot mol^{-1}}$
Methane/ natural gas ($x = 1$, $y = 4$, $z = 0$)	$\Delta H^0_{298} = +178\,\mathrm{kJ \cdot mol^{-1}}$	$\Delta H^0_{298} = -13\,\mathrm{kJ \cdot mol^{-1}}$
Methanol ($x = 1$, $y = 4$, $z = 1$)	$\Delta H^0_{298} = +49\,\mathrm{kJ \cdot mol^{-1}}$	$\Delta H^0_{298} = -92\,\mathrm{kJ \cdot mol^{-1}}$
Ethanol/ bio-ethanol ($x = 2$, $y = 6$, $z = 1$)	$\Delta H^0_{298} = +226\,\mathrm{kJ \cdot mol^{-1}}$	$\Delta H^0_{298} = -186\,\mathrm{kJ \cdot mol^{-1}}$
Propane ($x = 3$, $y = 8$, $z = 0$)	$\Delta H^0_{298} = +398\,\mathrm{kJ \cdot mol^{-1}}$	$\Delta H^0_{298} = -137\,\mathrm{kJ \cdot mol^{-1}}$
Glycerol ($x = 3$, $y = 8$, $z = 3$)	$\Delta H^0_{298} = +219\,\mathrm{kJ \cdot mol^{-1}}$	$\Delta H^0_{298} = -411\,\mathrm{kJ \cdot mol^{-1}}$
1-butanol ($x = 4$, $y = 10$, $z = 1$)	$\Delta H^0_{298} = +445\,\mathrm{kJ \cdot mol^{-1}}$	$\Delta H^0_{298} = -325\,\mathrm{kJ \cdot mol^{-1}}$

production by Oliveira *et al.* (2011). These processes are further detailed below.

The concept of SERP is not new; the first report of SMR in the presence of a calcium-based sorbent can be found in 1868 (Motay and Marechal, 1868), according to Harrison (2008). Up to date, most of the SERP researches have focused on the use of natural gas (methane) as feedstock for SR, also known as SE-SMR (Wang and Rodrigues, 2005). Besides, the sorption enhanced WGS (SE-WGS) process (Cunha *et al.*, 2015; Soria *et al.*, 2015) with carbon monoxide as feedstock for H_2 production in an Integrated Gasification Combined Cycle (IGCC) plant, and (SE-SRE) (bio-ethanol) (Wu *et al.*, 2016) for renewable H_2 production from biomass, have received a lot of research interest.

SE-SMR

Many publications available on SE-SMR process can be found in some recent reviews (Barelli *et al.*, 2008; Dou *et al.*, 2016; Harrison, 2008; Yancheshmeh *et al.*, 2016). Primary efforts were devoted to the development of high temperature carbon dioxide sorbents and multi-cycle operation performance of the catalyst-sorbent materials. Calcium-based sorbents and potassium promoted hydrotalcite (K-HTlc) sorbents have received more attention than lithium-based sorbents due to the relatively low adsorption rate and high cost of lithium-based materials (Kato *et al.*, 2005), even though they have excellent multi-cycle stability.

Calcium-based sorbents are employed in most of the SE-SMR studies. Johnsen *et al.* (2006) carried out an experimental study with dolomite as the carbon dioxide sorbent in a bubbling fluidized bed reactor. The operating temperature and pressure were 873 K and 101 kPa, respectively. The product gas with a hydrogen concentration over 98 mol% could be obtained. Wu *et al.* (2005) performed a SE-SMR process using calcium oxide as the carbon dioxide adsorbent. A 94 mol% hydrogen-rich gas was achieved, which is approximately equal to 96% of the theoretical equilibrium limit, and much higher than the equilibrium concentration of 67.5 mol% without carbon dioxide sorption under the same conditions at 773 K, 200 kPa pressure and a steam-to-carbon ratio $(R_{S/C})$ of 6. In addition, the residual mole fraction of carbon dioxide was less than 0.1%. Chanburanasiri *et al.* (2011) performed SE-SMR tests using Ni/CaO multifunctional material with a nickel metal content of 12.5 wt.% as catalyst and carbon dioxide sorbent, and a high hydrogen concentration (80 mol%) was obtained at atmospheric pressure with a $R_{S/C}$ of 3 and 873 K. In addition, SE-SMR tests with Ni-HTlc derived catalyst and two different CaO-based sorbents, Ca-based pellets and Rheinkalk limestone, have been carried out by

Broda *et al.* (2013). They found that high-purity hydrogen (99%) can be obtained at 823 K and $R_{S/C} = 4$, and the pellets showed better performance during the cyclic operation due to the stable nanostructured morphology.

On the other hand, Oliveira *et al.* (2011) conducted sorption enhanced experiments using a potassium promoted hydrotalcite as carbon dioxide adsorbent. It was found that during cyclic SE-SMR experiments, methane conversion and hydrogen purity were 43% and 75%, respectively, which is better than for a normal SMR reactor. Additionally, a mathematical model for this process was proposed and tested. It was found that this model was able to describe the SE-SMR experiments without any fitting parameters.

However, the performance of the SE-SMR process can be affected by several operating parameters, by design parameters as well as by physical and chemical parameters, while effects of these parameters are always coupled. As a result, it would be almost impossible to achieve the ideal conditions of SE-SMR process just by experimental tests alone. Modelling and simulation can significantly improve the performance of SE-SMR systems.

Different operating conditions were tested by Xiu *et al.* (2002a, 2002b, 2003) to achieve high hydrogen purities and low carbon monoxide concentrations in the effluent stream from SE-SMR. The simulations (Xiu *et al.*, 2002a) have shown that a hydrogen purity of 86.8 mol% together with 587 ppm of carbon dioxide and 50 ppm carbon monoxide can be achieved. Xiu *et al.* (2004) also applied a subsection-controlling strategy to design the SE-SMR reactor. The product gas had a concentration of hydrogen over 85 mol%, together with a carbon monoxide concentration below 30 ppm and a carbon dioxide concentration below 300 ppm.

Ochoa-Fernandez *et al.* (2005) studied SE-SMR with lithium zirconate as sorbent. A kinetic equation for the sorbent material, with carbon dioxide partial pressure and temperature as parameters, has been developed. Afterwards, the hydrogen production process by SE-SMR was simulated in a fixed bed reactor, where a Ni catalyst derived from a hydrotalcite-like precursor has been employed as SR catalyst. Simulation results show that the product gas with a hydrogen purity more than 95 mol%, and the concentration of carbon monoxide less than 0.2 mol%, can be produced in a single step.

Lee *et al.* (2004, 2006) simulated two different SE-SMR systems with a fixed bed reactor and a moving bed reactor, respectively. In the moving bed reactor, the nickel-based reforming catalyst and calcium oxide based carbon dioxide sorbent pellets are moved simultaneously with gaseous reactants. Effects of operating parameters such as temperature, pressure and $R_{S/C}$ on SE-SMR systems have been investigated. In the fixed bed reactor, at 923 K, a $R_{S/C}$ of 7 in the feed and a pressure higher than 100 bar, a product gas with less than 50 ppm of carbon monoxide can be obtained during the pre-breakthrough stage. In the moving bed reactor, by using the optimal feed rates of $1 \, \text{kg} \cdot \text{h}^{-1}$ of calcium oxide and $22.4 \, \text{NL} \cdot \text{h}^{-1}$ of methane at 973 K with a $R_{S/C}$ of 3, the purity of hydrogen can reach 94 mol% with 2.5 mol% of carbon monoxide in the gas phase.

SE-SRE

To date, most of the SERP researches are focused on the use of methane (natural gas) as feedstock, while only a small number of experimental studies have been carried out on SE-SRE.

Among these studies, calcium oxide-based materials have been widely used as carbon dioxide sorbents. Lysikov *et al.* (2008) compared different feedstocks for hydrogen production via SERP over a mixture of a commercial nickel-based catalyst and calcium oxide adsorbent. It was found that ethanol exhibits the best performance among all the feedstocks studied. A 98 vol.% purity of hydrogen-rich gas, with impurities of carbon monoxide and carbon dioxide less than 20 ppm, was obtained. With ethanol as feedstock, He *et al.* (2009) carried out SERP over a mixture of Co–Ni catalyst derived from hydrotalcite-like compounds and calcined dolomite as carbon dioxide sorbent. A good performance was observed at 823 K and the product gas contains 99 mol% hydrogen with only 0.1 mol% of carbon monoxide. Besides, a study on SE-SRE with Ni- and Co-incorporated MCM-41 catalysts, with calcium oxide as carbon dioxide sorbent, has been performed by Gunduz and Dogu (2012), where the yield of hydrogen can reach 94% (dry basis) at 873 K with a $R_{S/C}$ of 1.6 in the feed.

On the other hand, other high temperature carbon dioxide sorbents such as lithium-based materials and HTlc materials can also be used. Iwasaki *et al.* (2007) performed SE-SRE reaction with lithium silicate as carbon dioxide sorbent over 1 wt.% Rh/CeO$_2$ catalyst. A product gas with hydrogen purity 96 mol% was obtained at 823 K, atmospheric pressure and $R_{S/C}$ of 1.5 in the feed. In addition, Essaki *et al.* (2008) reported that the concentration of hydrogen can reach 99 mol% along with carbon monoxide below 0.12 mol% by SE-SRE over a commercial nickel-based catalyst, with lithium silicate as sorbent at similar conditions. Besides, a commercial nickel-based catalyst with an HTlc sorbent was employed by Cunha *et al.* (2012) as a hybrid system in a multi-layered pattern arrangement. The results of SE-SRE at 673 K with a $R_{S/C}$ of 5 in the feed showed that pure hydrogen was produced in the first 10 min of reaction.

Additionally, hybrid materials (Cunha *et al.*, 2013; Wu *et al.*, 2013b) with Cu or Ni as the active metal phase and HTlc sorbent as the support have also been employed for SE-SRE. These materials were found to have good catalytic performance to produce hydrogen from SE-SRE due to their relatively high surface areas and small crystal sizes.

2.3. Material Developments

2.3.1. Catalysts

2.3.1.1. *Active phases and supports*

A variety of catalytic reactions have been used for hydrogen production. Depending on the feedstock used, some selected active phases on supported catalysts have been widely used for hydrogen production as shown in Table 2.2.

Recently, feedstocks such as methanol and ethanol have been proposed for hydrogen production by SR reaction since they can be produced in considerable amounts from renewable sources and contribute to low carbon emission. Besides, SR of glycerol, butanol

Table 2.2   General active phases used for WGS, SR of methane, methanol and ethanol. Adapted from Wu *et al.* (2016).

Group	VIIIB			IB	IIB
	8	9	10	11	12
Active phase on	Fe^4	Co^3	$Ni^{1,2,3}$	$Cu^{2,3,4}$	Zn^2
supported	$Ru^{1,2}$	$Rh^{1,3}$	$Pd^{2,4}$	Ag	Cd
catalysts	Os	$Ir^{2,3}$	$Pt^{1,2,3,4}$	$Au^{2,4}$	Hg

Note: 1-methane; 2-methanol; 3-ethanol; 4-CO (WGS).

and other undesired by-products of bio-fuels has also been proposed as an alternative to reduce wastes and improve economic feasibility of bio-projects (Wang and Cao, 2011). It can be found from Table 2.2 that the most investigated active phases in hydrogen production can be classified as noble and non-noble metals mainly belonging to the VIIIB group. The supported noble metal catalysts with Rh, Ru, Pd, Pt, Re, Au and Ir as active phase have been extensively investigated for the SR of oxygenated hydrocarbons. These noble metal catalysts are found as very active within a wide range of operating conditions, e.g. temperatures from 623 K to 1073 K and space velocities from 5000 h^{-1} up to 300,000 h^{-1}. The excellent catalytic performance of noble metal catalysts might be related with their excellent capability in C–C and C–O bond cleavage (Koehle and Mhadeshwar, 2012). The catalytic activity order of noble metal active phases for SR has been summarized by Jones *et al.* (2008) experimentally and theoretically as Ru > Rh > Ir > Pt, while Wei and Iglesia (2004) showed that the activity of Pt is higher than that of Ir, Rh, and Ru in terms of C–H bond activation. Wang *et al.* (2009) found that Rh- and Ir-based catalysts have promising conversion efficiencies due to the lower decomposition barriers and the superior redox capabilities than other catalytic systems. In addition, noble-metal catalysts usually have very low carbon formation rates because carbon does not dissolve in these systems (Subramani *et al.*, 2009). However, comparing with metal loadings of non-noble metal samples (typically 10–25 wt.%), the supported noble metal catalysts have relatively low metal loadings (0.5–5 wt.%); the extremely high unit price may limit their industrial application in large scale. Besides, the activity of catalysts is rarely found as the limiting factor (Rostrup-Nielsen, 2008) and the limitation for SR reaction rate in conventional reformers is usually not the kinetics but the effectiveness factor, which is typically less than 5% due to severe mass and heat transfer limitations (Adris *et al.*, 1996). As a result, increasing research efforts have been focused on the development of non-noble metal based catalysts, where Cu, Co and especially Ni-based catalyst systems have received plenty of research interest in recent years.

It was found that copper is very effective for dehydrogenation during the SR of oxygenated hydrocarbons due to its ability to maintain the C–C bond (Cavallaro and Freni, 1996). Therefore, catalyst systems of copper in the presence of zinc oxide supported on alumina ($Cu/ZnO/Al_2O_3$) are the most frequently studied in the SR of methanol and WGS (cf. Chapter 3) due to the good selectivity and activity (Yong *et al.*, 2013), and especially low level of CO formation. However, for feedstocks containing C–C bonds, e.g. ethanol, glycerol, butanol, etc., the cleavage of C–C bond is essential for hydrogen production, which also requires much higher reforming temperatures (generally >773 K) comparing with methanol SR ($\sim$523 K) and low-temperature WGS ($\sim$450 K). Unfortunately, Cu-based catalysts always undergo a relative fast deactivation due to the aggregation of copper particles on the surface of the support materials under high temperature conditions.

The support material also has significant effects on the reaction performance of SR, because it is responsible for the active metal dispersion, which affects the sintering resistance; moreover, sometimes the support material can directly participate in the reaction by facilitating the reactants adsorption (Nieva *et al.*, 2014). Co-based catalysts with different supports have been studied by Llorca *et al.* (2002, 2003, 2004), where γ-Al_2O_3, SiO_2, TiO_2, V_2O_5, ZnO, Sm_2O_3, and CeO_2 have been tested as supports. Good activity and selectivity has been reported over the Co/ZnO sample, where nearly 5 mol of hydrogen can be produced per mol of ethanol converted (Llorca *et al.*, 2003). Besides, it was also found that the cobalt phase is very efficient in C–C bond cleavage even at temperatures as low as 673 K (Llorca *et al.*, 2004). But Co-based catalysts are found to have a high tendency towards coke formation (Haryanto *et al.*, 2005), which will lead to deactivation during SR.

As described in Section 2.2.2, SR of methane/natural gas (SMR) is the most commonly used hydrogen production technology. Many different metals are active for SMR (Table 2.2), but the nickel-based catalyst system is widely employed industrially due to the lower cost (Rostrup-Nielsen, 1984). Iulianelli *et al.* (2016) reported in a qualitative percentage distribution of the major catalysts (active phases) studied for SMR reaction that more than 65% of SMR catalysts are nickel-based samples. In commercial large-scale hydrogen production systems from SMR, the fixed bed SR operates with the Ni-based catalyst held on a ceramic pellet substrate bed with a temperature typically at 1073–1373 K and pressure at 30–45 bar. Among supported catalysts with non-noble metal active phase, Ni is much more active for SR than Cu and exhibits better catalytic performance than Co, since it is less prone to coke formation (Pana-giotopoulou *et al.*, 2014). The relatively high C–C bond breaking activity and the low production cost make the nickel-based catalysts suitable for reforming reactions (Auprêtre *et al.*, 2002; Breen *et al.*, 2002; Frusteri *et al.*, 2004). The reaction performance of Ni-based catalysts mainly depends on the support. The typical Al_2O_3 support is acidic and favours hydrocarbon cracking and polymerization, which may lead to coke formation. Therefore, many researches can be found related with alternative supports for Ni-based SMR catalysts. Nieva *et al.* (2014) found that the activity order of Ni-based catalysts with different supports is $Ni–MgAl_2O_4 > ZnAl_2O_4 > Al_2O_3 > SiO_2$, whereas $Ni/ZnAl_2O_4$ shows the highest ability to avoid carbon deposition and metal sintering. Additionally, the use of La_2O_3, ZrO_2 and CeO_2 as promoter or secondary support makes the Ni-catalysts more stable, with better coke resistance and allows higher methane conversion than other aforementioned single sup-ports (Iulianelli *et al.*, 2016). Up to date, Ni-based catalysts have been most extensively investigated among all transition active metals for SR, but the activity, sulphur poisoning, carbon formation and sintering are still regarded as the major challenges (Sehested, 2006).

Numerous developments on WGS catalysts are also taking place to remove/convert the CO content in syngas. Interested readers may refer to Chapter 3 and to some recent reviews for more detailed information (LeValley *et al.*, 2014; Ratnasamy and Wagner, 2009; Smith *et al.*, 2010).

2.3.1.2. *Reaction mechanism*

Numerous studies on the reaction mechanism of hydrogen production from SR with different feedstocks can be found in literature. A brief survey on the reaction mechanisms of SMR and SRE will be given in this section since they are the most widely employed and investigated systems.

Methane is a stable and highly symmetrical molecule. Reports in literature for SMR investigations indicate that the key-step for a successful hydrogen production must be the sufficient supply of activation energy to crack the C–H bonds in the methane molecule. It has also been reported that the theoretical "methane activation" in SMR should occur at temperatures higher than 573 K (Gritsch *et al.*, 2007), due to the high dissociation energy of methane ($\Delta H^{\circ}_{diss} = 435.3 \, kJ \cdot mol^{-1}$). Also the electronic configuration of methane is $1s^2 2s^2 2p^2$, with two unpaired electrons present in two 2p orbitals. The transfer of one electron from the 2s to the 2p orbitals results in four single filled orbital's ($1s^2 \, 2s^1 \, 2p^3$) for bonding. Thus, the arrangement of the four C–H bonds in the methane molecule is the result of a tetrahedral symmetry which minimizes electron repulsion.

As a result, the activation of C–H bond by dissociative adsorption of methane is generally regarded as the rate-determining step (RDS) for the SMR reaction. It is also in agreement with the assumption of first-order dependence on the methane concentration. For instance, Wei and Iglesia (2004) found that the reaction rates are only proportional to the partial pressure of CH_4, and the methane dissociation is kinetically limited by the initial activation of the C–H bonds, and unaffected by the H_2O concentration. This observation also leads to the conclusion that the dissociation of methane is the RDS for SMR.

In addition, Beebe *et al.* (1987) found that the activation energy for dissociative adsorption of methane is related with the crystal surface of nickel, being that CH_4 has a lower adsorption activation energy on Ni(100) than on Ni(110) and Ni(111).

Rostrup-Nielsen *et al.* (2002) assumed that the RDS is the irreversible dissociative adsorption of methane associated with substantial energy and entropy barriers, and requires at least a certain amount of active sites for carbon atoms formation, since methane is a stable and highly symmetrical molecule with low sticking coefficients for adsorption. Wei and Iglesia (2004) postulate that the reactivity of the active metal phase towards C–H bond breaking governs the overall reaction kinetics. It was found that reaction rates are proportional to methane partial pressure, but independent of carbon dioxide and water pressures, which led them to the conclusion of sole kinetic relevance of C–H bond activation steps. The catalyst surface may be uncovered by reactive intermediates, due to the fast steps of the activation of the co-reactant and methane activation by obtaining the chemisorbed carbon intermediates. Further, it is postulated that recombinative desorption steps of H atoms with OH species form hydrogen and water. The current understanding of methane dissociation (Rostrup-Nielsen *et al.*, 2002) leads to the conclusion that the reaction does not proceed via an adsorbed precursor state. It has been found that the adsorbed methyl species may decompose so far that an adsorbed carbon species is obtained (Cunha, 2009).

The first reaction step, dissociative adsorption of CH_4, can be written as:

$$CH_4 + 2^* \leftrightarrow {}^*CH_3 + {}^*H \quad (RDS) \tag{2.9}$$

The $^*CH_3$ species are subsequently dissociated by many possible surface reactions until chemisorbed carbon atoms are formed on the metal surface, as shown in Fig. 2.6(a). Bengaard *et al.* (2002) performed DFT calculations of the activation of methane on Ni(111) and Ni(211) surfaces, which are reproduced in an energy diagram given in Fig. 2.6(b). They proved that the energies of the CH_x species are higher than those for methane. While the adsorbed methyl

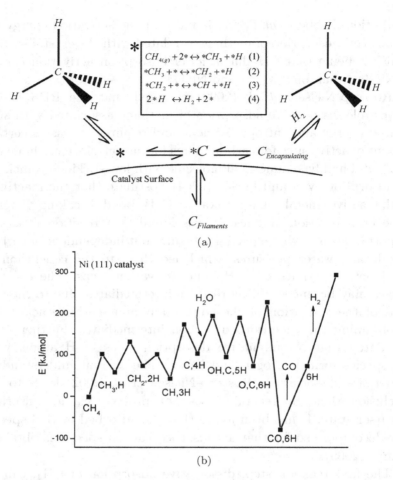

Figure 2.6 The mechanism of methane decomposing (a) and potential energy diagram (b). Reprinted with permission from Wu *et al.* (2014d). Copyright (2016), John Wiley and Sons.

species $-CH_3$ show a similar energy to $-CH_2$ and $-CH$ species, the energies of the chemisorbed carbon $^*C$ species are lower than for CH_x species and CH_4, with higher activation energy for decomposition of CH_x species to $^*C$ than to the hydrogenation of CH_x species to methane.

Afterwards, CO species can be produced from the surface reaction of $^*C$ with adsorbed O obtained from the dissociation of water. Therefore, the RDSs of the SMR reaction can be described as follows:

$$^*CH_3 + ^* \leftrightarrow {}^*CH_2 + {}^*H \tag{2.10}$$

$$^*CH_2 + ^* \leftrightarrow {}^*CH + {}^*H \tag{2.11}$$

$$^*CH + ^* \leftrightarrow {}^*C + {}^*H \tag{2.12}$$

$$H_2O + 2^* \leftrightarrow {}^*OH + {}^*H \tag{2.13}$$

$$^*OH + ^* \leftrightarrow {}^*O + {}^*H \tag{2.14}$$

$$^*C + {}^*O \leftrightarrow {}^*CO + {}^* \tag{2.15}$$

$$^*CO \leftrightarrow CO + {}^* \tag{2.16}$$

$$2^*H \leftrightarrow H_2 + 2^* \tag{2.17}$$

According to Wu *et al.* (2014d), if water is introduced in excess, the adsorbed $^*CH_3$ species may react with the adsorbed hydroxyl species or even intermolecular oxygen species, thus forming adsorbed CO and hydrogen:

$$^*CH_3 + {}^*OH \leftrightarrow {}^*CH_2OH + {}^*H \tag{2.18}$$

$$^*CH_2OH + ^* \leftrightarrow {}^*CH_2O + {}^*H \tag{2.19}$$

$$^*CH_2O + ^* \leftrightarrow {}^*CHO + {}^*H \tag{2.20}$$

$$^*CHO + ^* \leftrightarrow {}^*CO + {}^*H \tag{2.21}$$

On the other hand, the reaction system for steam reforming of ethanol (SRE) itself is much more complex than that for SMR. Depending on the reaction conditions and active phase used, different reaction pathways and species can be obtained during SRE, as given in Fig. 2.7.

In a first step, ethanol can go through dehydrogenation or dehydration reactions. Dehydration reaction is favoured when the catalytic system shows acidic features resulting in the formation of ethane, which is a precursor for coke formation through polymerization. On the other hand, the formation of acetaldehyde is preferred through dehydrogenation reaction when a basic catalyst support has been used. Acetaldehyde can decompose into CH_4 and

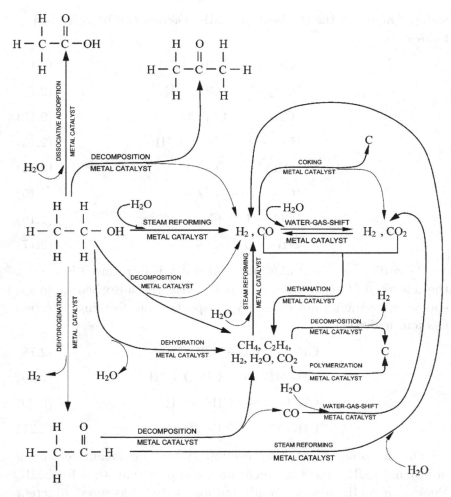

Figure 2.7 An overall reaction scheme involved for SRE. Reprinted with permission from Haryanto *et al.* (2005). Copyright (2016), American Chemical Society.

CO. The aforementioned two reactions occur in low temperature ranges, but it is also believed that ethanol can directly crack into H_2, CO and CH_4 according to Fig. 2.7. Afterwards, the WGS reaction takes place to convert CO into CO_2 and produce H_2 at the same time. However, there is a possibility that a latent WGS occurs, where acetaldehyde reacts directly with water to produce CO_2, H_2 and CH_4.

Table 2.3 Possible reactions during SRE. Adapted from Wu *et al.* (2016).

Reactions	Chemical description	ΔH^0_{298K} [kJ $\cdot$ mol^{-1}]
Dehydration	$CH_3CH_2OH \leftrightarrow C_2H_4 + H_2O$	+45.4
Dehydrogenation	$CH_3CH_2OH \leftrightarrow CH_3CHO + H_2$	+68.4
Acetaldehyde Decomposition	$CH_3CHO \leftrightarrow CH_4 + CO$	−18.8
Direct ethanol Decomposition	$CH_3CH_2OH \leftrightarrow H_2 + CH_4 + CO$	+49.7
WGS	$CO + H_2O \leftrightarrow CO_2 + H_2$	−41.1
Latent WGS	$CH_3CHO + H_2O \leftrightarrow CO_2 + H_2 + CH_4$	−59.9
Steam methane reforming	$CH_4 + H_2O \leftrightarrow CO + 3H_2$	+205.9

The remaining CH_4 has a critical aspect in SRE, as it limits the production of H_2, at low and intermediate temperatures. The SMR is required for methane conversion, following the mechanism discussed above in this section. The general reactions that can occur during SRE are summarized in Table 2.3.

The mechanism of ethanol decomposition on the Ni(111) was investigated by Gates *et al.* (1986). The sequence of bond scission steps occur as ethanol undergoes dissociative reactions. Bond activation occurs in the descending order –O–H, $-CH_2$, –C–C–, $-CH_3$. The products observed are CH_3CHO, CH_4, CO, H_2 and carbon adsorbed on the nickel surface. The latter species dissolves into the nickel bulk. Acetaldehyde and methane desorb, and the formation of both of these products is controlled by scission of the methylene (CH_2 group). The CH_3 group is cleaved from the intermediate surface CH_3– CHO species to form $*CH_3$. Similarly as the ethanol decomposition mechanism, the adsorption of ethanol is suggested to occur on the active metal phase during the SRE according to the study from Bshish *et al.* (2011), while water is chemisorbed on the support. A comprehensive review on SRE was given by Mattos *et al.* (2012), where experimental data from infrared spectroscopy during SRE has been analyzed. It is found that the dissociation of ethanol to ethoxy species can occur on the support, and the favoured supports should be sufficiently basic to inhibit acid-catalyzed dehydration of ethanol to ethylene, which leads to coke formation by polymerization reaction. The dehydrogenation of the ethoxy species into acetaldehyde is catalyzed by the active metal phases, and the presence of labile O adsorbed atoms or hydroxyl groups located on the support may also assist this step. Acetate can be obtained from the oxidization (nucleophilic attack) of acetaldehyde species with oxygen provided by the support. Finally, the metal particles are involved in the conversion of acetate to CH_4 and CO_x species, where steam can promote this process. As a result, the authors (Mattos *et al.*, 2012) believed that both the active phase and the support have critical effects on the reaction mechanism of SRE.

2.3.1.3. *Reforming kinetics*

Early work on kinetic studies of SMR were available in the literature during the 1950s. Afterwards, Bodrov *et al.* (1964, 1967, 1968) performed several comprehensive investigations on SMR kinetics with nickel-based catalysts. The following expression was proposed to describe SMR reaction rate, where methane adsorption was assumed as the RDS:

$$r_{SMR} = \frac{k_{SMR} \cdot p_{CH_4}}{1 + a \cdot p_{H_2O} \cdot p_{H_2}^{-1} + b \cdot p_{CO}} \qquad (2.22)$$

Allen *et al.* (1975) also carried out SMR experiments, using a commercial nickel catalyst and concluded that the RDS was the desorption of products. Later, De Deken *et al.* (1982) determined the SMR kinetics on a Ni/Al_2O_3 catalyst by adding hydrogen into the feed stream. The kinetic model was composed of two expressions that accounted for the formation of CO and CO_2.

The aforementioned expressions of SMR kinetics and the study on SMR, methanation and WGS with nickel-based catalyst from Xu and Froment (1989) were compared by Elnashaie *et al.* (1990). The most critical difference was the dependence for the partial pressure of steam: negative (Bodrov *et al.*, 1964, 1967, 1968), positive (De Deken *et al.*, 1982) and positive or negative depending on the temperature condition (Xu and Froment, 1989). The conclusion was that the work of Xu and Froment (1989) is the most general kinetic model, and the model has been extensively employed and tested till now.

According to the reaction kinetics of SMR using a nickel-based catalyst from Xu and Froment (1989), based on the Langmuir–Hinshelwood expression, the rate equations can be summarized as:

$$R_1 = \left(\frac{k_1}{P_{H_2}^{2.5}} \right) \frac{\left(P_{CH_4} P_{H_2O} - \frac{P_{CO} P_{H_2}^3}{K_{eq1}} \right)}{DEN^2} \tag{2.23}$$

$$R_2 = \left(\frac{k_2}{P_{H_2}^{3.5}} \right) \frac{\left(P_{CH_4} P_{H_2O}^2 - \frac{P_{CO_2} P_{H_2}^4}{K_{eq2}} \right)}{DEN^2} \tag{2.24}$$

$$R_3 = \left(\frac{k_3}{P_{H_2}} \right) \frac{\left(P_{CO} P_{H_2O} - \frac{P_{CO_2} P_{H_2}}{K_{eq3}} \right)}{DEN^2} \tag{2.25}$$

$$DEN = 1 + K_{CO} P_{CO} + K_{H_2} P_{H_2}$$
$$+ K_{CH_4} P_{CH_4} + K_{H_2O} \left(\frac{P_{H_2O}}{P_{H_2}} \right) \tag{2.26}$$

where $P_i (i = CH_4, H_2O, H_2, CO, CO_2)$ are the partial pressures of species i, R_j $(j = 1, 2, 3)$ are the reaction rates of the reaction $1(CH_4 + H_2O \leftrightarrow CO + 3H_2)$, $2(CH_4 + 2H_2O \leftrightarrow CO_2 + 4H_2)$ and $3(CO + H_2O \leftrightarrow CO_2 + H_2)$, respectively. k_j are the reaction rate constants defined by Arrhenius equation $k_j = k_{0j} \cdot \exp(-E_j/RT)$, K_{eq_j} $(j = 1, 2, 3)$ are the equilibrium constants for reactions 1, 2 and 3, respectively. K_i are the adsorption constants of species i defined as $K_i = K_{0i} \cdot \exp(-\Delta H_i/RT)$. The reaction between adsorbed species was assumed to be the RDS, and the activation energies for the three reactions were found to be $240.1 \, kJ \cdot mol^{-1}$, $243.9 \, kJ \cdot mol^{-1}$ and $67.13 \, kJ \cdot mol^{-1}$, respectively.

Researches on SMR later developed similar models based on different approaches (Power rate law, Langmuir–Hinshelwood, Temkin identity and other microkinetic analysis) to describe the kinetics of SMR. A brief survey of the reported SMR kinetic studies has been performed by Angeli *et al.* (2014) as shown in Table 2.4. We may find that the activation energies vary from one report to another; the reasons for these variations can be explained with the use of different catalysts (active phases, supports, particle sizes, morphologies, catalyst activation/pre-treatments) and/or operating conditions (temperature ranges, feed ratios, etc.).

The kinetic behaviour of the SRE reaction has also been extensively studied over various catalysts in recent years to describe the reaction rate by using a power law rate expression or expressions based on certain mechanistic assumptions. Similar as the kinetic studies on SMR, these SRE kinetic reports show significant differences. Kinetic analysis can be generally classified into two groups, according to the type of kinetic expressions used for analysis, the Power rate law expressions derived from data fitting or the use of a Langmuir–Hinshelwood and/or Eley–Rideal kinetic model.

On one hand, many of the SRE kinetics measured are given by power law expressions without considering the mechanism of elementary steps. Data from kinetic measurements of the SRE reaction are normally fitted into the following power law expression:

$$-r_{\text{SRE}} = k_0 \exp\left(\frac{-E_A}{RT}\right) p_{EtOH}^{\alpha} \cdot p_{H_2O}^{\beta} \qquad (2.27)$$

where the rate constant k_0, activation energy E_A, and reaction orders α and β vary between different studies. Power rate laws provide little information about the underlying processes. A survey of the SRE kinetics researches based on the power rate law expressions can be found in Table 2.5. However, one should recall that the power rate law kinetic model used in one study might not be suitable for other types of catalysts under different reaction conditions. Therefore, a better understanding of the chemical processes involved should be employed for the development of kinetic models. The establishment of a kinetic model with elementary steps requires a deep knowledge of the reactions involved in the overall SRE process. Kinetic models normally use the Langmuir–Hinshelwood (or sometimes Eley–Rideal)

Table 2.4   Kinetic expressions for steam methane reforming. Adapted from Angeli *et al.* (2014).

Catalyst	$T(K)$	Reaction rate expression	Kinetic parameters	RDS	Reference
Ni/MgAl$_2$O$_4$	773–943	$R_1 = \left(\dfrac{k_1}{P_{H_2}^{2.5}}\right)\dfrac{\left(P_{CH_4}P_{H_2O} - P_{CO}P_{H_2}^3/K_{eq1}\right)}{DEN^2}$ $R_2 = \left(\dfrac{k_2}{P_{H_2}^{3.5}}\right)\dfrac{\left(P_{CH_4}P_{H_2O}^2 - P_{CO_2}P_{H_2}^4/K_{eq2}\right)}{DEN^2}$ $R_3 = \left(\dfrac{k_3}{P_{H_2}}\right)\dfrac{\left(P_{CO}P_{H_2O} - P_{CO_2}P_{H_2}/K_{eq3}\right)}{DEN^2}$ $DEN = 1 + K_{CO}P_{CO} + K_{H_2}P_{H_2} +$ $K_{CH_4}P_{CH_4} + K_{H_2O}P_{H_2O}/P_{H_2}$	$k_{01} = 1.16 \times 10^{16}$ mol $\cdot$ kPa$^{0.5} \cdot$ kg$_{cat}^{-1} \cdot$ s^{-1}, $E_1 = 240.1\,$kJ $\cdot$ mol^{-1} $k_{02} = 2.79 \times 10^{15}$ mol $\cdot$ kPa$^{0.5}$ kg$_{cat}^{-1} \cdot$ s^{-1}, $E_2 = 243.9\,$kJ $\cdot$ mol^{-1} $k_{03} = 5.41 \times 10^{6}$ mol $\cdot$ kPa$^{-1} \cdot$ kg$_{cat}^{-1} \cdot$ s^{-1}, $E_3 = 67.13\,$kJ $\cdot$ mol^{-1}	1. Formation of CO from CH$_4$ 2. Formation of CO$_2$ from CH$_4$ 3. Formation of CO$_2$ from CO via WGS	Xu and Froment (1989)
Ni/MgO	823–1023	$R = kP_{CH_4}$	$k_0 = 5.5 \times 10^{3}$ mol $\cdot$ kPa$^{-1} \cdot$ s^{-1} $E = 102\,$kJ $\cdot$ mol^{-1}	Dissociative methane adsorption	Wei and Iglesia (2004)
Ni/Al$_2$O$_3$	733–890	Same as Xu and Froment (1989)	$k_{01} = 5.79 \times 10^{13}$ mol $\cdot$ kPa$^{0.5} \cdot$ kg$_{cat}^{-1} \cdot$ s^{-1} $E_1 = 217.01\,$kJ $\cdot$ mol^{-1}		

(Continued)

Sorption Enhanced Reaction Processes

Table 2.4 *(Continued)*

Catalyst	$T(K)$	Reaction rate expression	Kinetic parameters	RDS	Reference
			$k_{02} = 1.29 \times 10^{14}$ mol $\cdot$ kPa$^{0.5} \cdot$ kg$_{cat}^{-1} \cdot$ s^{-1} $E_2 = 215.84$ kJ $\cdot$ mol^{-1} $k_{03} = 9.33 \times 10^4$ mol $\cdot$ kPa$^{-1} \cdot$ kg$_{cat}^{-1} \cdot$ s^{-1}, $E_3 = 68.20$ kJ $\cdot$ mol^{-1}	Same as Xu and Froment (1989)	Oliveira *et al.* (2009)
Ni/CaAl$_2$O$_4$	748–823	$R_1 = \left(\dfrac{k_1}{P_{H_2}^{2.5}}\right)\left(P_{CO}P_{H_2O} - P_{CO_2}P_{H_2}/K_1\right)/DEN^2$ $R_2 = \left(\dfrac{k_2}{P_{H_2}^{3.5}}\right)\left(P_{CH_4}P_{H_2O}^2 - P_{CO_2}P_{H_2}^4/K_1 K_2\right)/DEN^2$ $DEN = 1 + K_{CO}P_{CO} + K_{H_2O}P_{H_2O}/P_{H_2}$	$k_{01} = 1.49 \times 10^5$ mol $\cdot$ kPa$^{0.5} \cdot$ kg$_{cat}^{-1} \cdot$ s^{-1} $k_{02} = 4.3 \times 10^{12}$ mol $\cdot$ kPa$^{0.5}$ kg$_{cat}^{-1}$ s^{-1},	$E_1 = 32.6$ kJ mol^{-1}, $E_2 = 186$ kJ mol^{-1}	Same as Xu and Froment (1989)	Soliman *et al.* (1992)
Rh/Al$_2$O$_3$	673, 773, 873	$R_{SR} = \dfrac{k_{55}C_{CH_4}}{\left(1 + \frac{k_{56}}{k_{57}}\sqrt{\frac{k_1}{k_2}C_{H_2}}\right)\left(1 + \sqrt{\frac{k_1}{k_2}}C_{H_2} + \frac{k_{19}}{k_{20}}C_{CO}\right)^2}$ $\times (1 - \eta_{SR})$		Methane activation	Maestri *et al.* (2008)

Rh/Ce$_\alpha$ Zr$_{1-\alpha}$ O$_2$	748–848	$$R_{WGS} = \dfrac{k_7 \frac{k_{13}}{k_{14}} C_{H_2O}}{\left(1 + \sqrt{\frac{k_1}{k_2}C_{H_2} + \frac{k_{19}}{k_{20}}C_{CO}}\right)^2}(1 - \eta_{WGS})$$ ηi is the equilibrium factor of reaction i $$R_1 = \dfrac{k_1}{P_{H_2}^{2.5}}\left(P_{CH_4}P_{H_2O} - \dfrac{P_{H_2}^3 P_{CO}}{K_I}\right) \times \Omega_I \Omega_S$$ $$R_2 = \dfrac{k_2}{P_{H_2}}\left(P_{CO}P_{H_2O} - \dfrac{P_{H_2}P_{CO_2}}{K_{II}}\right) \times \Omega_I \Omega_S$$ $$R_3 = \dfrac{k_3}{P_{H_2}^{3.5}}\left(P_{CH_4}P_{H_2O}^2 - \dfrac{P_{H_2}^4 P_{CO_2}}{K_{III}}\right) \times \Omega_I \Omega_S$$ $$\Omega_I = \dfrac{1}{1 + k_{CH_4}P_{CH_4}/P_{H_2}^{0.5} + k_{CO}P_{CO} + k_{CO_2}P_{CO_2}+k_{H_2}P_{H_2}}$$ $$\Omega_S = \dfrac{1}{1 + k_{H_2O}P_{H_2O}/P_{H_2} + k_{H_2}P_{H_2}}$$	Redox surface reactions	Halabi *et al.* (2010)
Rh-based	773	$R_{overall} = k \cdot P_{CH_4}P_{H_2O}^0$	Dissociative methane adsorption	Ligthart *et al.* (2011)

Table 2.5   The kinetics of steam reforming of ethanol based on the power rate law expressions reported by Wu et al. (2014d).

Expression	E_A (kJ·mol^{-1})	Active metal	Support/ Precursor	T (K)	$R_{S/C}$ (mol/mol)	Reference
$r = k \cdot C_{EtOH}$	7.0	Ni (20.6 wt.%)	Y$_2$O$_3$	403	0	Sun et al. (2005, 2008)
	5.9	after heat treated at 773 K				
	16.9	Ni (16.1 wt.%)	Al$_2$O$_3$			
	1.9	Ni (15.3 wt.%)	La$_2$O$_3$			
$r = k \cdot p_{EtOH}^{2.52} \cdot p_{H_2O}^{7.0}$	n.a.	Ni	Al$_2$O$_3$	673	2.15	Therdthianwong et al. (2001)
$r = k \cdot p_{EtOH}^{0.8}$	144	Ni (43.6 wt.%)	Ni–Al–LDH	823–923	2.75	Mas et al. (2008)
$r = k \cdot C_{EtOH}$	149	Raney-Type 68.9 wt.% Ni- 28.2 wt.% Cu- 2.9 wt.% Al$_2$O$_3$		523–573	0.5	Morgenstern and Fornango (2005)
$r = k \cdot p_{EtOH}^{0.711} \cdot p_{H_2O}^{2.71}$	23	Ni (25 wt.%)	MgO-Al$_2$O$_3$	673–873	1.5–9	Mathure et al. (2007)

$r = k \cdot N_{EtOH}^{0.43}$	4.4	Ni (15 wt.%)	Al_2O_3	593–793	2.15	Akande et al. (2006)
$r = k \cdot p_{EtOH}$	96	Ru (5 wt.%)	Al_2O_3	873–973	5	Vaidya and Rodrigues (2006)
$r = k \cdot N_{EtOH}^{3.64}$	51	Ni (20.6 wt.%)	n/a	673–863	2.15	Akpan et al. (2007)
$r = k \cdot C_{EtOH}^{\alpha}, \alpha = 0.2/1.2$	85	Rh-Pt (~4 wt.%)	monolith (BASF: SR10D)	837–973	1.5–5	Simson et al. (2009)
$r = k \cdot p_{EtOH}^{\alpha} \cdot p_{H_2O}^{\beta}, \alpha = 0.5 \pm 0.05, \beta = 0 \pm 0.015$	18.4	Pt (1-5 wt.%)	CeO_2	573–723	0.75–3	Ciambelli et al. (2010)

mechanism. It assumes that all reaction species are adsorbed on the catalyst surface before they take part in network reactions. However, the mechanisms proposed are based on assumptions and simplifications (Chorkendorff and Niemantsverdriet, 2003), such as RDS, dominating species on surface, quasi-equilibrium, steady state, etc.

Four RDSs are proposed for the SRE by Akande *et al.* (2006) based on an Eley–Rideal mechanism: adsorption of ethanol on active sites, dissociative adsorption of ethanol, surface reaction of the adsorbed oxygenated hydrocarbon fraction with no adsorbed steam, surface reaction of the adsorbed hydrocarbon fraction with non-adsorbed steam. The kinetics obtained was found to have a better fitting and description of reactions involved in SRE than the work from Mathure *et al.* (2007), where they used acquired experimental data from SRE to fit them in the power law kinetic model with a deviation of 10.2%. One mechanistic kinetic model based on the Langmuir–Hinshelwood approach was proposed by Sahoo *et al.* (2007). The surface reactions of SRE, WGS and ethanol decomposition have been considered as the RDS in the mechanism. The fitted data indicate that the formation of acetaldehyde from ethoxy is the RDS for reforming, and the correlation coefficient R^2 value for all cases was above 0.95. Mas *et al.* (2008) reported kinetic experiments on SRE with the formation of H_2, CO, CO_2 and traces of CH_4. Experiments with different methane concentrations showed that ethanol conversion decreases as methane concentrations increase, revealing the existence of competitiveness between both ethanol and methane for adsorption on the same active site. It was reported by Görke *et al.* (2009) that the RDS steps in the model were CO_2 desorption, dissociative ethanol adsorption and reaction of adsorbed methane with steam from the gas phase for WGS, ethanol decomposition and SMR, respectively. Since a micro reactor has been used, neither heat nor mass transfer limitations were detected.

In a kinetic study of SRE by Llera *et al.* (2012), CO was found as a final product and they found that both CO_2 and CH_4 are intermediate products. A kinetic model involving two reaction steps where ethanol reacts irreversibly, and CH_4 and CO_2 can react reversibly, has been proposed. They postulate that the surface reactions are the RDS steps, and that the dissociative adsorption of methane is a reversible step in quasi-equilibrium, as well as all remaining reactions. 18 reaction steps with 4 RDS steps were used making the model quite complex. The model parameters have shown acceptable fit with a goodness of R^2 higher than 0.95.

In order to have a comprehensive investigation of SRE behaviour, Wu *et al.* (2014d) proposed a simplified kinetics for SRE on a nickel-based catalyst in a large temperature range (473–873 K), where ethanol reacts with a nickel site directly converting into adsorbed acetaldehyde and releases molecular hydrogen simultaneously in the first step. The adsorbed acetaldehyde is released from the free active nickel site, or it further reacts to form methane and adsorbed *CO simultaneously. Afterwards, CO can be released, while the produced methane and two free active sites can go through the reversible dissociative adsorption reaction, which is the RDS. The kinetic model is based on the reaction pathway shown in Fig. 2.8 marked with bold line. Both the power rate law and Langmuir–Hinshelwood

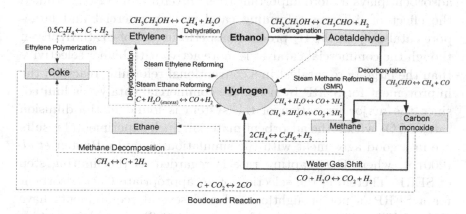

Figure 2.8 The reaction pathway used in the kinetic study for SRE. Reprinted with permission from Wu *et al.* (2014d). Copyright (2016), John Wiley and Sons.

kinetic models have been employed and found able to describe the catalytic SRE process in a fixed bed reactor over a large temperature range; the activation energy for SRE reaction was found to be $31.8\,kJ \cdot mol^{-1}$.

2.3.2. High temperature CO_2 adsorbents

2.3.2.1. *Candidates*

Carbon dioxide is known as a GHG, which is present in the atmosphere and absorbs a part of the thermal radiation reflected from the surface of the earth. The global CO_2 concentration increased from about 280 ppm to 380 ppm in 150 years, since the industrial revolution. To avoid irreversible changes in global climate caused by the CO_2 concentration increase, innovative materials and processes have been developed for CO_2 capture. Gas-solid adsorption is one of the most promising strategies for post-combustion CO_2 capture applications (Wang *et al.*, 2012a). On the other hand, CO_2 adsorbents can be used to enhance the hydrogen production performance of the conventional reforming processes, which can be regarded as an attractive pre-combustion CO_2 capture technology. Comparing with the effect of a catalyst on the performance of SERP, the adsorbent plays a more important role. Oliveira *et al.* (2011) studied the effect of different reforming catalysts (commercial and large-pore catalysts) on the H_2 production performance of SE-SMR. Even though the commercial catalyst is more active with higher selectivity than the large-pore catalyst in a conventional reforming reaction, the improvement for SERP by using the commercial catalyst is limited, since the major limitation of the system is found in the diffusion rate of CO_2 within the adsorbent material. Such experimental results are in a good agreement with the simulation study by Rusten *et al.* (2007b), where CO_2 capture rate is regarded as the limiting step of SERP. Therefore, the selection of an appropriate CO_2 adsorbent for a SERP is not straightforward and several requirements have to be fulfilled. According to Yong *et al.* (2002), several important aspects should be taken into account in order to find an appropriate adsorbent for SERP:

(1) high selectivity and adsorption capacity for CO_2 at high temperatures;

(2) high adsorption/regeneration rates for CO_2 at the operating conditions;

(3) stable adsorption capacity of CO_2 after repeated adsorption/regeneration cycles;

(4) adequate mechanical strength of adsorbent particles after repeated adsorption/regeneration cycles.

In fact, SERP requires regeneration of the adsorbent after saturation with CO_2. Regeneration can be carried out by temperature swing and/or pressure swing depending on the nature of adsorbent and energy efficiency.

Depending on the interaction energy between the adsorbate and the surface of the adsorbent, adsorption can be categorized into physical adsorption (physisorption) or chemical adsorption (chemisorption). A brief comparison of the features for physisorption and chemisorption can be found in Table 2.6. Due to the weak bonds formed in physisorption by van der Waals and coulombic forces, a physisorbent has the advantage of being easily regenerated. Whereas

Table 2.6 Basic features of physical adsorption and chemical adsorption. Adapted from Ruthven (1984).

Property	Physical adsorption	Chemical adsorption
Selectivity	Non-specific	Highly specific
Effects of pressure and temperature	Favoured at high pressure and low temperature	Possible over a wide range of pressures and temperature conditions
Kinetics	Rapid	May be slow
Desorption	Reversible	Irreversible/reversible
Heat of adsorption	Low heat of adsorption (usually $<40\,\mathrm{kJ \cdot mol^{-1}}$)	High heat of adsorption (chemical bond, tens to hundreds $\mathrm{kJ \cdot mol^{-1}}$)
Adsorbate–adsorbent interaction	Monolayer or multilayer	Monolayer only, dissociation may be involved

chemisorption is a chemical reaction and involves electron transfer, chemisorbents are useable at high temperatures, but have the disadvantage that regeneration requires relatively large amounts of energy.

Adsorbent materials used for CO_2 adsorption under low temperature conditions (<473 K) by physisorption are generally the physisorbents, e.g. activated carbons, zeolites, alumina and amine-based adsorbents. These physisorbents are sensitive to temperature changes and have relatively low CO_2 selectivity due to the weak physical adsorption force (Yang, 2003). However, SERP are usually carried out under intermediate- (473–673 K) and high-temperature (>673 K) conditions. Clearly, conventional physisorbents cannot be used for CO_2 adsorption under operation conditions employed for SERP due to the limitations discussed above. As a result, chemisorbents, which have much higher selectivity towards CO_2, relatively higher adsorption capacities at intermediate- and high-temperature conditions comparing with physisorbents, become ideal candidates for CO_2 adsorption in SERP (Wang *et al.*, 2011a). The general properties of the most widely used CO_2 adsorbents for SERP applications, calcium oxide-based, lithium-based and hydrotalcite-like materials, can be found in Table 2.7.

2.3.2.2. *Calcium oxide-based materials*

Calcium oxide-based materials are readily available in several combinations on earth. The most important ones are the limestone, which is a naturally occurring mineral consisting mainly of $CaCO_3$, and the dolomite, which is a combination of calcium and magnesium

Table 2.7 A comparison of the general properties of calcium oxide, lithium-based and hydrotalcite-like materials for SERP applications.

Material	Capacity	Stability	Kinetics
CaO-based	Good ($\sim2 - \sim17\,\text{mol} \cdot \text{kg}^{-1}$)	Poor	Fair
Li_2ZrO_3-based	Fair ($\sim2 - 8\,\text{mol} \cdot \text{kg}^{-1}$)	Good	Poor
Hydrotalcite-like (HTlc)	Poor ($\sim1\,\text{mol} \cdot \text{kg}^{-1}$)	Good	Good

carbonates, $CaMg(CO_3)_2$. As a result, the use of calcium oxide is environmental and economically attractive since calcium-based materials such as limestone and dolomite are abundant and low cost (Abanades *et al.*, 2007). Besides, due to the high sorption capacities at high temperatures, calcium oxide has been extensively employed for the selective adsorption of CO_2 in SERP by the carbonation reaction:

$$CaO(s) + CO_2(g) \rightarrow CaCO_3(s) \quad (\Delta H^0_{873K} = -171.3\,kJ \cdot mol^{-1})$$

$$(2.28)$$

In addition to the theoretical high adsorption capacity of $17.8\,mol \cdot kg^{-1}$, acceptable carbonation (adsorption) rates in the range of $0.08-0.4\,mmol \cdot s^{-1}$ can be achieved at temperatures higher than 773 K, while the decarbonation reaction (regeneration) at elevated temperatures ($>1073\,K$) will take place following the reverse carbonation reaction, releasing the adsorbed CO_2 and regenerating the calcium-based adsorbent materials.

The carbonation reaction of CaO (Eq. (2.28)) is characterized by a rapid initial reaction rate (surface reaction-controlled step) followed by a second stage of product layer diffusion-controlled process to a quite slow adsorption rate (Albrecht *et al.*, 2008). This change in reaction rate can be attributed to the formation of a layer of carbonate around the adsorbent so that the mass transfer resistance increases for the diffusion of CO_2 to reach the internal CaO particle for further carbonation (Bhatia and Perlmutter, 1983). An illustration of the $CaCO_3$ production through carbonation reaction of CaO in the presence of water vapour can be found in Fig. 2.9.

Cyclic studies suggested that the thickness of the $CaCO_3$ layer formed in the carbonation cycle is 22 nm (Barker, 1973). The $CaCO_3$ layer also influences the activation energy, which is initially low but increases with conversion. The relatively low adsorption rate in the latter step (layer diffusion-controlled process) has a very limited effect on the enhancement of hydrogen production during SERP. Therefore, the initial fast step is considered to be critical in determining the reactor size, conversion and hydrogen purity (Cobden *et al.*, 2009).

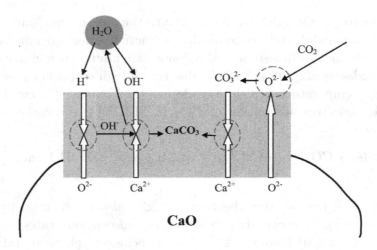

Figure 2.9 CaO carbonation reaction with steam. Reprinted with permission from Dou *et al.* (2016). Copyright (2016), Elsevier.

Calcium oxide-based adsorbents are also significantly deactivated by the high temperature cyclic carbonation/calcination treatment. The loss of adsorption capacity for calcium oxide-based adsorbents during cyclic carbonation/calcination was attributed mainly to pore blockage (Alvarez and Abanades, 2005) and sintering (Kuramoto *et al.*, 2003) and was accompanied by textural and surface area changes (Abanades and Alvarez, 2003). Abanades *et al.* (2004) found that the conversion of calcium oxide decreased sharply from 70% in the first cycle to 20% after 10 cycles when tested in a fluidized bed, which is believed to be related with the formation of a bottleneck in the limestone pore network (Alvarez and Abanades, 2005). To describe the adsorption capacity evolution of calcium oxide during extended cyclic adsorption/regeneration operations, Abanades (2002) fitted several series of experimental data to one equation:

$$X_N = f^{N+1} + b \qquad (2.29)$$

where X_N is the conversion of calcium oxide to calcium carbonate at the Nth cycle, with the constants $f = 0.782$ and $b = 0.174$.

A direct consequence of these deactivation problems is that both capacity and kinetics of CO_2 are respectively smaller and slower as the number of calcination/carbonation cycles increases, which is not desirable for cyclic operation. Therefore, the deactivation is a major concern against the application of calcium oxide-based adsorbents in SERP for hydrogen production. It was found by Wang *et al.* (2013) that steam can be introduced during the adsorption process to prevent the sintering of adsorbents and facilitate the diffusion of CO_2 within the adsorbent. When dolomite was used, the rate of sintering was less pronounced due to the excess pore volume created by decomposition of $MgCO_3$ (Kuramoto *et al.*, 2003). The work from Barker (1973) showed that if 10 nm CaO particles were used, the conversion was maintained at 93% over 30 cycles at 902 K. Analysis based on simulation results for SE-SMR showed that calcium oxide must maintain at least 15% of its thermodynamic capacity in the cyclic operation to avoid large loss in the system efficiency (Cobden *et al.*, 2009). The other major drawback associated with calcium oxide based materials is the high temperatures required for regeneration. A large temperature gap between adsorption and regeneration (TSA) may increase the energy consumption.

2.3.2.3. *Lithium-based adsorbents*

Lithium-based adsorbents are another kind of adsorbent material for selective CO_2 removal at high temperatures. The interest in lithium salts as CO_2 high temperature adsorbents can be seen from the international patents registered by Toshiba corporation (Kato *et al.*, 2003a; Nakagawa *et al.*, 2002) and by the NTNU group in Norway (Ronning *et al.*, 2006). Among different lithium salts, lithium zirconate (Li_2ZrO_3) was first found able to adsorb CO_2 at intermediate and high temperatures with a adsorption capacity of $4.54 \, mol \cdot kg^{-1}$ for CO_2 (Nakagawa and Ohashi, 1998) by using potassium carbonate promoted lithium zirconate. The carbonation reaction of lithium zirconate is:

$$Li_2ZrO_3(s) + CO_2(g) \leftrightarrow Li_2CO_3(s) + ZrO_2(s)$$

$$(\Delta H^0_{873K} = -81.0 \, kJ \cdot mol^{-1}) \tag{2.30}$$

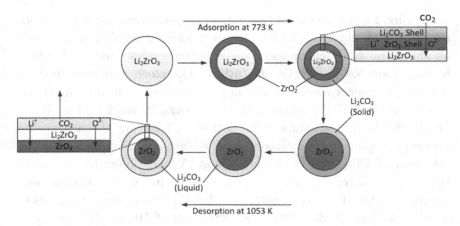

Figure 2.10 Mechanism for CO_2 adsorption and desorption on lithium zirconate. Adapted from Ida and Lin (2003). Copyright (2016), with permission from American Chemical Society.

It was found that only little CO_2 can be adsorbed when using the pure lithium zirconate, and the sorption kinetics is extremely slow, approximately 250 min being required to achieve the maximum adsorption capacity (Xiong *et al.*, 2003). A double-shell mechanism for CO_2 adsorption/desorption on lithium zirconate was proposed by Ida and Lin (2003) to explain such behaviour, as demonstrated in Fig. 2.10.

It can be found that during the sorption process, CO_2 diffuses to the surface of Li_2ZrO_3 and reacts with Li^+ and O^{2-} on the surface to form lithium oxide and lithium carbonate nuclei. When lithium oxide nuclei grow to form a shell covering unreacted Li_2ZrO_3, the sorption rate decreases because Li^+ and O^{2-} have to diffuse through the lithium oxide shell to react with CO_2. Worse still, lithium carbonate nuclei form another shell outside of lithium oxide shell. The lithium carbonate shell can also decrease the adsorption rate, because CO_2 molecules have to diffuse through this shell before reaction. It was found that the diffusion of CO_2 in solid lithium carbonate is the rate limiting step for CO_2 adsorption over Li_2ZrO_3 materials (Ida and Lin, 2003).

By doping lithium zirconate with potassium promoter, the sorption kinetics can be improved by 40 times when compared to the pure sample, as the CO_2 diffuses faster to reach the lithium zirconate (Nakagawa and Ohashi, 1998). Besides, sodium carbonates and oxides have also been reported to show promoting effects on Li_2ZrO_3 (Kato *et al.*, 2003a), so that the adsorption capacity of lithium zirconate can be increased up to $5.0 \, mol \cdot kg^{-1}$ at 673 K with a promising adsorption kinetics (saturated within 10 min) (Ochoa-Fernandez *et al.*, 2005). The improvement in the characteristics of the material was credited to a new preparation method that was able to produce high purity lithium zirconate crystals (Ochoa-Fernández *et al.*, 2006). Regeneration of the new material was achieved at 923 K.

In addition to Li_2ZrO_3, lithium orthosilicate (Li_4SiO_4) also shows promising results in the adsorption of CO_2. The adsorption is achieved by a carbonation reaction:

$$Li_4SiO_4(s) + CO_2(g) \leftrightarrow Li_2CO_3(s) + Li_2SiO_3(s)$$

$$\Delta H_{298K} = -142.0 \, kJ \cdot mol^{-1} \qquad (2.31)$$

Lithium orthosilicate is cheaper than lithium zirconate and has a sorption rate 30 times faster than pure lithium zirconate at 773 K with a CO_2 concentration of 20%, and the sorption capacity can reach $6.13 \, mol \cdot kg^{-1}$ (Kato *et al.*, 2005). Regeneration can be performed at temperatures above 973 K, which is much lower than that required for CaO. Another feature of Li_4SiO_4 is the capacity to remove CO_2 even from diluted streams. It was reported that Li_4SiO_4 can adsorb up to $6.14 \, mol \cdot kg^{-1}$ of CO_2 from the diluted stream ($\sim$4%) in 2 h (Seggiani *et al.*, 2011). Bretado *et al.* (2005) proposed the impregnated suspension method to prepare Li_4SiO_4 with higher purity, and the new material obtained showed a CO_2 adsorption capacity of $8.2 \, mol \cdot kg^{-1}$. When Li_4SiO_4 crystals were prepared into a pellet, mass transfer limitations and sintering were observed. Essaki *et al.* (2006) reported a decrease of 25% in the capacity at 873 K between the powder and pellet samples. Kato *et al.* (2005) reported a 27% decrease in the capacity of the pellets during cyclic tests. The decrease in the capacity could be reduced by doping the orthosilicate with 5% Li_2ZrO_3.

Besides Li_2ZrO_3 and Li_4SiO_4, Kato *et al.* (2005) examined CO_2 adsorption performances over several different lithium-based adsorbents including lithium ferrite ($LiFeO_2$), lithium nickel oxide ($LiNiO_2$), lithium titanate (Li_2TiO_3) and lithium methasilicate (Li_2SiO_3), etc. Despite of the large number of lithium salts investigated, Li_4SiO_4 was found to have the highest adsorption capacity ($6.13\,mol \cdot kg^{-1}$). The lithium-based materials have great potential due to their excellent CO_2 sorption capacity as well as stability after many adsorption/regeneration cycles (Ochoa-Fernández *et al.*, 2006). However, the main obstacle for the practical application for SERP of lithium-based adsorbents is still the kinetic limitation (Wang *et al.*, 2011a). Although Li_4SiO_4 can achieve higher conversion and production capacity at lower steam-to-carbon ratio ($R_{S/C}$) compared to that of Li_2ZrO_3 as CO_2 adsorbent, the CO_2 adsorption kinetics of Li_4SiO_4 is found to be the limiting step for SERP (Rusten *et al.*, 2007b). Besides, similar double-shell adsorption mechanisms have also been found by Zhang *et al.* (2013) for the lithium silicate material and high temperature ($\sim973\,K$) is required for desorption. To improve the adsorption kinetics, some authors (Wang *et al.*, 2014) synthesized several Li_2ZrO_3-based adsorbents by the solid-state reaction method from mixtures of Li_2CO_3, K_2CO_3 and ZrO_2 with different compositions. They found that the Li_2CO_3/K_2CO_3-doped Li_2ZrO_3 adsorbent is able to achieve excellent CO_2 uptake capability and presented the maximum adsorption rate at $798\,K$. Recently, the use of LiOH is found able to decrease the synthesis temperature of Li_4SiO_4 to as low as $873\,K$ and undesirable small crystallites can be avoided, which leads to the formation of macroporous framework (Kim *et al.*, 2015). As a result, CO_2 adsorption rate ($56.1\,mg \cdot g^{-1} \cdot min^{-1}$) has been improved significantly comparing with the conventional Li_4SiO_4($<3\,mg \cdot g^{-1} \cdot min^{-1}$).

2.3.2.4. *Layered double hydroxides/hydrotalcite-like compounds*

Layered double hydroxides (LDHs), also known as hydrotalcite-like compounds (HTlc), are widely used as a kind of chemisorbent, which belong to a large class of anionic clay minerals having the general formula:

$$\left[M_{1-x}^{II} M_x^{III} \left(OH \right)_2 \right] \left[A^{n-} \right]_{x/n} \cdot z H_2 O \qquad (2.32)$$

where M^{II} and M^{III} are divalent (Mg^{2+}, Mn^{2+}, Fe^{2+}, Co^{2+}, Ni^{2+}, Cu^{2+}, Zn^{2+}, etc.) and trivalent (Al^{3+}, Mn^{3+}, Fe^{3+}, etc.) metal ions, respectively, and A^{n-} is the anion (CO_3^{2-}, Cl^-, SO_4^{2-}, etc.). The divalent and trivalent ions form a crystalline structure that is positively charged due to the presence of the trivalent ion, and x is usually between 0.17 and 0.33 (Ulibarri *et al.*, 2001). Besides, the excess of positive charge can be balanced by the anions and water molecules within the interlayer space between the two positive layers (Ding and Alpay, 2001; Ram Reddy *et al.*, 2006). Due to the diversity of compositions and the unique characteristics related, they have been widely used as catalysts, precursors, ion exchangers, etc. for both gas phase and liquid phase applications (Auerbach *et al.*, 2004). In the natural mineral of HTlc, M^{II} is magnesium (Mg) and M^{III} is aluminium (Al), and the compensating anions are CO_3^{2-} and OH^- with a formula as $Mg_6Al_2(OH)_{16}CO_3 \cdot 4H_2O$. The CO_2 can be adsorbed in the form of the carbonate ion (CO_3^{2-}) in the interstitial layer (San Román *et al.*, 2008; Yong and Rodrigues, 2002). A representation of the structure of a hydrotalcite can be seen in Fig. 2.11.

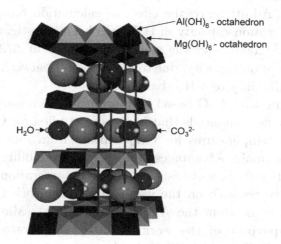

Figure 2.11 Structure of hydrotalcite. Reprinted with permission from Reijers *et al.* (2005). Copyright (2016) American Chemical Society.

However, the fresh HTlc materials usually do not have a promising adsorption performance due to their relatively low basicity, which is partially caused by the presence of adsorbed water that hinders the access of CO_2 to the basic sites located on the surface and on the interstitial layer. Therefore, HTlc often need to be activated by a thermal treatment before being used, the calcination process being performed between 473 K and 823 K in a nitrogen atmosphere in most cases. Upon thermal treatment, the interlayer spacing of the hydrotalcite decreases due to the loss of adsorbed water at around 473 K, but the layered structure remains. The material will be completely dehydrated and partially dehydroxylated after heating to 673 K, and the CO_3^{2-} located in the interlayer is decomposed. Such obtained material can be referred as layered double oxides (LDO) (Hutson *et al.*, 2004). Comparing with the original HTlc, LDO have an amorphous 3D structure with higher surface area and pore volume (Hutson *et al.*, 2004). It has been found that samples calcinated at temperatures <773 K can be transformed back to the original layered structure by contacting with a carbonate solution. The phenomenon is the so-called "memory effect" of HTlc materials (Reijers *et al.*, 2005). However, the structure of HTlc may change irreversibly and form a spinel phase when heated to temperatures above 773 K. Additionally, the effect of calcination temperature on the CO_2 adsorption capacity of HTlc has been investigated by Ram Reddy *et al.* (2006), and the sample calcinated at 673 K had the highest adsorption capacity, due to the trade-off between the surface area and availability of active basic sites.

Comparing with CaO-based high temperature adsorbents, HTlc are relatively new materials that have been studied for CO_2 adsorption at high temperatures for CO_2 capture from flue gas and for SERP. Their main advantages are the high stability, promising kinetics and resistance to steam in cyclic operation. The first comprehensive research on the use of HTlc to adsorb CO_2 at high temperature came from the Air Products (USA) (Mayorga *et al.*, 1997). The purpose of the work was to demonstrate a SE-SMR process with HTlc as the CO_2 adsorbent, and pressure swing was used to regenerate the saturated HTlc. The adsorbent was modified by using potassium promoter and the CO_2 equilibrium capacity was determined at three different temperatures (573 K, 673 K and 773 K). It was found that the capacity decreased with increasing temperature from around $1 \, mol \cdot kg^{-1}$ at 573 K to $0.6 \, mol \cdot kg^{-1}$ at 773 K in the presence of steam ($p_{CO_2} = 152 \, kPa$). Additionally, after 15 adsorption/desorption cycles ($T = 673 \, K, p_{CO_2} = 30 \, kPa$), the K-promoted HTlc was found to have a working capacity of $0.45 \, mol \cdot kg^{-1}$.

Afterwards, a CO_2 adsorption capacity of $0.62 \, mol \cdot kg^{-1}$ at $p_{CO_2} = 44 \, kPa$ and 673 K in the presence of steam was reported using K-promoted HTlc (Ding and Alpay, 2000a). The adsorption capacity can be maintained after treatment with steam, which indicated that HTlc materials are suitable for cyclic operation in PSA. The Langmuir model was used to describe the adsorption equilibrium data at 673 K and 753 K and the linear driving force (LDF) model was employed to describe the adsorption kinetics. However, they found that the desorption kinetics could not be described by the simple LDF model. As a result, the model was modified by calculating the effective diffusivity from the mass balance to the extrudate, assuming that pore diffusion is dominant, and the modified LDF was able to give a good fit of the experimental results. Simulation results indicated that it is important to understand the mass transfer kinetics as well as the mechanism of adsorption of the HTlc adsorbents. Ebner *et al.* (2006) conducted a series of non-equilibrium experiments in K-promoted HTlc and proposed a new mechanism combining three coupled terms: temperature-dependent parameters, reactions with different kinetics and CO_2 adsorption capacities. In addition, the same group (Ebner *et al.*, 2007) later developed a non-equilibrium kinetic model to describe the adsorption mechanism based on three reversible reactions and four species, as shown in Fig. 2.12.

Generally speaking, the interaction between the adsorbed CO_2 and the basic sites of HTlc is stronger than for a zeolite but weaker than for alkali metal oxide (Wang *et al.*, 2011a). As a result, the adsorption (for zeolite <473 K) and regeneration (for CaO > 1073 K) process can be carried out at intermediate-temperatures (473−773 K)

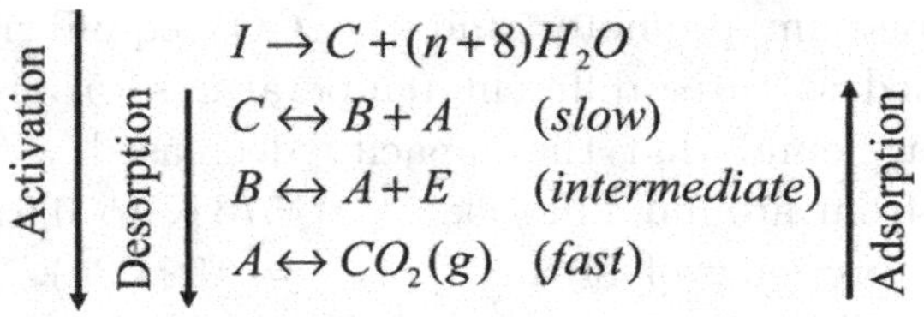
Activation
Desorption
I → C + (n+8)H2O
C ↔ B + A (slow)
B ↔ A + E (intermediate)
A ↔ CO2(g) (fast)
Adsorption

$$I \equiv [Mg_3Al(OH)_8]_2(CO_3)\cdot K_2CO_3 \cdot nH_2O$$
$$A \equiv CO_2(ad)$$
$$E \equiv Mg_6Al_2K_2O_{10}$$
$$B \equiv Mg_6Al_2K_2O_9(CO_3)$$
$$C \equiv Mg_6Al_2K_2O_8(CO_3)_2$$

Figure 2.12 Reaction pathway of CO_2 adsorption/desorption in a K-promoted HTlc. Reprinted with permission from Ebner *et al.* (2007). Copyright (2016), American Chemical Society.

for SERP with HTlc as the adsorbent. Many studies can be found regarding the SERP with HTlc materials as selective CO_2 adsorbents. It was found that HTlc doped with K_2CO_3 is able to have a promising adsorption performance within a temperature range 573–773 K, and the CO concentrations in the product gas from SR reaction can be restrained within a ppm level (Hufton *et al.*, 1999a; Xiu *et al.*, 2002b). But the relatively low CO_2 adsorption capacity of HTlc becomes the biggest obstacle for SERP in industrial applications. Intense efforts have been devoted to improve the adsorption capacity of HTlc for CO_2 in recent years, mainly including the following aspects: synthesis conditions, operating conditions (temperature/pressure/steam presence), particle sizes and alkali promoter doping, organic anions intercalation as well as activated carbon/carbon nanotube/carbon nanofiber support. A brief survey on the adsorption performance of HTlc materials can be found in Table 2.8.

2.3.2.5. *Other candidates*

Magnesium oxide (MgO) materials have a moderate CO_2 adsorption capacity by the formation of carbonates from ambient

Table 2.8 Adsorption performance of CO_2 on HTlc materials from literature.

Material	Adsorption temperature (K)	p_{CO_2} (bar)	Regeneration temperature (K)	Cycles	Capacity (mol. kg^{-1})	References
HTlc($Mg_6.Al_2(CO_3)-(OH)_{16} \cdot 4H_2O$)	573	1 (dry)			0.25	
	673	0.153 (dry)			1.18	
	673	0.153 (humid)			1.07	Ficicilar and Dogu (2006)
	773	0.161 (dry)	n.a.	n.a.	0.97	
	773	0.161 (humid)			0.87	
	800	0.155 (dry)			0.80	
	800	0.155 (humid)			0.66	
K-HTlc (PURALOX MG70)	673	5.6 (humid)	673	1400	0.66	van Selow et al. (2009)
HTlc($Mg-Al-CO_3$)	473	~1 (dry)	673	4	0.48	Ram Reddy et al. (2006)
HTlc (EXM696)	573	1 (humid)	n.a.	n.a.	0.52	Yong et al. (2000a)
HTlc($Ca_{0.75}Al_{0.25}(OH)_2-(CO_3)_{0.125}$)	603	1 (dry)	n.a.	n.a.	0.90	Hutson and Attwood (2008)
HTlc($Mg_{0.75}Al_{0.25}(OH)_2-(ClO_4)_{0.25}$)					1.03	
K-HTlc (MG78, 30.9 wt.% K_2CO_3)	723	0.70 (dry)	723	~6000	0.45	Mayorga et al. (1997)
K-HTlc (commercial)	673	0.45 (dry)	673		0.63	Ding and Alpay (2000a)
	673	0.44 (humid)		4	0.65	
	753	0.58 (dry)	753		0.52	
	753	0.55 (humid)			0.58	
K-HTlc (commercial)	481	1 (humid)	n.a.	n.a.	0.90	Ding and Alpay (2001)
	575				0.80	

(Continued)

Table 2.8　(*Continued*)

Material	Adsorption temperature [K]	p_{CO_2} [bar]	Regeneration temperature [K]	Cycles	Capacity [mol·kg^{-1}]	References
K-HTlc (commercial)	673	0.5 (dry)			0.36	Lee et al. (2007b)
		1 (dry)			0.79	
		3 (dry)			0.88	
	793	0.5 (dry)	n.a.	n.a.	0.30	
		1 (dry)			0.50	
		3 (dry)			0.59	
K-HTlc-[[Mg$_3$Al(OH)$_8$]$_2$ CO$_3$ · K$_2$CO$_3$ · nH$_2$O]	523	0.20 (dry)	n.a.	n.a.	0.11	Ebner et al. (2006)
	723	0.93 (dry)			0.55	
	773	1.13 (dry)			0.46	
	673	1 (dry)	673	4	0.60	Ebner et al. (2007)
K-HTlc (MG70, 22 wt.% K$_2$CO$_3$)	673	0.05 (humid)	673	20	0.51	Reijers et al. (2005)
K-HTlc (MG30, 22 wt.% K$_2$CO$_3$)	673	0.05 (humid)	673	20	0.62	
K-HTlc (MG30, 20 wt.% K$_2$CO$_3$)	673	0.4 (humid)	673	75	0.71	Oliveira et al. (2008)
Cs-HTlc (MG30, 20 wt.% Cs$_2$CO$_3$)	673	0.4 (humid)			0.41	
K-Na-HTlc (18.5 wt.% K$_2$CO$_3$-1.5 wt.% Na$_2$CO$_3$)	573	0.4 (dry)	573	5	1.21	Martunus et al. (2011)
	673		673		1.04	

Sorbent						Reference
K-HTlc (MG70, 20 wt.% K_2CO_3)	653	0.05 (dry)	653	4	0.15	Du et al. (2010)
K-HTlc (Mg_3Al_1-CO_3, 20 wt.% K_2CO_3)	773	1.01 (dry)	773	n.a.	0.20	Wang et al. (2011b)
		0.05 (dry)			0.22	
		1.01 (dry)			0.40	
		1.01 (dry)			0.85	
HTlc (MG30)	673	1.01 (dry)	n.a.	n.a.	0.098	Lee et al. (2010)
K-HTlc (MG30, 26 wt.% K_2CO_3)					0.94	
HTlc (MG50)	673	1.01 (dry)	n.a.		0.091	
K-HTlc (MG50, 26 wt.% K_2CO_3)					0.89	
HTlc (MG70)	673				0.13	
K-HTlc (MG70, 26 wt.% K_2CO_3)					0.68	
K-HTlc (MG61, 22 wt.% K_2CO_3)	673	0.85 (humid)	673	30	0.89	Halabi et al. (2012a)
Mg_3Al_1-stearate Hydrotalcite	573	0.2 (dry)	n.a.	n.a.	1.25	Wang et al. (2012b)
K-HTlc (MG30, 20 wt.% KNO_3)	473	0.2 (dry)	673	6	1.18	Wu et al. (2013a)
	656	0.5 (humid)	656 and 708	10	1.13	
Ga-K-HTlc (HTC-10Ga-20K)	573	1.08 (dry)	573	0-5	1.82	Miguel et al. (2014)

temperature to $\sim$773 K. Comparing with CaO-based adsorbents, the major advantage of using MgO as CO_2 adsorbent for SERP is the relatively low energy requirement for regeneration. However, practical applications of MgO for SERP are limited. Similar as the CaO-based adsorbents, the $MgCO_3$ may form an impermeable layer around the unreacted MgO particle during the carbonation process hindering the diffusion of CO_2 molecules. Furthermore, poor thermal stability of MgO during cyclic adsorption/regeneration is usually reported. As a result, highly durable, reactive, and mechanically strong MgO adsorbent candidates are required to minimize attrition losses during the SERP cyclic operation (Dou *et al.*, 2016).

Double salts are solid adsorbents containing MgO and alkali metal salts. Double salts were generally tested in the temperature range of 573–773 K. At 673 K and $p_{CO_2} = 70$ kPa, these adsorbents showed a CO_2 adsorption capacity above 1 mol $\cdot$ kg^{-1} (Mayorga *et al.*, 2001), but further research on the kinetics and stability of double salts is required (Reijers *et al.*, 2005).

Basic alumina has been tested for the adsorption of CO_2 at temperatures up to 573 K. The commercial basic alumina sample (98AX316) showed a CO_2 adsorption capacity $>$0.3 mol $\cdot$ kg^{-1} at a CO_2 partial pressure of 100 kPa at 573 K (Yong *et al.*, 2000b). By using K-promoted γ-alumina adsorbent, Li *et al.* (2014) found that the CO_2 adsorption capacity can be improved to 0.67 mol $\cdot$ kg^{-1} at 573 K with $p_{CO_2} = 40$ kPa in the presence of steam.

Generally, the adsorbents mentioned as CO_2 uptake candidates for SERP reported up to date either suffer from low adsorption capacities, slow adsorption rates or poor adsorption/regeneration cyclic stability. There is still much to do in the development of promising CO_2 adsorbents for SERP. More detailed information regarding the high temperature CO_2 adsorbents developed for SERP applications can be found from some recent reviews (Dou *et al.*, 2016; Shokrollahi Yancheshmeh *et al.*, 2016).

2.3.3. Multifunctional materials

In some cases, alternated layers of catalyst and adsorbent configurations have been employed for SERP by packing the adsorptive reactor with layers of catalyst and adsorbent sequentially. Consequently, the column becomes a combination of several reforming reactors and CO_2 adsorbers in series. Such arrangement could be useful when there is a need to remove and replace the deactivated catalyst and/or adsorbent for external regeneration purposes (Agar, 2005). On the other hand, a uniform mixture of the catalyst and adsorbent materials is widely used in most SERP studies. The effects of pairs of catalyst–adsorbent (well-mixed and layered configurations) on the overall performance of hydrogen production from SE-SMR have been investigated by Rawadieh and Gomes (2009) numerically. They found that even though the CO_2 concentration at the outlet stream from the well-mixed catalyst–adsorbent column was higher than that from layered system, the H_2 purity can be enhanced as the number of layers increases under cyclic steady state conditions, and well-mixed configuration was able to produce the hydrogen product with highest purity and the lowest CO content. A similar behaviour was also reported by other authors (Lu *et al.*, 1993), where both adsorptive reactor systems have been simulated. However, due to the differences from the particle size, shape and density of both materials, it is very difficult to obtain such an ideal well-mixed catalyst-adsorbent system in practice.

Alternatively, by incorporating reforming catalyst (as active phase) into high temperature CO_2 adsorbent material, which is also known as bifunctional catalyst or multifunctional catalyst or hybrid material in various publications, researchers have found promising H_2 production from SERP. In addition to the advantage of homogenous catalyst/adsorbent mixture, gradients (both temperature and concentration) can be eliminated, while the mass and heat transfer can be enhanced in microscale for SERP comparing with the conventional well-mixed configuration. Besides, Agar (2005) believed that the development of hybrid materials may also provide an additional degree of freedom in the design of adsorptive reactors to further improve the performance for hydrogen production by SERP. The illustration of possible benefits by using multifunctional materials for SERP can be found in Fig. 2.13.

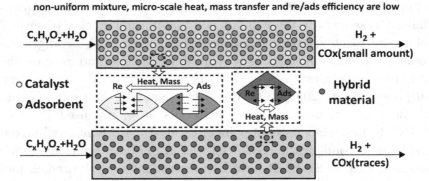

Figure 2.13 A comparison between the well-mixed catalyst–adsorbent system and the multifunctional material for SERP.

The first study on H_2 production from SERP with multifunctional material was reported by Satrio *et al.* (2005) where they prepared a "core-in-shell" type pellet with a lime/dolomite core and a porous shell made of Ni/Al_2O_3. The H_2 product obtained at 873 K with methane as the feedstock had a high purity (96%, dry basis) with a yield of 95%. Afterwards, a series of $NiO–CaO–Ca_{12}Al_{14}O_{33}$ multifunctional materials with different Ca: Al: Ni ratios have been synthesized by Martavaltzi and Lemonidou (2010) for SE-SMR. The hybrid sample with 16% of Ni content can produce a stream rich in H_2 (>90%) during the pre-breakthrough period (>100 min). Besides, Ni/CaO hybrid material was prepared by Chanburanasiri *et al.* (2011) through the impregnation method, but the Ni crystallites were found to have a non-uniform distribution on the adsorbent support. Wu *et al.* (2012a) used a HTlc as the precursor of the $Ni/CaO–Al_2O_3$ multifunctional materials where Ni metal active phase appeared as spherical particles embedded in the surrounding CaO/Al_2O_3 matrix. As a result, better Ni dispersion can be achieved on CO_2 adsorbent. Besides, Feng *et al.* (2012) studied the effect of La_2O_3 promoter contents on the H_2 production performance from SE-SMR with $Ni/CaO/Al_2O_3$ hybrid samples; they found that the Ni particle and surface area can be kept stable after several 873–1073 K temperature cycles because the formation of $NiAl_2O_4$ can be inhibited by La_2O_3.

In addition to $Ni/CaO/Al_2O_3$-based hybrid materials for SERP, a series of multifunctional materials have been developed for SERP with ethanol as the feedstock at LSRE (U. Porto, Portugal) with Cu, Ni, and Cu–Ni alloy metals as the active phase(s), while HTlc or K-promoted HTlc CO_2 adsorbents were used as support. For K–Ni–Cu–HTlc hybrid sample, it was found that Cu can preferentially catalyze ethanol dehydrogenation and WGS reactions, while Ni favours acetaldehyde decomposition and SMR (Cunha *et al.*, 2014). Consequently, a high-purity hydrogen stream (99.8 mol%, dry basis) can be obtained during the pre-breakthrough period at 773 K with a $R_{S/C} = 5$ in the feed. However, the specific surface area of these hybrid samples decreases compared with the original HTlc CO_2 adsorbent (MG30), which leads to a reduction on the CO_2 adsorption capacity. Recently, Dewoolkar and Vaidya (2015) prepared two multifunctional materials, Ni–HTlc and Ni–CaO/Al_2O_3, and compared the performances of these samples for SERP with methane as the feedstock. Both multifunctional materials have shown longer breakthrough times than powdered mixtures of commercial Ni/Al_2O_3 catalyst and respective adsorbent (HTlc or CaO). Additionally, Ni-HTlc has lower adsorption capacity ($1.1\,\text{mol}\cdot\text{kg}^{-1}$) than Ni-CaO/$Al_2O_3$ ($12.3\,\text{mol}\cdot\text{kg}^{-1}$) but high cyclic stability and ease of regeneration, 673 K for Ni-HTlc and 1173 K for Ni-CaO/Al_2O_3.

2.3.4. Pellet modelling

Transient 1D reactor models with axial dispersion were widely used to simulate the SERP within the adsorptive reactor. For the modelling of catalyst and adsorbent pellets, a two-particle heterogeneous model can be used to describe the mass and heat balances within catalyst and adsorbent materials, and a set of boundary conditions at the particle surface and particle centre are required to solve the equations. For instance, a two-particle heterogeneous model in spherical coordinates developed for the Ni-based methane reforming catalyst and K-HTlc CO_2 adsorbent by Oliveira *et al.* (2011) can be found in Table 2.9. The reaction kinetics of the catalyst can be found from Xu and Froment (1989),

Table 2.9 Mathematical model for catalyst and adsorbent pellets employed by Oliveira *et al.* (2011).

Catalyst

Mass balance

$$\varepsilon_{p,\,\text{cat}}\frac{\partial C_{\text{cat},\,i}}{\partial t} = \varepsilon_{p,\,\text{cat}}D_{p,\,\text{cat}}\left(\frac{\partial^2 C_{\text{cat},\,i}}{\partial r_{\text{cat}}^2} + \frac{1}{R_{\text{cat}}}\frac{\partial C_{\text{cat},\,i}}{\partial r_{\text{cat}}} + \frac{\partial^2 C_{\text{cat},\,i}}{\partial z_{\text{cat}}^2}\right)$$

$$- v_{\text{cat}}\frac{\partial C_{\text{cat},\,i}}{\partial z_{\text{cat}}} + \rho_{\text{cat}}\sum_{j=1}^{3} v_{j,\,i}R_j \tag{2.33}$$

Heat balance

$$[\varepsilon_p \hat{C}_{vg}\bar{C}_{T,\text{cat}} + (1 - \varepsilon_{p,cat})\rho_{\text{solid,cat}}\hat{C}_{ps,\text{cat}}]\frac{\partial T_{\text{cat}}}{\partial t}$$

$$= -v_{\text{cat}}\hat{C}_{\text{pg,cat}}\bar{C}_{T,\text{cat}}\frac{\partial T_{\text{cat}}}{\partial z_{\text{cat}}} + \lambda_{\text{cat}}\frac{\partial^2 T_{\text{cat}}}{\partial z_{\text{cat}}^2} + a_{p,\text{cat}}h_{f,\text{cat}}(T - T_{\text{cat}})$$

$$+ \varepsilon_p R T_s \frac{\partial \bar{C}_{t,\text{cat}}}{\partial t} + \rho_{\text{cat}}\sum_{j=1}^{3} R_j(-\Delta H_j) \tag{2.34}$$

Adsorbent

Mass balance to the macropores of the adsorbent (LDF)

$$\frac{\partial \bar{C}_{\text{sorb},\,i}}{\partial t} = \frac{15 D_{p,\,\text{sorb}}}{R_{\text{sorb}}^2}\frac{Bi}{Bi+1}(C_i - \bar{C}_{\text{sorb},i}) - \frac{\rho_{\text{sorb}}}{\varepsilon_{p,\,\text{sorb}}}\frac{\partial \bar{q}_i}{\partial t} \tag{2.35}$$

Mass balance to the micropores of the adsorbent (LDF)

$$\frac{\partial q_{\text{CO}_2}}{\partial t} = k_{\text{CO}_2}(q_{\text{eq, CO}_2} - \bar{q}_{\text{CO}_2}) \tag{2.36}$$

Heat balance to the adsorbent extrudate

$$\left[\varepsilon_{p,\text{sorb}}\hat{C}_{vg}\bar{C}_{T,\text{sorb}} + (1 - \varepsilon_{p,\text{sorb}})\rho_{\text{solid,sorb}}\hat{C}_{ps,\text{sorb}} + (1 - \varepsilon_{p,sorb})\frac{\rho_{\text{sorb}}}{1 - \varepsilon_{p,\text{sorb}}}\right.$$

$$\left.\times \left(\sum_{j=1}^{5}\bar{q}_i\right)\hat{C}_{vg}\right](1 - \varepsilon_c)\frac{\partial T_{\text{sorb}}}{\partial t} = (1 - \varepsilon_c)a_{p,\text{sorb}}h_{f,\text{sorb}}(T - T_{\text{sorb}})$$

$$+ (1 - \varepsilon_c)\varepsilon_{p,\text{sorb}}R_g T_{\text{sorb}}\frac{\partial \bar{C}_{T,\text{sorb}}}{\partial t}$$

$$+ (1 - \varepsilon_c)\rho_{\text{sorb}}\sum_{j=1}^{5}\frac{\partial \bar{q}_i}{\partial t}(\beta(-\Delta H_1) + (1 - \beta)E_2) \tag{2.37}$$

---

and the adsorption equilibrium isotherms of CO_2 for the adsorbent can be found from Oliveira *et al.* (2008). Generally, the mass transfer within the pellet can be described by several diffusion mechanisms and equations: local equilibrium (LEQ), linear driving force (LDF), surface diffusion, and pore diffusion. The first two mechanisms are achieved by using certain assumptions to simplify the mass transfer processes and the need to solve the mass balance at the particle scale can be eliminated, which is able to save the computation time to solve the equations to a great extent. Ding and Alpay (2000b) compared the first two mechanisms (LEQ and LDF) for the description of CO_2 adsorption/desorption process over the K-HTlc material; they found that the LDF model based on pore diffusion, and accounting for the nonlinearity of the isotherm can give a good prediction. The last two mechanisms lead to more rigorous models that take into account mass balance at both scales (interstitial fluid and adsorbent particle), but the computation time required to simulate the cyclic SERP operation may increase a lot compared with the simulation with the first two approaches.

On the other hand, numerical studies on the comparison of H_2 production performance with combined catalyst/adsorbent hybrid pellet (multifunctional material) and with conventional catalyst-adsorbent two-pellet design have also been carried out (Solsvik and Jakobsen, 2011, 2012). To elucidate the effectiveness factor with different mathematical modelling assumptions, both parallel pore and random pore models have been used for effective diffusivities calculation. According to the simulation results, the performance of combined pellet is found to be more promising compared to the conventional mixed catalyst and adsorbent configuration, due to the relatively high effectiveness factor values associated with the hybrid materials. This is in a good agreement with the numerical investigation performed by others (Rout and Jakobsen, 2013). Besides, Lawrence and Grünewald (2006) suggested that the transport limitations within hybrid material might be exploited to enhance the process performance by optimal distribution of the functionalities within a particle based on the results obtained by Morbidelli *et al.* (1982). Consequently, a recent study (Lugo and Wilhite, 2016)

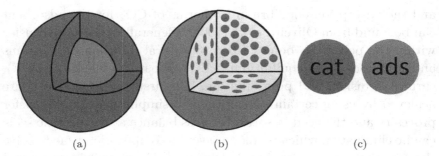

(a)   (b)   (c)

Figure 2.14 Schematic of catalyst–adsorbent configurations: (a) core–shell design; (b) uniform-distributed design; and (c) two-pellet design. Adapted from Lugo and Wilhite (2016). Copyright (2016), Elsevier.

exploited several of the most popular adsorbent–catalyst designs by numerical simulation for SE-WGS applications; the different catalyst–adsorbent configurations can be found in Fig. 2.14.

Lugo and Wilhite (2016) found that the one-pellet designs (multifunctional material) always have greater adsorbent utilization than a two-pellet design. The comparison between the two multifunctional designs shows that the uniform-distributed design is recommended over core–shell for better adsorbent utilization. Finally, the core–shell design could be helpful in an adiabatic SE-WGS reactor, because hot-spot temperature can be reduced by ~40 K compared to the uniform-distributed.

As a result, further numerical studies on the pellet modelling and simulations are required, as they are proving to be more and more important for the improvement of SERP performance at the particle level.

2.4. Reactor Design

The essential of SERP is the packing of adsorbent and catalyst into a reactor, which design will significantly affect the mass and heat transfer processes, and consequently the hydrogen production performance. Therefore, researches related with the adsorptive reactor design, including experimental studies of SERP with fixed bed reactors, experimental studies with fluidized bed reactors, mathematical modelling and simulations regarding the configuration of adsorbent and catalyst, operating conditions, etc. will be addressed along this section.

2.4.1. Experimental studies with fixed bed reactors

A comprehensive research on SERP with methane as the feedstock in a fixed bed reactor was performed by Balasubramanian *et al.* (1999) using commercial Ni-based catalyst and CaO-based material as the CO_2 adsorbent. The adsorbent material was prepared from $CaCO_3$ with a high purity (99.97%) by carrying out calcination at 1023 K, under nitrogen atmosphere for 4 h. The obtained CaO particles with a diameter from 45 to 210 μm were used in the tests. The NiO content of the reforming catalyst was $\sim$22 wt.%, and the catalyst particles used were crushed and sieved to 150 μm. The production distribution at the reactor outlet stream as a function of reaction time can be found in Fig. 2.15, being the reactor packed with 6.56 g of CaO and 7.0 g of catalyst at 923 K, 15 atm, $R_{S/C} = 4$ and with 70% of N_2 in the feed.

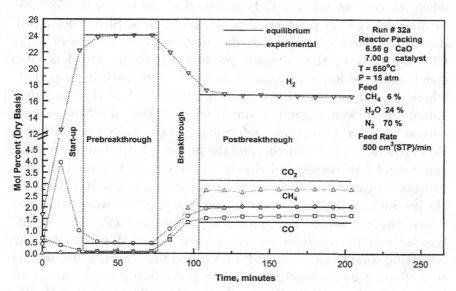

Figure 2.15   Reactor response curves during SERP. Reprinted with permission from Balasubramanian *et al.* (1999). Copyright (2016), Elsevier.

The complete reactor response curves during SERP can be divided into four periods: start-up, pre-breakthrough, breakthrough and post-breakthrough. The unsteady state start-up period can be related with two factors. On one hand, the dead volume in the reactor system causes a delay between opening the valve to feed reactive gas and the time that product gas needs to reach the sampling valve. On the other hand, in this particular example the reforming catalyst was not pre-reduced and the H_2 produced may be consumed to reduce NiO to Ni to produce active phases. During the pre-breakthrough period the SMR, WGS, and CO_2 adsorption process occurred at maximum efficiency. Therefore, H_2 content was maximum and the undesired products were minimum during this period. According to the experimental data, we may find that simultaneous reaction equilibrium can be closely approached as the experimental hydrogen content of ~23.9 mol% was in a good consistency to the equilibrium value of 24.1 mol% from the thermodynamic calculation indicated by the respective horizontal lines in the pre-breakthrough. The breakthrough period started when the concentration of CO_2 increased at the exit of the column. Meanwhile, the H_2 production performance from SERP decreased due to the lower CO_2 removal efficiency, and the contents of CH_4, CO and CO_2 in the obtained product increased. The following steady-state period is designated as the post-breakthrough period, where the CO_2 adsorbent became saturated and the CO_2 adsorption enhancement was negligibly small, so that the SE-SMR process turned into the conventional reforming process. Only the SMR and WGS reactions were occurring and the product distributions from the experiment test correspond closely to the equilibrium of reforming process indicated by the respective horizontal lines. The H_2 content in the product gas during the post-breakthrough period was 16.6 mol%, which was 30% lower than that during the pre-breakthrough period. Similar reactor response curves can be obtained when different operating conditions are tested, by varying the temperature, steam-to-methane ratio, acceptor-to-catalyst ratio, feed gas flow rate, etc. The same authors (Balasubramanian *et al.*, 1999) found that the H_2 product was able to achieve a purity over 95% when N_2 was not used as the carrier gas in the feed, which was close to the equilibrium limitation of SERP with CaO-based material as the CO_2 acceptor. In addition, increasing the ratio of acceptor-to-catalyst from 1 to 3.3 extended the duration of the pre-breakthrough period by $\sim 1/3$ and provided additional flexibility to increase the CH_4 content of the feed gas and/or the feed gas flow rate, which was also in a good agreement with the results from Wang and Rodrigues (2005) who found that the operation performance of SERP can be affected by the ratio between the catalyst-to-adsorbent to a large extent.

Cunha *et al.* (2012) carried out an experimental investigation of hydrogen production from SERP with ethanol as the feedstock in a fixed bed reactor. A commercial 15 wt.% Ni/γ-Al_2O_3 catalyst and HTlc CO_2 adsorbent material have been used and packed in the reactor with a multilayer pattern, as shown in Fig. 2.16. Hydrogen

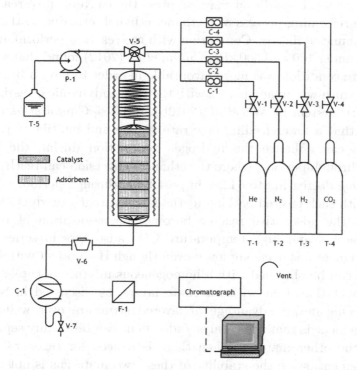

Figure 2.16 Schematic diagram of the experimental unit for SERP with ethanol as feedstock, and where catalyst and adsorbent were packed in the reactor with a multilayer pattern. Reprinted with permission from Cunha *et al.* (2012). Copyright (2016), John Wiley and Sons.

can be generated in the temperature range between 373 K and 873 K from conventional SRE. However, both the reaction kinetics as well as the adsorbent capacity are strongly dependent on the operational temperature. Generally, higher temperature favours the reaction kinetics, while the adsorbent capacity may decrease under high temperature condition. Therefore, the optimal operating temperature in which there is adsorptive enhancement should be determined. The authors (Cunha *et al.*, 2012) found that HTlc material was a relative promising CO_2 adsorbent for SERP at 673 K to produce CO-free H_2 product. On the other hand, the uptake of the adsorbent is raised by increasing the pressure of the adsorbate due to the pressure dependency of the adsorption isotherm. Therefore, higher pressure favours the CO_2 adsorption process during SERP. However, higher pressure condition may suppress the conversion of reactants for the reforming process of methane, ethanol, etc. due to the thermodynamic equilibria. Comparing with the reaction performances of SERP under 100 and 500 kPa, Cunha *et al.* (2012) found that a lower pressure condition was more favourable in order to have a better H_2 yield, which was in agreement with the thermodynamic investigation of SERP system by Wu *et al.* (2012b). Besides, Cunha *et al.* (2012) found that a lower feeding flow rate regime and multilayer pattern system can enhance the hydrogen production during the initial breakthrough periods, where the ethanol conversion can reach 100%, the value decreasing to 64.2% in post-breakthrough period.

With such layered packing method for hydrogen production from SERP, the adsorptive reactor becomes an association of reforming reactors and high temperature CO_2 adsorbers in series. This arrangement has been employed even though H_2 product with higher purity can be obtained with a homogeneous mixture of catalyst and adsorbent than that with alternate layers, as discussed in Section 2.3.3. One major advantage of layered structure over well-mixed arrangement is that the catalyst/adsorbent can be readily separated from the other material when there is a need for regeneration or replacement, since the stability of these two materials is not always the same. Besides, the composition of each layer can be carefully controlled to ensure the equilibrium is not reached within the catalyst layers and CO_2 breakthrough is not premature (Oliveira *et al.*, 2011).

Reijers *et al.* (2011a) performed an experimental study by using two different CO_2 adsorbents in a fixed bed adsorptive reactor for SE-SMR with layered pattern. The CaO adsorbent was used in the first layer for the bulk adsorption due to the high capacity and good kinetics, while the Ba_2TiO_4 material was packed in the second layer for the final polishing of H_2 product, because it can remove CO_2 even under very low pressures. Better reaction performance can be obtained from the reactor with two kinds of adsorbents than with a CaO-only bed; the CO_2 concentration in the obtained product was below 0.03% during a 17 min pre-breakthrough period with the CaO–Ba_2TiO_4 configuration, while the concentration of CO_2 was around 0.25% for the CaO-only bed during the 10 min pre-breakthrough period.

In addition to the bi-adsorbent configuration, Vaidya and Rodrigues (2006) developed another layered configuration with two different catalytic systems for SERP with ethanol as the feedstock in order to avoid carbon deposit. As shown in Eq. (2.38), ethanol can be converted into acetaldehyde and hydrogen in a first layer on the copper-based catalyst, which is very effective for dehydrogenation reaction due to the ability to maintain the C–C bond at low temperatures:

$$C_2H_5OH \xrightarrow{\text{Cu-basedcatalyst}} CH_3CHO + H_2 \qquad (2.38)$$

As a result, ethylene formation from dehydration reaction of ethanol on Ni-based catalyst can be prevented and thereby reducing catalyst coking (carbon deposition from ethylene polymerization). Acetaldehyde formed can decompose into CH_4 and CO. The resulting mixture, consisting of hydrogen, acetaldehyde, methane, CO_2 and water is then fed into a second layer of a nickel-based catalyst, coupled with CO_2 adsorbent. Such concept is also applicable for other feedstocks to avoid the carbon deposition during SERP. Besides, the effect of pellet (catalyst and adsorbent) size on the H_2 production performance from SE-SMR was tested by Xie *et al.* (2012); it was found that small catalyst and adsorbent pellets with a diameter of 0.16 mm exhibit higher conversion than large ones ($d_p = 1.42$ mm), as expected.

More recently, hybrid materials with different catalyst-to-adsorbent ratio have been prepared and used to further improve the concept of multi-section fixed bed adsorptive reactor for SE-WGS (Lee and Lee, 2014) and SE-SMR (Lee *et al.*, 2015). It has been found that the concentration of impurity can be decreased further in the produced gas when hybrid pellets with high catalyst content are packed in the first section and pellets with high adsorbent content placed in the second section. In both studies, high purity hydrogen (CO content <10 ppm) can be produced by applying the multi-section packing concept where the performance of SERP is not limited by the reaction kinetics. These experimental results are consistent with the simulation results from Wu *et al.* (2014c), where a similar concept has been employed in a homogeneously packed column with two subsections for hydrogen production by SERP with ethanol as the feedstock.

2.4.2. Experimental studies with fluidized bed reactors

Gorin and Retallick (1963) filled a patent on H_2 production from SERP with methane as feedstock in a fluidized bed reactor. In 1986, Brun-Tsekhovoi *et al.* (1986) performed an experimental study of SE-SMR with a fluidized bed reactor, where the purity of hydrogen product was found to reach up to 96 vol.%. Calcined dolomite was used as the CO_2 adsorbent material, and the diameter of adsorbent pellets was much larger than that of the Ni-based microsphere catalyst. Consequently, the pellets can descend and be collected at the bottom of the reactor during SERP, then the used adsorbent was regenerated in a separate vessel and recycled to the reactor.

Johnsen *et al.* (2006) performed a comprehensive experimental study on SERP using a bubbling fluidized bed reactor under atmospheric pressure condition; the experimental set-up can be found in Fig. 2.17, where the lower part of the major component was a pre-heater and the upper part was the fluidized bed reactor with an expanded freeboard. A filtration unit and a gas cooler unit can be found in Fig. 2.17 on the right side. Methane was fed to the upper

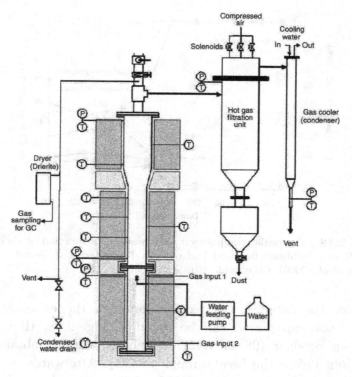

Figure 2.17 Schematic diagram of the experimental unit for SE-SMR in a bubbling fluidized bed reactor. Reprinted with permission from Johnsen *et al.* (2006). Copyright (2016), Elsevier.

part of the pre-heater through Gas input 1 stream where it was mixed with steam fed from the water feeding pump. Within the reactor, Ni-based SR catalyst and CaO-based adsorbent material obtained from dolomite have been used for SE-SMR. Dolomite was calcined at 1123 K under N_2 atmosphere and then used as the CO_2 acceptor for SERP; the material contained 32 wt.% of CaO, 20.3 wt.% of MgO and traces of SiO_2, Al_2O_3, Fe_2O_3, Na_2O, TiO_2 and K_2O. Additionally, the catalyst was reduced by a H_2/N_2 mixture for 1–2 h to ensure the activation prior to each SE-SMR test. A typical product distribution curve during SERP is given in Fig. 2.18; the reaction was performed at 873 K with $R_{S/C} = 3$ and the superficial gas velocity within the reactor was $0.032\,\mathrm{m\cdot s^{-1}}$.

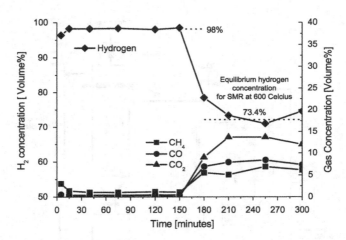

Figure 2.18 The outlet composition (dry basis) as a function of time during SE-SMR in a bubbling fluidized bed reactor. Reprinted with permission from Johnsen *et al.* (2006). Copyright (2016), Elsevier.

Since the catalyst was reduced prior to the experiment, no obvious start-up period can be found in Fig. 2.18. High purity hydrogen product (98–99 vol.%, dry basis) can be obtained for ~150 min before the breakthrough of CO_2. Afterwards, a sudden drop in concentration of H_2 to ~73 vol.% can be observed. Meanwhile, the CO_2 content in the product increased from 0.3 vol.% to 13–14 vol.%. The precise time of the breakthrough (between 150 and 180 min) cannot be determined due to the limited number of sampling points, but the shape of the curve was found to have a good agreement with the calculated time for complete carbonation of 170 min (assuming 100% calcium utilization). The fluidized bed adsorptive reactor periodically alternates between SERP and high-temperature calcination process for regeneration. High purity hydrogen product (98–99 vol.%, dry basis) can be obtained during the pre-breakthrough period for four reaction/regeneration cycles, and the H_2 concentration during the post-breakthrough period was also kept at equilibrium after four cycles, which led to the conclusion that the activation of Ni-based SR catalyst can be maintained during the cyclic operation. However, reduction in the duration of the pre-breakthrough period was found due to the loss of CO_2 adsorption capacity of the CaO-based adsorbent used. By using a fluidized bed reactor, uniform temperature distribution within the bed (temperature difference <4 K) can be achieved by rapid mixing of the solids, which could be helpful for processes where temperature uniformity is important (Johnsen *et al.*, 2006). Besides, the overall reaction and adsorption rate was found to be sufficiently fast to produce high purity hydrogen (>95 vol.%) under a superficial gas velocity up to $0.096\,\mathrm{m \cdot s^{-1}}$ (still lower than expected commercial velocities) at 873 K.

In addition to fluidized bed reactor configuration, Dou *et al.* (2013) developed a continuous SERP system for high-purity H_2 production from glycerol by using two moving bed reactors (Fig. 2.19). The moving velocity of catalyst ($NiO/NiAl_2O_4$) and adsorbent (lime) within the reforming reactor and regenerator was kept at $\sim 0.1\,\mathrm{m \cdot min^{-1}}$. Needle valves were used to seal and adjust pressure balance to prevent gas mixing between the two moving bed reactors, and the riser was used for particles transport by N_2 flow. Extended time of operation (over 60 min) was performed by using the moving bed reactor configuration. It was found that H_2-rich products with a H_2 content of 93.9% and 96.1% can be obtained with a $R_{S/C} = 3$ at reforming temperature of 773 K and 873 K, respectively. Very small amounts of CO_2, CH_4 and CO were formed during the process. It was observed that the maximum conversion of CaO was only 15.3%, and no obvious activity loss of the adsorbent was found after several carbonation/calcination cycles when the regeneration temperature was kept at 1173 K.

On the other hand, instead of using catalysts for methane reforming in fluidized bed adsorptive reactors, the concept of sorption enhanced chemical reforming loops, which evolved from chemical looping reforming, has been proposed in recent years. Schematics of these two processes for H_2 production can be found in Fig. 2.20. By using chemical looping reforming, the feedstock is oxidized to H_2 by a metal oxide (NiO, Fe_3O_4, etc.) and the produced CO_2 can be further captured during the process to enhance the H_2 production.

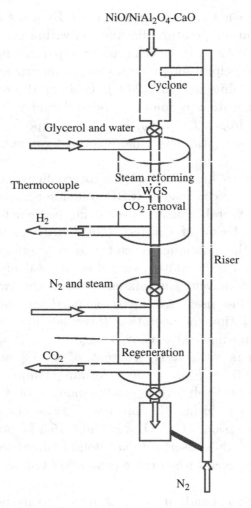

Figure 2.19 Experimental setup for continuous high-purity hydrogen production by sorption enhanced SR of glycerol with two moving bed reactors. Reprinted with permission from Dou *et al.* (2013). Copyright (2016), Elsevier.

Recently, Dou *et al.* (2014b) demonstrated the high purity H_2 production from glycerol as the feedstock by sorption enhanced chemical reforming loops with NiO as metal oxide and CaO as adsorbent; the experimental setup is similar as that shown

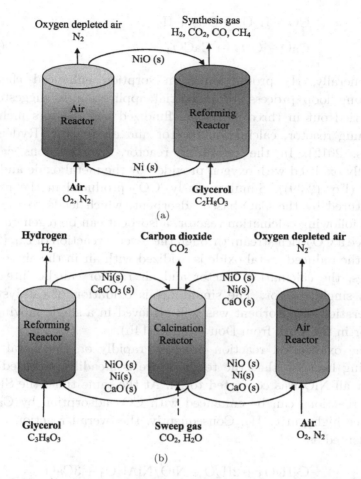

Figure 2.20 A comparison between chemical-looping reforming (a) and sorption enhanced chemical reforming loops (b) for H_2 production. Reprinted with permission from Dou *et al.* (2014a). Copyright (2016), Elsevier.

in Fig. 2.20, where the whole process involves the following reactions:

$$C_3H_8O_3 + H_2O + NiO/NiAl_2O_4$$
$$\rightarrow CO + 2CO_2 + 5H_2 + Ni/NiAl_2O_4 \quad (2.39)$$

$$C_3H_8O_3 \rightarrow 3CO + 4H_2 \quad (2.40)$$

$$CO + H_2O \leftrightarrow CO_2 + H_2 \tag{2.41}$$

$$CaO + CO_2 \leftrightarrow CaCO_3 \tag{2.42}$$

Generally, H_2 production from sorption enhanced chemical reforming loop process for industrial application is suggested to be carried out in three individual fluidized bed reactors including reforming reactor, calcination reactor and air reactor (Rydén and Ramos, 2012). In the reforming reactor, hydrocarbons can be partially oxidized with oxygen provided by the metal oxide and with steam (Eq. (2.39)). Simultaneously, CO_2 produced in the process is captured by the CaO-based adsorbent, which leads to a SERP. In the following calcination reactor, adsorbent can be regenerated to produce a CO_2-rich stream through the reverse reaction of Eq. (2.42), while the reduced metal oxide is oxidized with air in the air reactor. Besides, the calcination reactor and air reactor can be integrated into a single reactor. The simultaneous oxidation of catalyst and regeneration of adsorbent was well achieved in a single moving bed reactor in the work from Dou *et al.* (2014b).

The oxidization reaction proceeds rapidly on the metal oxide releasing heat, so that the temperature is steadily increased until almost all NiO was converted to Ni. At the same time, the SR and WGS reactions can be enhanced with CO_2 adsorption by CaO to produce high purity H_2. Consequently, the overall reaction can be represented as:

$$C_3H_8O_3 + 2H_2O + NiO/NiAl_2O_4 + 3CaO$$

$$\rightarrow 6H_2 + 3CaCO_3 + Ni/NiAl_2O_4 \tag{2.43}$$

The reaction (Eq. (2.43)) is a highly exothermic reaction with $\Delta H_{873K} = -359.14 \, kJ \cdot mol^{-1}$. Therefore, the endothermic and exothermic reactions can achieve overall heat balance and the reforming reactor was continuously operated at auto-thermal state when $R_{S/C} \leq 3$, according to the temperature profiles at the bottom of the reforming reactor for different initial temperatures, as shown in Fig. 2.21. Upon further increase of $R_{S/C}$ to 4.5, the temperature within the reactor may decrease with time-on-stream, which indicates

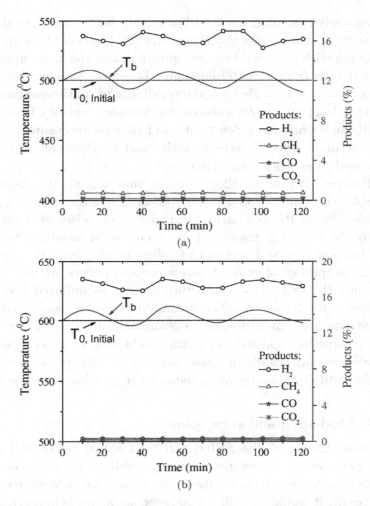

Figure 2.21 Temperature profiles at the bottom of the reforming reactor and product distributions during the sorption enhanced chemical reforming loop process with glycerol as feedstock, $R_{S/C} = 3$ at the different initial temperatures: (a) 773 K; (b) 873 K. Reprinted with permission from Dou *et al.* (2014b). Copyright (2016), Elsevier.

the energy requirement is greatly increased in the excess of steam. As a result, high steam concentration is not desirable for sorption enhanced chemical reforming loop process since it lowers the energy efficiency of the process (Dou *et al.*, 2014b).

Generally speaking, fluidized bed adsorptive reactors are able to offer a much more uniform temperature distribution along the bed comparing with the fixed bed configuration, and the heat supply for fluidized beds from an external source is usually simpler than that for fixed bed reactors. Besides, the possibility of continuous addition of fresh adsorbent and/or catalyst to overcome loss of CO_2-uptake capacity of adsorbent or deactivation of catalyst to ensure a steady flow rate of H_2 product stream with constant composition can be considered as another major advantage.

However, Wu *et al.* (2016) found that among the literature available, the fluidized bed configuration is not the general option of choice for SERP comparing with fixed bed adsorptive reactors due to the following reasons: (a) difficult solid handling for gas-tight introduction and removal of solids to and from the reactor, and the properties of catalyst and adsorbent pellets are not always the same; (b) well-mixed contacting pattern in fluidized bed does not favour SERP as equilibrium-limited reactions, while counter-current flow pattern is desired to optimize utilization of the cyclic CO_2 adsorption capacity; (c) extra mechanical stability demands for catalyst and adsorbent materials, the possibility of blockages, abrasion and electrostatic phenomena must be taken into account.

2.4.3. Modelling and simulation

The performance of the SERP can be affected by several operating parameters, design parameters as well as physical/chemical parameters, while effects of these parameters are always coupled. As a result, it would be difficult to ascertain suitable operating and design parameters for SERP by experimental investigations alone. Modelling and numerical simulation can be employed to improve the performance of SERP systems. In this section, the development and design of fixed bed adsorptive reactor for SERP by modelling and simulation have been included.

As shown in Fig. 2.22, an adsorptive reactor model must take into account the simultaneous mass, heat, and momentum balances of the bulk gas phase. Besides, the transport and thermo-physical

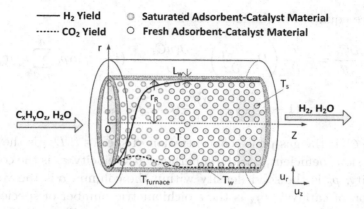

Figure 2.22 Schematic diagram of an adsorptive reactor model.

properties of the gas mixture, the transport between the bulk gas phase and particle surface, the diffusion, adsorption and reaction processes within the particle (described in Section 2.3.4), and a set of boundary conditions should all be included in order to perform a rigorous numerical simulation.

An 1D, non-isothermal model developed by Wu *et al.* (2013b, 2014a) for SERP with ethanol as the feedstock will be used to demonstrate the numerical simulation study in a fixed bed adsorptive reactor with the following assumptions:

- Axial dispersed plug flow, momentum balance coefficient simplified by the Ergun equation;
- Ideal gas behaviour;
- No mass, velocity or heat variations in the radial direction of the column;
- The column porosity, and cross sectional area, are constant along the column;
- CO_2 is considered as the only adsorbed species on the adsorbent.

Based on the above assumptions, the following governing equations and the corresponding initial conditions and general boundary conditions for the SERP process can be derived. The overall mass balance equation in the fixed bed is

$$\varepsilon_c \frac{\partial C_i}{\partial t} = \varepsilon_c \frac{\partial}{\partial z}\left(D_{\text{ax}}\frac{\partial C_i}{\partial z}\right) - \frac{\partial(uC_i)}{\partial z} + (1-\varepsilon_c)\alpha\rho_c \sum_{j=1}^{n} \nu_{j,i}r_j$$

$$-(1-\varepsilon_c)\frac{a(1-\alpha)k_f}{Bi+1}(C_i - \bar{C}_{i,\text{sorb}}) \qquad (2.44)$$

where C_i is the gas-phase concentration of species i, D_{ax} is the axial dispersion coefficient, u is the superficial gas velocity, ε_c is the column porosity, ρ_c is the solid density within the column. α is the volume fraction of catalyst, $\nu_{j,i}$ is the stoichiometric number of species i in reaction j, r_j is the rate of reaction j, a is the specific area of pellet, k_f is the film mass transfer coefficient, Bi is the Biot number, $\bar{C}_{i,\text{sorb}}$ is the averaged concentration in the macropores of the adsorbent, t is the time, and z is the axial coordinate.

The mass transfer from the gas phase to the macropores in the adsorbent is given by

$$\frac{\partial \bar{C}_{i,\text{sorb}}}{\partial t} = \frac{15D_p}{r_p^2}\frac{Bi}{Bi+1}(C_i - \bar{C}_{i,\text{sorb}}) - \frac{\rho_c}{\varepsilon_p}\frac{\partial \bar{q}_i}{\partial t} \qquad (2.45)$$

where $\bar{q}_i$ represents the average adsorbed species concentration in the pellet, D_p is the pore diffusivity, r_p is the radius of pellet, and ε_p is the porosity of the adsorbent.

Since CO_2 is the only species adsorbed, the mass balance for the adsorbed CO_2 in the adsorbent can be expressed by

$$\frac{\partial \bar{q}_{CO_2}}{\partial t} = \frac{15D_{\mu,CO_2}}{r_{\text{crystal}}^2}(q_{\text{eq},CO_2} - \bar{q}_{CO_2}) \qquad (2.46)$$

where D_{μ,CO_2} is the crystal diffusivity, r_{crystal} is the crystal radius and $q_{CO_2,\text{eq}}$ is the CO_2 adsorbed concentration in the equilibrium state.

The momentum balance in the fixed bed is given by Ergun equation as

$$\frac{\partial P}{\partial z} = -\frac{150\,\mu_g(1-\varepsilon_c)^2}{\varepsilon_c^3 d_p^2}u + \frac{1.75(1-\varepsilon_c)\rho_g}{\varepsilon_c^3 d_p}|u|u \qquad (2.47)$$

where d_p is the diameter of the adsorbent, μ_g and ρ_g are the bulk gas mixture viscosity and density, respectively. The total pressure P is given by

$$P = C_t R T \tag{2.48}$$

where $C_t = \sum C_i$ is the total concentration, T is the temperature of gas phase and R is the ideal gas constant. The energy balance includes three phases (gas, solid, and column wall). The energy balance for gas phase in the fixed bed is given by:

$$\varepsilon_c C_t C_v \frac{\partial T}{\partial t} = \frac{\partial}{\partial z} \left(\lambda \frac{\partial T}{\partial z} \right) - u C_t C_p \frac{\partial T}{\partial z} + \varepsilon_c R T \frac{\partial C_t}{\partial t}$$
$$- (1 - \varepsilon_c) a h_f (T - T_s) - \frac{2 h_w}{r_c} (T - T_w) \tag{2.49}$$

where C_v and C_p are the specific molar heat at constant volume and the specific molar heat at constant pressure of the gas mixture, respectively. λ is the effective thermal conductivity, h_f is the film heat transfer coefficient, h_w is the film heat transfer coefficient between the gas phase and the wall. r_c is the radius of the column and T_w is the temperature of the wall. T_s stands for the temperature of the solid.

The solid-phase energy balance in the fixed bed is expressed by

$$[\varepsilon_p C_{v,\text{ads}} \bar{C}_{t,\text{sorb}} + (1 - \varepsilon_p) \rho_c C_{ps}] \frac{\partial T_s}{\partial t}$$
$$= a h_f (T - T_s) + \varepsilon_p R T_s \frac{\partial \bar{C}_{t,\text{sorb}}}{\partial t}$$
$$+ \rho_c \left[\sum_{j=1}^{4} r_j (-\Delta H_j) + \frac{\partial \bar{q}_{CO_2}}{\partial t} (-\Delta H_{\text{ads}}) \right] \tag{2.50}$$

where $C_{v,\text{ads}}$ is the molar specific heat in the adsorbed phase at constant volume, $\bar{C}_{t,\text{sorb}}$ is the average total gas concentration within the adsorbent, C_{ps} is the particle specific heat at constant pressure, r_j and ΔH_j are the reaction rate and reaction enthalpy of reaction j, respectively. ΔH_{ads} is the isosteric heat of adsorption.

Finally, the energy balance for the column wall is given by

$$\rho_w C_{pw} \frac{\partial T_w}{\partial t} = \frac{2r_c}{l_w(2r_c + l_w)} h_w(T - T_w) - \frac{2U(T_w - T_{\text{furnace}})}{\left[(2r_c + l_w)\ln\left(\frac{2r_c + 2l_w}{2r_c}\right)\right]} \tag{2.51}$$

where ρ_w is the density of the column wall, C_{pw} is the specific heat at constant pressure of the column wall and l_w is the thickness of the column wall. T_{furnace} is the temperature of the furnace and U is the overall heat transfer coefficient between the column wall and the external air in the furnace.

The values of parameters for transport and thermo-physical properties of the gas mixture can be calculated employing frequently used correlations given in Table 2.10.

A reaction kinetic model based on the Langmuir–Hinshelwood–Hougen–Watson rate expression has been used to predict the reaction performance, where four reactions are considered in the reaction scheme shown in Table 2.11. Among these reactions, ethanol and water are the reactants, and carbon monoxide, carbon dioxide, hydrogen, acetaldehyde and methane are the products involved.

The rate expressions of reactions shown in Table 2.11 are:

Ethanol dehydrogenation

$$r_{\text{ETD}} = \frac{k_{\text{ETD}} p_{C_2H_6O}}{\text{DEN}} \left(1 - \frac{1}{K_{\text{ETD}}} \frac{p_{C_2H_4O} p_{H_2}}{p_{C_2H_6O}}\right) \tag{2.52}$$

Acetaldehyde decomposition

$$r_{\text{ACD}} = \frac{k_{\text{ACD}} p_{C_2H_4O}}{\text{DEN}} \left(1 - \frac{1}{K_{\text{ACD}}} \frac{p_{CO} p_{CH_4}}{p_{C_2H_4O}}\right) \tag{2.53}$$

Methane SR

$$r_{\text{SMR}} = \frac{k_{\text{SMR}} p_{H_2O} p_{CH_4}}{\text{DEN}^2} \left(1 - \frac{1}{K_{\text{SMR}}} \frac{p_{CO} p_{H_2}^3}{p_{H_2O} p_{CH_4}}\right) \tag{2.54}$$

Table 2.10   Calculation of transport parameters and thermo-physical properties of the gas mixture.

Axial mass and heat dispersion coefficients by Wakao and Funazkri correlations (Wakao and Funazkri, 1978):

$$\frac{\varepsilon_c D_{\text{ax}}}{D_m} = 20 + 0.5\, Sc\text{Re}, \ \frac{\lambda}{k_g} = 7 + 0.5\, \text{PrRe}$$

Schmidt, Reynold, Prandtl and Biot numbers are calculated as follows:

$$Sc = \frac{\mu_g}{\rho_g D_m}, \ \text{Re} = \frac{\rho_g u d_p}{\mu_g}, \ \text{Pr} = \frac{\mu_g C_{p,m}}{k_g}, \ \text{Bi} = \frac{r_p k_{\text{cat}}}{5\varepsilon_p D_p}$$

Heat capacity

Molar heat capacity at constant pressure of component i can be obtained with Shomate equation (Poling *et al.*, 2001):

$$C_{p,i} = A + Bt + Ct^2 + Dt^3 + \frac{E}{t^2}, \ t = \frac{T}{1000}$$

Molar heat capacity at constant pressure of the gas mixture is calculated as

$$C_p = \sum_{i=1}^{n} y_i C_{p,i}$$

where y_i is the molar fraction of component i.

The mass heat capacity at constant pressure of component i is calculated as:

$$C_{p,m,i} = \frac{C_{p,i}}{M_i}$$

where M_i is the molecular weight of component i.

The mass heat capacity at constant pressure of the gas mixture was calculated as

$$C_{p,m} = \frac{\sum_{i=1}^{n} y_i C_{p,i}}{\sum_{i=1}^{n} y_i M_i}$$

Viscosity

Viscosity of the gas component i was calculated according to the first order Chapman–Enskog equation (Poling *et al.*, 2001):

$$\mu_i = 2.67 \times 10^{-6} \frac{(M_i T)^{0.5}}{\varepsilon_i \Omega_\mu}, \quad \Omega_\mu = 1.16 \left(\frac{\varepsilon_i}{kT}\right)^{0.15} + 0.52 e^{\frac{-0.77kT}{\varepsilon_i}} + 2.16 e^{\frac{-2.44kT}{\varepsilon_i}}$$

where ε_i/k is the characteristic Lennard–Jones energy of component i.

The viscosity of the gas mixture is calculated with Wilke method (Poling *et al.*, 2001):

$$\mu_g = \sum_{i=1}^{n} \frac{y_i \mu_i}{\sum_{j=1}^{n} y_i \Phi_{ij}}, \quad \Phi_{ij} = \left[8\left(1 + \frac{M_i}{M_j}\right)\right]^{-0.5} \left[1 + \sqrt{\frac{\mu_i}{\mu_j}}\left(\frac{M_i}{M_j}\right)^{-0.25}\right]^2$$

(Continued)

Table 2.10   (*Continued*)

Thermal conductivity

The thermal conductivity of the gas component i is calculated according to the
  following equations proposed by Eucken (Poling *et al.*, 2001):

$$k_i = \left(C_{p,m,i} + 1.25\frac{R}{M_i} \right) \mu_i$$

Thermal conductivity of the gas mixture is calculated with Wassiljewa method
  (Poling *et al.*, 2001):

$$k_g = \sum_{i=1}^{n} \frac{y_i k_i}{\sum_{j=1}^{n} y_i \Phi_{ij}}$$

Density

The density of the gas mixture is obtained from

$$\rho_g = \frac{P}{RT} \left(\sum_{i=1}^{n} y_i M_i \right)$$

Molecular diffusivity

The molecular diffusivity of the mixture gas is calculated follows:

$$D_m = \sum_{i=1}^{n} D_{m,i} y_i$$

The molecular diffusivity of the gas component i is calculated as follows:

$$D_{m,i} = \frac{1 - y_i}{\sum_{j=1}^{n} \frac{y_j}{D_{ij}}}$$

where D_{ij} is obtained from the Chapman–Enskog equation (Poling *et al.*, 2001):

$$D_{ij} = \frac{1.18809 \times 10^{-7}}{P\sigma_{ij}^2 \Omega_{ij}} \sqrt{T^3 \left(\frac{1}{M_i} + \frac{1}{M_j} \right)}, \quad \sigma_{ij} = (\sigma_i + \sigma_j)/2$$

where σ_i is the characteristic Lennard–Jones length of component i. Ω_{ij} is
  calculated by

$$\Omega_{ij} = 1.06 \left(\frac{\varepsilon_{ij}}{kT} \right)^{0.1561} + 0.19 e^{-0.476\frac{kT}{\varepsilon_{ij}}} + 1.04 e^{-1.53\frac{kT}{\varepsilon_{ij}}} + 1.76 e^{-3.89\frac{kT}{\varepsilon_{ij}}}, \varepsilon_{ij} = \sqrt{\varepsilon_i \varepsilon_j}$$

WGS

$$r_{\text{WGS}} = \frac{k_{\text{WGS}} p_{H_2O} p_{CO}}{\text{DEN}^2} \left(1 - \frac{1}{K_{\text{WGS}}} \frac{p_{CO_2} p_{H_2}}{p_{H_2O} p_{CO}} \right) \tag{2.55}$$

where p_i is the partial pressure of the species i in the gas phase, and
k_{ETD}, k_{ACD}, k_{SMR} and k_{WGS} are the reaction rate constants. K_{ETD},
K_{ACD}, K_{SMR} and K_{WGS} are the equilibrium constants, which can
be obtained from thermodynamic data. The denominator (DEN) is

Table 2.11   Reaction scheme of the ethanol reforming process.

Reactions	Chemical description	ΔH_{298K} [kJ · mol^{-1}]
Ethanol dehydrogenation	$C_2H_6O \leftrightarrow C_2H_4O + H_2$	+68.4
Acetaldehyde decomposition	$C_2H_4O \leftrightarrow CH_4 + CO$	−18.8
Methane SR	$CH_4 + H_2O \leftrightarrow CO + 3H_2$	+205.9
WGS	$CO + H_2O \leftrightarrow CO_2 + H_2$	−41.4

given by

$$\text{DEN} = 1 + K^*_{\text{EtOH}}y_{\text{EtOH}} + K^*_{\text{H}_2\text{O}}y_{\text{H}_2\text{O}} + K^*_{\text{CH}_4}y_{\text{CH}_4}$$
$$+ K^*_{\text{OH}}y_{\text{H}_2\text{O}}y_{\text{H}_2}^{-0.5} + K^*_{\text{CH}_3}y_{\text{CH}_4}y_{\text{H}_2}^{-0.5} + K^*_{\text{CO}}y_{\text{CO}} + K^*_{\text{H}}y_{\text{H}_2}^{0.5}$$
$$(2.56)$$

where each K^*_i is the adsorption equilibrium constant of the adsorbed species.

On the other hand, the CO_2 adsorption performance of the adsorbent (HTlc) is usually described by the Langmuir isotherm according to the following equation:

$$q_{\text{CO}_2} = q_{\max}\frac{b_{\text{CO}_2}p_{\text{CO}_2}}{1 + b_{\text{CO}_2}p_{\text{CO}_2}}, \quad b_{\text{CO}_2} = b_{\text{CO}_2,0}e^{\frac{-\Delta H_{\text{ads}}}{RT}} \quad (2.57)$$

where $q_{\max}$ is the maximum capacity, $b_{\text{CO}_2,0}$ is the pre-exponential factor and ΔH_{ads} is the adsorption enthalpy. In addition, bi-Langmuir model can also be found for K-HTlc materials (Oliveira *et al.*, 2011).

The partial differential equations (PDEs) can be converted into ordinary differential equations (ODEs) and solved by discretisation methods, e.g. central finite difference scheme, backward finite difference scheme, forward finite difference scheme, orthogonal collocation on finite elements scheme, etc. together with the algebraic equations of the system to obtain the pressure, concentration and temperature profiles within the reactor during SERP.

The validity of the mathematical model proposed for SERP simulations should be checked with experimental data, as shown in

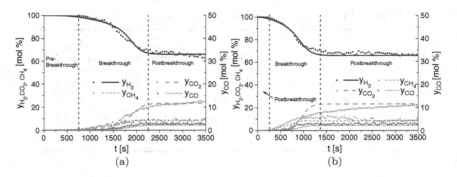

Figure 2.23 Product distributions (a), (b) as a function of time from the fixed bed adsorptive reactor under different feeding rates with ethanol as the feedstock. Reprinted with permission from Wu *et al.* (2014b). Copyright (2016), Elsevier.

Figs. 2.23(a) and 2.23(b), where symbols are the experimental values and the lines in the graph correspond to the simulated values at the outlet of the reactor. Besides, simulations are also helpful to give the information inside the reactor column which is difficult to obtain by SERP experiments.

It can be found that the experimental points during the whole breakthrough period can be successfully described by the numerical simulations. Consequently, such mathematical model can be helpful to probe feasible operation regions to fulfill CO-free and high purity H_2 production. For instance, the effects of $R_{S/C}$ and adsorbent mass fraction on the H_2 production performance can be found in Fig. 2.24. Wu *et al.* (2013b) found that the H_2 concentrations at both the pre-breakthrough period and steady state increase with increasing $R_{S/C}$, and thus a high $R_{S/C}$ is favoured for high purity H_2 production during SERP. However, the breakthrough times of CO_2 in Fig. 2.24(a) decrease with the $R_{S/C}$ increase, and the concentration of H_2 with $R_{S/C} = 10$ after pre-breakthrough (540 s) is even lower than that with $R_{S/C} = 5$ in the feed, which is caused by the lower CO_2 adsorption capacity due to the decrease of p_{CO_2} within the reactor. As a result, a feasible $R_{S/C}$ can be determined through numerical simulations instead of repeated experimental tests. Another important design parameter is the ratio of the catalyst and adsorbent within the adsorbent bed. A compromise between

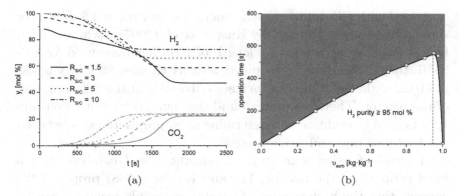

Figure 2.24 Simulation results for molar fractions of H_2 and CO_2 at different $R_{S/C}$ conditions (a) and the pre-breakthrough time for H_2 production with different adsorbent mass ratios (b). Reprinted with permission from Wu *et al.* (2013b). Copyright (2016), Elsevier.

catalytic activity (catalyst mass) and adsorbent capacity (adsorbent mass) has to be found. The increase of adsorbent mass ratio from 0 to $\sim$0.95 favours the extension of pre-breakthrough period according to Fig. 2.24(b), but further increase of the adsorbent fraction may decrease the operation time drastically due to the reduction of catalytic efficiency. As a result, further improvements in the H_2 production performance can be achieved by adjusting the ratio of the two solid phases according to simulation results.

The residence time is another important parameter that may affect the H_2 production performance of SERP. Comparing with the decrease of feed flow rate, Wang and Rodrigues (2005) found that the use of a prolonged reactor under the same residence time conditions can produce H_2 with higher productivity (according to the simulation results). Xiu *et al.* (2002a) investigated the effect of the residence time by varying the length of reactor column, and found that a hydrogen product with higher purity can be obtained during the SE-SMR in a fixed bed adsorptive reactor of 6 m length compared with reactors of 2 and 4 m under the same feed flow rate. However, the use of a fixed bed adsorptive reactor with extended length may decrease the efficiency (productivity) of SERP. According to the simulation results (Xiu *et al.*, 2002a), the ratio of unused bed (where adsorbent is not saturated with CO_2) during cyclic operation may reach 0.8. In a later simulation study performed by Rusten *et al.* (2007a), where a fixed bed adsorptive reactor with a length of 20 m has been used for SE-SMR, they found that the H_2 purity already reached 86% (dry basis) at 10 m, while the value can increase only $\sim$2% in the following half of the reactor. Therefore, we can find that numerical simulations are very helpful to evaluate and determine an appropriate ratio between the feed flow rate and reactor length.

Instead of using a mixture of adsorbent and catalyst with a fixed ratio along the reactor, Lee and Kadlec (1988) proposed the concept that the performance of parallel reversible reactions can be improved by optimizing the distribution of catalyst and adsorbent in a fixed bed adsorptive reactor. The idea has been adopted and examined by Xiu *et al.* (2003) for SE-SMR with numerical simulations, with a Ni/Al_2O_3 reforming catalyst and a HTlc material as the CO_2 adsorbent. Catalysts and adsorbents are distributed into three subsections within the reactor, and the wall-temperature is controlled at different values as indicated in Fig. 2.25, which is referred as SERP with subsection-controlling strategy. Catalytic reactions mainly occur in the inlet (subsection-I) for reforming reaction and in the outlet (subsection-III) for methanation and WGS reactions. Therefore, the catalyst-to-adsorbent ratios in inlet and outlet zones (0.5) are higher than that in the subsection-II (0.25), where CO_2

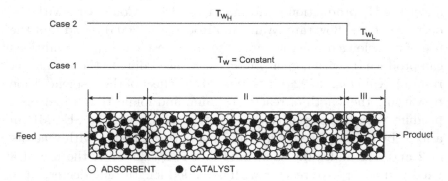

Figure 2.25 Subsection-controlling strategy in a fixed bed adsorptive reactor. Reprinted with permission from Xiu *et al.* (2003). Copyright (2016), Elsevier.

adsorption process and SERP mainly occurred. In addition, two configurations for the wall-temperature of subsection-III have been tested. In the first configuration (Case 1), the wall-temperature for subsection-III is the same as the other two subsections (723 K). While in Case 2, a wall-temperature of 673 K is used for subsection-III, since lower temperature favours the equilibrium to shift towards the product side of methanation and WGS reactions, which leads to a reduced CO content in the H_2 product. Consequently, the produced H_2 product has a purity of 87.7% with a CO content of 30 ppm in Case 2, while the CO content reaches 342 ppm in Case 1.

In a more recent work (Jang *et al.*, 2013), a multi-section column strategy has been tested for SE-WGS, where the fixed bed adsorptive reactor is divided into two sections. Investigations on the effects of catalyst-to-adsorbent ratio in each section, length ratio between each section, and total ratio between catalyst and adsorbent have been performed by numerical simulations. Similarly to the results from Xiu *et al.* (2003), Jang *et al.* (2013) found that the H_2 production performance can be improved by increasing both the catalyst content at the inlet zone and the adsorbent content at the outlet zone. Additionally, the use of a short first section where WGS reaction dominantly occurs and a long second section where SE-WGS dominantly takes place can improve the overall reaction performance ~4.5% in the H_2 productivity comparing with results obtained from a single section arrangement. Besides, H_2 productivity can be further enhanced as the total adsorbent fraction raises, which agrees with the results from the numerical research by Wang and Rodrigues (2005) for SERP with methane as the feedstock and the results obtained by Wu *et al.* (2014a) for SE-SRE with a similar two subsections strategy.

2.5. Regeneration and Cyclic Operation

Investigations of SERP for hydrogen production have been mainly focused on the hydrogen performance during the reaction step; however, due to the limited adsorption capacity, one major technological challenge of SERP is its cyclic nature. The CO_2 adsorbent, when saturated, has to be periodically regenerated (desorbed) under a certain operating condition which is different from that for SERP without affecting both the performances of the catalyst and the adsorbent during the regeneration period. On one hand, for the continuous SERP process using fluidized bed adsorptive reactors, the hydrogen product and CO_2 by-product are usually obtained with a steady flow rate and of constant composition from the reactor and the regenerator, respectively. Therefore, the regeneration step for fluidized bed adsorptive reactors is typically carried out in a regenerator at elevated temperature and under the same pressure condition of reaction step. On the other hand, for fixed bed adsorptive reactor configuration, each column with adsorbent and catalyst swings between at least two different states: reaction, where reactant is fed to the reactor to produce hydrogen; and regeneration under a different operating condition to regenerate the saturated CO_2 adsorbent. The cyclic operation of these steps is mainly determined by the adsorptive capacity and the regeneration time required. For continuous hydrogen production, at least two fixed bed adsorptive reactors are required to operate in parallel, and the total number of reactors used in a practical application depends on the durations of productive and regenerative periods as well as several intermediate rinse, purge, etc. steps that might be necessary. In this section, the regeneration and cyclic operation will be discussed based on the methodology used for the desorption of CO_2 from the adsorbent.

2.5.1. Pressure swing

Pressure swing processes have been developed for decades with considerable experience available, which can be helpful for the design of a sorption enhanced H_2 production process with cyclic regeneration operation. Therefore, pressure swing reactor or pressure swing regeneration (PSR) were the first to be considered when Vaporciyan and Kadlec (1987) proposed the earliest adsorptive reactor concept with cyclic operation for SERP. Generally speaking, PSR or a similar regeneration method, concentration swing regeneration (CSR), can be carried out by decreasing the total pressure within the reactor column or inlet purge (usually with N_2/He or steam) to decrease the partial pressure of CO_2, and consequently decrease the CO_2 adsorption capacity. Additionally, PSR or CSR can be completed in a period of time ranging from seconds to minutes, which is the most commonly used regeneration method for SERP with K-HTlc type material as CO_2 adsorbent. Many variations of SERP with PSR method can be found in literature, as shown in Table 2.12, and these configurations are mainly adapted from the PSA process for gas separation.

Consequently, the cyclic operation of a pressure swing reactor can be performed with many methods for H_2 production from SERP; the number and arrangement of reactors in a network also has variations, as given in Fig. 2.26: (a) two steps, (b) three steps with a delivery step, (c) four steps with an added purge or (d) with added backfill and (e) six steps with two purges and repressurization. Besides, the duration for each step varies with particle size of the packing material, column length and CO_2 adsorption properties on the adsorbent to be separated.

In Fig. 2.26(a), the first feed step where the reaction and *in situ* CO_2 capture process are carried out under high pressure is shown. The regeneration of the CO_2 adsorbent can be achieved by steam and/or N_2 purge at low pressure condition as an exhaust or depressurization step when the outlet is opened; the pressure at the outlet is constant and is the lowest pressure during the cycle. The use of steam is recommended when CO_2 is captured as a valuable product (Wright *et al.*, 2011), since the separation between CO_2/H_2O is relatively easy to implement and heat of steam can be recycled (Jeong *et al.*, 2012). For Fig. 2.26(b), after the adsorbent has been saturated by CO_2, the depressurization step (or delivery step) can be performed co-currently and the pressure within the column is reduced to a lower pressure condition (atmospheric pressure in most cases). On the other hand, a product purge step can be added after the depressurization step, as can be seen in Fig. 2.26(c); in this case the outlet port is opened and the product is fed back into the reactor, being such purge step used to ensure the purity of product

Table 2.12 PSR and CSR strategies developed for hydrogen production by SERP.

Operation Strategy	Conditions					Catalyst	Product		References
	T (K)	P (kPa)	$R_{S/C}$ (mol/mol)	Feedstock	Adsorbent		y_{H_2} (mol%)	y_{CO} (ppm)	
PSR (4 steps)	823	20–2027	3	CH_4	K-HTlc	Commercial Ni-based	90~95	n.a.	Hufton et al. (1999b); Mayorga et al. (1997)
PSR (4 steps)	763	179–455	6	CH_4	K-HTlc	Noble metal/Al_2O_3	94.4	<30	Waldron et al. (2001)
PSR (4 steps)	673	126–446	6	CH_4	K-HTlc	$Ni/MgAl_2O_4$	83.9	82	Xiu et al. (2002a)
PSR with reactive regeneration (5 steps)							87.9	33	Xiu et al. (2002a)
PSR (5 steps)	723	126–446	6	CH_4	K-HTlc	$Ni/MgAl_2O_4$	~81	<30	Xiu et al. (2002b)
PSR with subsection-controlling strategy (4 steps)	673–763	126–446	6	CH_4	K-HTlc	$Ni/MgAl_2O_4$	93	18	Xiu et al. (2003)
	673–733	126–446	6	CH_4	K-HTlc	$Ni/MgAl_2O_4$	~86	<30	Xiu et al. (2004)
PSR (10 steps)	623–723	111–3500	~2	CO	K-HTlc	HTS catalyst	96.2	0.6%	Allam et al. (2005)
PSR (5 steps)	673	100–2800	1.9	CO	K-HTlc	Commercial FeCr-based	n.a.	<15	van Selow et al. (2009)
PSR (4 steps)	723	126–446	6	CH_4	K-HTlc	$Ni/MgAl_2O_4$	81	24	Rawadieh and Gomes (2009)

CSR with pulsing methane feed (4 steps)	803	129	3	CH_4	K-HTlc	Rh/CeO_2	97.3	11	Duraiswamy et al. (2010)
CSR (5 steps)	673			C_2H_5OH			95.2	3502	Duraiswamy et al. (2010)
PSR (8 steps)	673	200–2360	6.9	CO	K-HTlc	No catalyst needed		carbon capture ratio 90.4%	Reijers et al. (2011b)
PSR (11 steps)	1013–1033	115–3000	2.2					carbon capture ratio 95%	Wright et al. (2011)
PSR (4 steps)	778	101–202	4	CH_4	$Li_2CO_3 - CaO$	Ni-based	99.8	130	Derevschikov et al. (2011)
PSR (4 steps)	773	101–400	4	CH_4	K-HTlc	Ni/Al_2O_3	76.4	n.a.	Oliveira et al. (2011)
CSR (2 steps)	673	450	6	CH_4	K-HTlc	$Rh/Ce_aZr_{1-a}O_2$	>99	<95	Halabi et al. (2012b)
PSR (11 steps)	~673	n.a.–3000	n.a.	CO	ALKASORB Class Sorbents	No catalyst needed	90–95	n.a.	Jansen et al. (2013)
PSR (11 steps)	773	110–2689	~5	CO	K-HTlc	FeCr-based	90	544	Najimi et al. (2013)
PSR with multi-section strategy (4 steps)	773	101–304	4	C_2H_5OH	K-HTlc	Ni-HTlc	>99	30	Wu et al. (2014c)
PSR (7 steps)	673			CO	K-HTlc	K-Cu-Ni/HTlc	99.1	25	Wu et al. (2014b)
PSR (11 steps)	673	111–2400	~3.9	CO	K-HTlc	No catalyst needed		carbon capture ratio 95%	Boon et al. (2015)
PSR (4 steps)	948–1048 (axial gradient)	101–507	6.3	C_2H_5OH	CaO	Ni-based	97–98	8000	Lysikov et al. (2015)

Notes: PSR: Pressure swing regeneration; CSR: concentration swing regeneration.

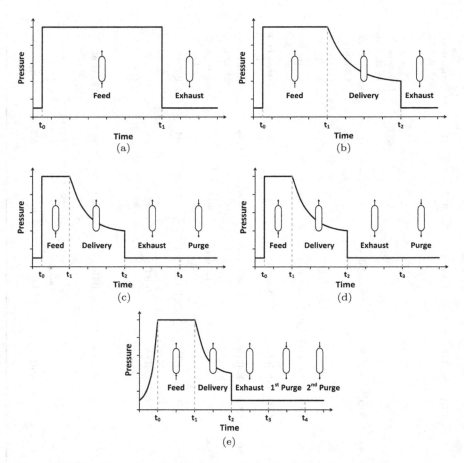

Figure 2.26 Several cyclic operating strategies for a pressure swing reactor. Adapted from Vaporciyan and Kadlec (1987).

in the following cycle (but the H_2 yield may decrease). A backfill step is illustrated in Fig. 2.26(d), where the exhaust port is closed after depressurization and the product stream will pressurize the column. This step can also improve the purity. Besides, a backfill step may employ the product or a non-adsorbing feed component to pressurize the reactor, as shown in Fig. 2.26(e) where a second purge (purge 2) with a composition different from purge 1 has been used.

A comprehensive and pilot-scale SERP with cyclic PSR research was developed by Mayorga *et al.* (1997) from Air Products company, where a fuel-cell grade H_2 product can be directly produced from methane feedstock by SERP at 673–773 K without sacrificing the conversion of CH_4. A fixed bed adsorptive reactor packed with a mixture of a commercial Ni-based catalyst and the K-HTlc as the CO_2 adsorbent has been employed. The saturated adsorbent was periodically regenerated by steam purging under a isothermal condition at the same temperature as the reaction under a lower pressure. The adsorbent was found to be very stable, and maintained CO_2 adsorption capacity of $0.3–0.45\,\text{mol}\cdot\text{kg}^{-1}$ over nearly 6000 cycles. The whole cyclic process consisted of four operating steps:

1. **SERP Step**: Initially, the column is pre-saturated with a mixture of steam and H_2 at the desired reaction temperature and pressure. Afterwards, steam and methane at a prescribed ratio are fed to the reactor and the H_2 product is collected at the outlet. When the H_2 purity in the product decreases to a preset level, the SERP step will be switched to the depressurization step.
2. **Depressurization Step**: The reactor is counter-currently depressurized, and the effluent gas can be recycled as feed to another reactor or used as heat source.
3. **Purge Step**: The reactor is counter-currently purged with a mixture of H_2 with steam to regenerate the CO_2 adsorbent. The desorption pressure may range between 20 and 110 kPa. The desorbed gas consists of CH_4, CO_2, H_2 and steam and is used as fuel after condensing the water content.
4. **Pressurization Step**: The reactor is counter-currently pressurized to the reaction pressure with a steam/H_2 mixture.

The main downside of PSR is that the length of time period for operating cycle and each step as well as the arrangement of fixed bed reactors have to be carefully tailored in parallel to achieve a continuous hydrogen production process. This is especially important when pressure equalization step(s) is(are) used in the cyclic operation to save energy (Boon *et al.*, 2015), which gives

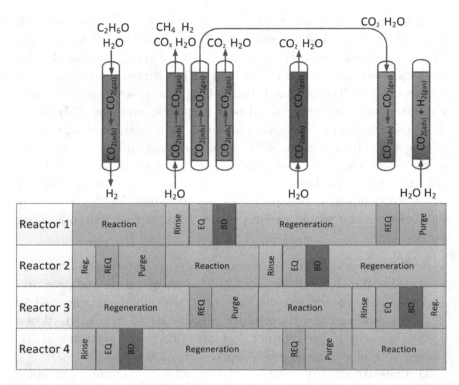

Figure 2.27 Four-reactor schemes and cyclic configurations employed in SERP. BD: blowdown (delivery), EQ: pressure equalization, REQ: received pressure equalization. Reprinted with permission from Wu *et al.* (2014b). Copyright (2016), Elsevier.

additional complexity comparing with operating strategies shown in Fig. 2.26. In addition, PSR fixed bed adsorptive reactors must be designed to ensure that the pressure drop does not affect the efficiency of regeneration. A demonstration of the complexity of SERP with PSR for H_2 production from ethanol as the feedstock can be found in Fig. 2.27, where a cyclic operating configuration for four fixed bed adsorptive reactors with seven steps has been developed by Wu *et al.* (2014b), considering the following stages:

- Reaction (co-currently to feed). Sorption enhanced reaction with reactants in the feed at high pressure;

- Rinse (counter-currently to feed). Rinse the column with steam ($y_{H_2O} = 100\%$) to remove the residual reactants and products within the gas phase at high pressure;
- Pressure equalization (EQ, counter-currently to feed). This step is performed by connecting two reactors at different pressure levels to save energy. The high pressure of the reactor can be reduced;
- Blowdown (BD, counter-currently to feed). After the pressure equalization step, the pressure of the column is reduced to atmospheric pressure;
- Regeneration (counter-currently to feed). Regenerating the CO_2 sorbent by steam ($y_{H_2O} = 100\%$) at low pressure;
- Received pressure equalization (REQ, co-currently to feed). The high pressure stream from one reactor on the pressure equalization step is recycled to another reactor to increase pressure and reduce compression energy;
- Purge (counter-currently to feed). Purging the column with H_2 and steam gas-mixture ($y_{H_2} = y_{H_2O} = 50\%$) with a pressure increase to high pressure (p_H) before the next cycle.

In such a study (Wu *et al.*, 2014b), the number of reactors should be more than two to achieve continuous hydrogen production since the pressure equalization step has been considered, and the use of four reactors (Fig. 2.27) can reduce the complexity of connections instead of using three columns. As a result, the duration of each step can be set taking into consideration the extension of the cycle to a process with four reactors, $t_{cycle} = 4 \times t_{reaction}, t_{reaction} = 4 \times t_{rinse} = 4 \times t_{EQ} = 4 \times t_{BD} = 4 \times t_{REQ} = 2 \times t_{purge} = 2/3 \times t_{regeneration}$. The low pressure value of 101 kPa for regeneration step was used; lower pressure has not been used because the use of a vacuum pump and sub-atmospheric steam can be cost intensive (Lee *et al.*, 2007a).

2.5.2. Temperature swing

Temperature/thermal swing regeneration (TSR) can be carried out by increasing the temperature of the adsorbent, this way decreasing the CO_2 adsorption capacity of the material and consequently releasing the adsorbed CO_2. A general survey of the H_2 production studies by SERP with cyclic TSR method is given in Table 2.13. As we can find from the previous researches, the number of operating steps required for TSR is generally less than that of PSR. For instance, Lee *et al.* (2007a) proposed a cyclic TSR operation for fuel-cell grade H_2 production through a SERP with methane as the feedstock using the K-HTlc as the CO_2 adsorbent; the cyclic process consisted of two steps:

(a) During the SERP step, a mixture of methane and steam is fed at 763 K with a pressure from ~150 kPa to 200 kPa into a fixed bed adsorptive reactor. The reactor is preheated to a temperature of ~863 K and filled with steam at the desired reaction pressure and temperature conditions. The product obtained from the reactor during this step is fuel-cell grade H_2 with CO content less than 20 ppm.

(b) During the thermal regeneration step, the adsorptive reactor is simultaneously depressurized to near-ambient pressure and counter-currently purged with superheated steam at ambient pressure and at ~863 K. Afterwards, a counter-current pressurization step was carried out with steam at ~863 K to the feed pressure. The effluent from the reactor during this step is a CO_2-rich waste gas.

According to the authors (Lee *et al.*, 2007a), several advantages of the proposed TSR concept over the general cyclic PSR process for SERP can be summarized as: (a) elimination of the usually expensive, sub-atmospheric steam purge step for CO_2 desorption and the use of an energy intensive vacuum pump during the process; (b) direct supply of the energy for endothermic SMR reaction before the SERP step; (c) higher utilization of the specific CO_2 uptake capacity of the adsorbent in the cycle due to more stringent regeneration; (d) higher yield and purity of H_2 product; and (e) less purge steam requirement per unit amount of H_2 produced. Besides, we can find that a CO_2 stream with high pressure can be obtained during the regeneration step, which may reduce the cost for CO_2 compression and capture (Manzolini *et al.*, 2015). Recently, Lysikov *et al.* (2015) performed SERP with a dual fixed bed adsorptive reactor configuration for continuous hydrogen production

Table 2.13 Temperature/thermal swing regeneration strategies developed for hydrogen production by SERP.

Operation strategy	Conditions						Product		Reference
	T (K)	p (kPa)	$R_{S/C}$ (mol/mol)	Feedstock	Adsorbent	Catalyst	y_{H_2} (mol%)	y_{CO} (ppm)	
TSR (2 steps)	775–1023	500	3–5	CH_4	lime	Commercial NiO-catalyst	98	<10	Harrison and Peng (2003)
TSR (2 steps)	823–1023	1	4	CH_4	lime	Supported Ni-material	~90	~5%	Kato et al. (2003b)
TSR (2 steps)	903–1123	101	5	CH_4	$CaO/Ca_{12}Al_{14}O_{33}$	Ni-based	>90	n.a.	Li et al. (2006)
TSR (2 steps)	763–863	150	4	CH_4	K-HTlc	$Ni/MgAl_2O_4$	>99	<10	Lee et al. (2007a)
Hybrid PSR-TSR with multi-section strategy (5 steps)	673–823	1520–2179	4	CO	K-HTlc/Na-alumina	$Cu/ZnO/Al_2O_3$	99	~10	Jang et al. (2012); Lee et al. (2008a)
TSR (2 steps)	1023–1123	689	~0.5	$CO - CO_2 - H_2 - CH_4$ synthesis gas	CaO	Fe_2O_3	~95	n.a.	Wiltowski et al. (2008)
TSR (2 steps)	848–1043	101	3	C_2H_5OH	dolomite	Co-Ni/HTls	>99	1000	He et al. (2009)
TSR (2 steps)	923–1173	101	2	CH_4	CaO-based	$NiO/NiAl_2O_4$	88	<10%	García-Lario et al. (2015)
TSR (4 steps)	923–1048	507	6.3	C_2H_5OH	CaO	Ni-based	~99	3000	Lysikov et al. (2015)

Note: TSR: Temperature/thermal swing regeneration.

with bio-ethanol as the feedstock, where both TSR and PSR methods have been used for the regeneration of the CaO adsorbent. The highest purity of H_2 product obtained by TSR cyclic operating method is able to reach approximately 99% while a H_2 product with a purity ~97% can be obtained from PSR configuration.

Another benefit of TSR is its simplicity, where the continuous H_2 production process can be performed with two parallel fixed bed adsorptive reactors operated in a two-step cyclic manner (Li *et al.*, 2006), as shown in Fig. 2.28. Three three-way valves are used to switch different feeds between two reactors. SERP with methane as the feedstock has been carried out in one reactor at 903 K while the calcination of saturated $CaO/Ca_{12}Al_{14}O_{33}$ adsorbent is performed in the other reactor at 1123 K with Ar flow as the inert purge gas. During the multi-cycle SERP tests, Li *et al.* (2006) found that a hydrogen stream with purity higher than 90% can be obtained continuously from the outlet with such arrangement.

In addition, a SE-WGS process with TSR was developed by the research group of Lee *et al.* (2008b) for the simultaneous production of H_2 and CO_2 from synthesis gas, where the fuel cell grade H_2 and compressed CO_2 as a by-product gas can be directly produced by reacting the CO and H_2O feedstock from a synthesis gas produced by gasification of coal in a fixed bed adsorptive reactor. The cyclic TSR process consists of five operating steps including (a) SERP step, (b) high pressure CO_2 rinse step, (c) batch heating step, (d) high pressure steam purge step, and (e) multitasking regeneration step consisting of multiple sub-steps such as depressurization, cooling, low pressure steam purge and pressurization. A schematic drawing of the process flow sheet with four parallel adsorptive columns can be found from Fig. 2.29. Shell and tube-type heat exchangers have been employed, where the WGS catalyst and the CO_2 adsorbent are packed in the tube sides and shell sides are used for flowing the heating and cooling streams for indirect energy exchange of the reactors.

SERP occurs at relatively lower temperature conditions and the saturated adsorbent is then flushed with a hot carrier stream (N_2 or steam) for regeneration. The main drawback of TSR technology is the

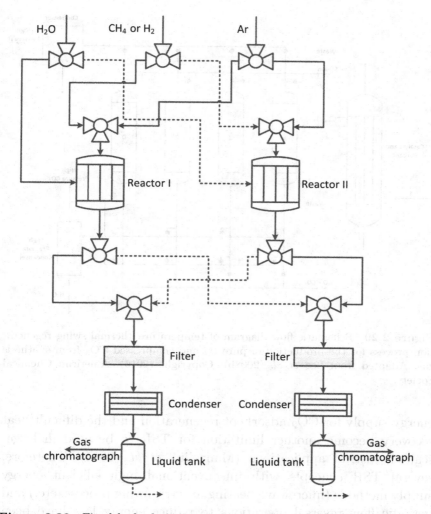

Figure 2.28 Fixed bed adsorptive reactors operated in a cyclic manner with TSR method. Adapted from Li *et al.* (2006). Copyright (2016), American Chemical Society.

extended period of time, which is usually much longer than the PSR or CSR, due to the thermal inertia of the reactor column, catalyst and adsorbent materials (Reßler *et al.*, 2006). Besides, the intermediate-(∼673–773 K) or high-temperature (>773 K) heat sources used for adsorptive reactors are not always available. Additionally, the intense

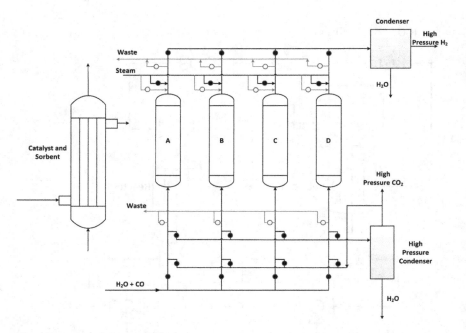

Figure 2.29   Schematic flow diagram of temperature/thermal swing regeneration process for the production of pure H_2 and compressed CO_2 from synthesis gas. Adapted from Lee *et al.* (2008b). Copyright (2016), American Chemical Society.

energy supply for CO_2 adsorbent regeneration and the difficult heat recovery become another limitation for TSR to be used in large-scale industrial applications (Manzolini *et al.*, 2015). Therefore, several TSR concepts with unconventional high efficient energy supply methods (microwave heating, electric swing process, etc.) and periodic flow reversal operations to reduce energy loss have been proposed in recent years.

On the other hand, a hybrid method with both PSR and TSR has been developed by increasing the temperature and decreasing the pressure during the regeneration step to improve the efficiency of the regeneration process and the CO_2 uptake capacity of adsorbent can therefore be further reduced. Besides, in addition to the hybrid TSR–PSR operation, Xiu *et al.* (2002a) proposed an even more efficient hybrid regeneration strategy — so-called "reactive regeneration",

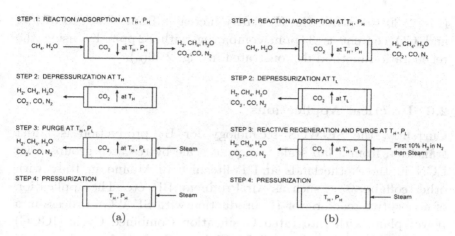

Figure 2.30 Operating steps for SERP with (a) the hybrid pressure–temperature swing regeneration process (TSR–PSR), and (b) TSR–PSR with reactive regeneration. Adapted from Xiu *et al.* (2002a). Copyright (2016), Elsevier.

also known as reaction enhanced desorption process for SE-SMR, where 10% H_2/N_2 gas mixture is fed into the reactor during the low pressure/high temperature regeneration step, as illustrated in Fig. 2.30. The hybrid TSR–PSR operation is given in Figs. 2.30(a) and 2.30(b) shows the operation of TSR–PSR-reactive regeneration. The reaction involved during the regeneration step is given in Eq. (2.58):

$$CO_2 \cdot \text{adsorbent} + 4H_2 \rightarrow \text{adsorbent} + CH_4 + 2H_2O$$

$$\Delta H_{298\,K} = -164.9\,\text{kJ} \cdot \text{mol}^{-1} \qquad (2.58)$$

Xiu *et al.* (2002a) found that by reacting with H_2, the adsorbed CO_2 can be converted into CH_4 through the methanation reaction (reversed methane reforming reaction). The use of a low pressure/high temperature condition with N_2 inert flow during regeneration step favours the desorption of CO_2 from the K-HTlc adsorbent. Besides, methanation reaction is an exothermic process, which can helps the decrease of CO_2 uptake capacity. Finally, under cyclic steady state, the conversion of methane can be improved from 56.6% to 67.5% by employing reactive regeneration strategy

(Fig. 2.30(b)), while the H_2 purity increases from 83.9% to 87.9%, and CO content is 33 ppm comparing with 82 ppm by using the regeneration method demonstrated in Fig. 2.30(a).

2.6. Practical Applications

Currently, the SE-WGS technology for H_2 production is under transition from a lab scale to a pilot scale by research groups from ECN in the Netherlands and Politécnico di Milano in Italy with other collaborators such as Air Products, BP, etc. The application of adsorptive reactors for H_2 production with SE-WGS process in a power plant with Integrated Gasification Combined Cycle (IGCC) technology by Gazzani *et al.* (2013a) can be found in Chapter 1. On the other hand, the application of SE-WGS for carbon capture in integrated steelworks was also evaluated (Gazzani *et al.*, 2013b), where SE-WGS was used for H_2 production and simultaneous CO_2 capture with blast furnace gas as the feedstock. The application of the SE-WGS process was investigated and compared with a conventional monoethanolamine (MEA)-based post-combustion absorption option as a reference. It was found that SE-WGS is able to reach a CO_2 avoidance as high as ~85%, while the MEA post combustion configuration has a carbon capture rate lower than 50%.

Besides, Gazzani *et al.* (2015) performed a more comprehensive simulation study to assess the potential of SE-WGS for pre-combustion CO_2 capture from steel mill off-gas, and compared with two commercially available CO_2 capture technologies, MEA-based post-combustion and methyldiethanolamine (MDEA)-based pre-combustion absorption processes, respectively. The results can be found in Fig. 2.31. The conventional post-combustion MEA process is able to achieve a CO_2 avoidance of 65% with a specific primary energy consumption for CO_2 avoided (SPECCA) of ~3.8 MJ $\cdot$ kg$_{CO_2}^{-1}$, which corresponds to an overall ~25–45% CO_2 emissions reduction since the power plant usually accounts for 40–70% of the total emissions from an integrated steel mill. By using the MDEA-based pre-combustion technology, the CO_2 capture rate can increase up to ~90% of CO_2 avoidance with a SPECCA of 3 MJ $\cdot$ kg$_{CO_2}^{-1}$. Finally,

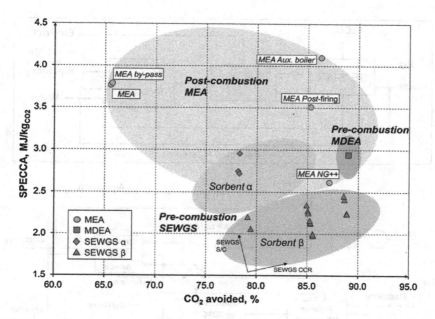

Figure 2.31 A comparison of the different plant solutions proposed in terms of CO_2 avoided and SPECCA (specific primary energy consumption for CO_2 avoided). Reprinted with permission from Gazzani *et al.* (2015). Copyright (2016), Elsevier.

the adoption of the SE-WGS process allows both decreasing the SPECCA to a range from $3.0\,\mathrm{MJ \cdot kg_{CO_2}^{-1}}$ to $2.0\,\mathrm{MJ \cdot kg_{CO_2}^{-1}}$, and increasing the CO_2 avoidance close to 90%. The authors (Gazzani *et al.*, 2015) also found that the SE-WGS performance is mostly affected by the performance of the CO_2 adsorbent employed (adsorbent α and β shown in Fig. 2.31). Because of the encouraging results, their work of SE-WGS will be demonstrated on pilot scale adjacent to the SSAB steel plant in Sweden in the Horizon 2020 project STEPWISE, which started on May 1, 2015 (Voldsund *et al.*, 2016).

In addition to SE-WGS, the zero emission gas (ZEG) power concept was developed by a Norwegian research group from the Institute for Energy Technology, which is a hybrid technology by using SERP with hydrocarbon (mainly methane) fuels as the feedstock and a high temperature Solid Oxide Fuel Cell (SOFC) for electricity production from H_2. The schematic diagram and of ZEG concept can be found

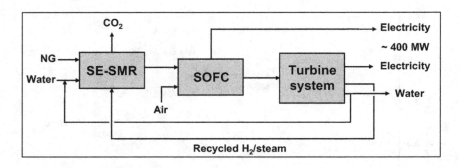

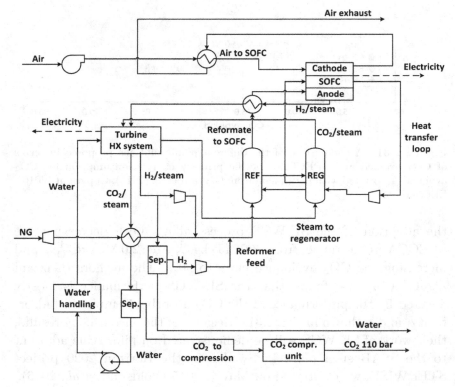

Figure 2.32 A schematic layout and process flow diagram of the ZEG power concept. Adapted from Meyer *et al.* (2011). Copyright (2016), Elsevier.

in Fig. 2.32, where natural gas (NG) is fed into the adsorptive reactor to produce high purity H_2 stream by SE-SMR reforming technology. SERP is carried out in the presence of a mixture of a nickel-based reforming catalyst (Haldor Topsøe) and a high temperature CaO-based adsorbent material (Arctic Dolomite, Franzefoss) for CO_2 *in situ* uptake operated at $\sim$873 K. The produced H_2 concentration is typically around 95 mol% (dry basis), and the H_2-rich stream is then fed to a SOFC for electricity production. The SERP was performed within a dual bubbling fluidized bed (DBFB) reactor system, where a reformer is employed for H_2 production and CO_2 capture, a regenerator to desorb CO_2 from the saturated adsorbent, and a piping loop that carries the adsorbent and reforming catalyst material between the two bubbling fluidized beds.

On one hand, high purity CO_2 (>96% after condensation) can be obtained during the adsorbent regeneration process using temperature swing with steam as the carrier gas, which may eliminate the costly separation steps downstream. On the other hand, the heat required to regenerate the saturated adsorbent can be provided by the SOFC waste heat through an internal heat transfer loop. The SOFC has a nominal operating temperature of 1103 K; an afterburner was employed in order to heat the anode gas temperature to >1273 K before it can be used to supply the energy requirement for the adsorbent regenerator. A techno-economical study performed using Aspen HYSYS simulation by Meyer *et al.* (2011) shows that, even with no income for the CO_2 captured and a quite moderate natural gas price of 19 EUR $\cdot$ MWh^{-1}, the ZEG-case shows profitability for an electric power price of 50 EUR $\cdot$ MWh^{-1} or higher.

With such promising simulation results, a prototype plant that demonstrates the ZEG-technology (Zero Emission Gas technology) has been built at Hynor Lillestrøm in Norway, with a hydrogen production capacity of 30 kW or approximately 1 kg $\cdot$ h^{-1} and a 20 kW SOFC module. Instead of using natural gas as the feedstock, a methane-rich gas stream was supplied from a municipal waste landfill site nearby through a pipeline, which contains about 41% CH_4, 30% CO_2, 27% N_2, and 2% other gases. Therefore, a small-scale gas upgrading system was installed to purify (desulphurization) and upgrade the landfill gas to a CH_4 content of 85–90%. The BioZEG technology has been tested in a 50 kW plant, where biogas derived methane feedstock was converted into hydrogen and electricity with integrated CO_2 capture process. Andresen *et al.* (2014) found that by using thermal integration between the SOFC module and the SERP system, a 70% overall energy efficiency can be reached. Additionally, the BioZEG concept can provide a carbon negative energy production when the CO_2 captured during the H_2 production process is sequestered or otherwise used.

2.7. Conclusions and Future Perspectives

By using SERPs for hydrogen production, the hydrogen yield and purity are expected to be improved by shifting thermodynamic equilibrium of the reaction towards product side through *in situ* carbon dioxide removal. This offers promising potential to simplify the hydrogen separation and purification steps, reduce or possibly eliminate carbon deposition in the reforming reactor, allow less expensive construction materials for adsorptive reactors due to the decreased reaction temperature, and the size of heat exchanger equipment may also decrease. Besides, an individual CO_2 capture step is also not required in some cases since a high-purity stream of CO_2 can be directly obtained during the adsorbent regeneration process.

In recent decades, much progress on SERP for H_2 production has been achieved in materials development, which includes the reforming and WGS catalysts, CO_2 adsorbents that can be used under SERP conditions with promising uptake capacity, and particularly hybrid materials (multifunctional catalyst–adsorbent). Comparatively speaking, more attention can be devoted to the development of CO_2 adsorbents, since a high and multi-cycle stable CO_2 adsorption capacity remains a major challenge. Hybrid materials can ensure a uniform distribution of active phase on adsorbent to achieve a better reaction performance than that with conventional catalyst/adsorbent mixture. They can also offer additional degrees of freedom (core–shell arrangement, catalyst-to-adsorbent ratio, etc.) for materials development. Besides, numerical simulation and optimization can be employed to achieve a rational hybrid material design for better SERP performance on the particle/pellet level.

On the other hand, both fluidized bed and fixed bed adsorptive reactor configurations can be found. Ideally, counter-current reactant–adsorbent flow pattern can be used in a fluidized bed adsorptive reactor to improve the utilization of the adsorptive capacity and H_2 purity. But difficult solid handling for SERP with a fluidized bed reactor remains the major obstacle. On the other hand, a low column utilization is generally found for H_2 production with fixed bed adsorptive reactors. Therefore, the simulated moving bed (SMB) reactor arrangement, where the countercurrent movement of adsorbent can be achieved by switching inlet and outlet ports along a sequence of fixed bed adsorptive reactors that has already been extensively used for SERP in liquid phase (Chapter 4), might be considered in the future for H_2 production with SERP in gas phase since water (steam) can be used as the reactant and desorbent stream at the same time.

The process design of the SERP is developing rapidly, but most cyclic operating strategies are adopted from conventional gas adsorption separation process (pressure swing, temperature swing, concentration swing, etc.) developed in the last century. Some latest approaches of the operating arrangements for gas adsorption separation process, such as electric swing adsorption, dual-reflux PSA, gas phase SMB adsorption, rapid cycle PSA, rotary-bed adsorption process might be tested to improve the cyclic efficiency of SERP.

Finally, it is believed that the future of H_2 production from SERP (sometimes together with a H_2 fuel cell system) relies on successful and practical applications such as SE-WGS for IGCC power plant and steel mill, SE-SMR for H_2 and electricity production with the commercialization as the ultimate goal.

Nomenclature

a_p	Area-to-volume ratio of the pellet	m^{-1}
Bi	Biot number	
C_i	Concentration of component i in the gas phase	$mol \cdot m^{-3}$
C_p	Specific molar heat at constant pressure	$J \cdot mol^{-1} \cdot K^{-1}$

C_{ps}	Specific heat of the solid	$J \cdot kg^{-1} \cdot K^{-1}$
C_{pw}	Specific heat of the column wall	$J \cdot kg^{-1} \cdot K^{-1}$
$\bar{C}_i$	Average concentration of compound i inside particle pores	$mol \cdot m^{-3}$
C_t	Total gas concentration	$mol \cdot m^{-3}$
$\bar{C}_{t,p}$	Total concentration in the pellet	$mol \cdot m^{-3}$
C_v	Specific molar heat at constant volume	$J \cdot mol^{-1} \cdot K^{-1}$
$C_{v,\text{ads}}$	Specific molar heat in the adsorbed phase at constant volume	$J \cdot mol^{-1} \cdot K^{-1}$
d_p	Pellet diameter	m
D_p	Pore diffusion coefficient	$m^2 \cdot s^{-1}$
D_{ax}	Axial dispersion coefficient	$m^2 \cdot s^{-1}$
E_a	Activation energy	$J \cdot mol^{-1}$
E_i	Activation energy of reaction i	$J \cdot mol^{-1}$
h_f	Film heat transfer coefficient between the gas and the pellet phase	$W \cdot m^{-2} \cdot K^{-1}$
h_w	Film heat transfer coefficient between the gas phase and the column wall	$W \cdot m^{-2} \cdot K^{-1}$
k_f	Film mass transfer coefficient	$m \cdot s^{-1}$
k_g	Thermal conductivity of the gas mixture	$W \cdot m^{-1} \cdot K^{-1}$
k_j	Rate constant of reaction j	$mol \cdot kg^{-1} \cdot s^{-1}$
K_i^*	Adsorption equilibrium constant of component i	
K_i	Adsorption equilibrium constant of component i	
K_j	Thermodynamic equilibrium constant of reaction j	
l_w	Wall thickness	m
N	Molar flow	$mol \cdot s^{-1}$
P	Total pressure	Pa
p_i	Partial pressure of component i	Pa
$\bar{q}_{CO_2}$	Average adsorbed CO_2 concentration in the sorbent	$mol \cdot kg^{-1}$
$q_{CO_2,eq}$	CO_2 concentration in the pellet at the equilibrium state	$mol \cdot kg^{-1}$
R_j	Reaction rate of component j	
r_j	Reaction rate of reaction j	$mol \cdot kg^{-1} \cdot s^{-1}$
r_{crystal}	Radius of the crystal	m
r_p	Radius of the pellet	m
R	Gas constant	$J \cdot mol^{-1} \cdot K^{-1}$
$R_{S/C}$	Steam-to-carbon molar ratio	
t	Time	s
T	Temperature of the gas phase	K
T_w	Wall temperature	K
T_{furnace}	Temperature in the furnace	K
u	Superficial velocity	$m \cdot s^{-1}$
U	Global external heat transfer coefficient	$W \cdot m^{-2} \cdot K^{-1}$
y_i	Molar fraction of component i	
z	Axial direction	m

Greek symbols

α	Volume fraction of catalyst	
λ	Effective thermal conductivity	$\text{W} \cdot \text{m}^{-1} \cdot \text{K}^{-1}$
ρ_c	Density of the column	$\text{kg} \cdot \text{m}^{-3}$
ρ_g	Gas density	$\text{kg} \cdot \text{m}^{-3}$
ρ_p	Density of the pellet	$\text{kg} \cdot \text{m}^{-3}$
ρ_w	Column wall density	$\text{kg} \cdot \text{m}^{-3}$
ΔH_{ads}	Adsorption enthalpy	$\text{J} \cdot \text{mol}^{-1}$
ΔH_j	Enthalpy of reaction j	$\text{J} \cdot \text{mol}^{-1}$
ε_c	Porosity of the column	
ε_p	Porosity of the pellet	
μ_g	Gas viscosity	$\text{Pa} \cdot \text{s}$
$\upsilon_{j,i}$	Stoichiometric number of component i in reaction j	

Subscripts

cat	Catalyst phase
eq	Equilibrium
sorb	Adsorbent phase

Acronyms

ATR	Auto thermal reforming
BD	Blowdown
CSR	Concentration swing regeneration
DEN	Denominator
EQ	Pressure equalization
GHG	Greenhouse gases
HTlc	Hydrotalcite like compound
IEO	International Energy Outlook
IGCC	Integrated Gasification Combined Cycle
LDF	Linear driving force
LDH	Layered double hydroxides
LDO	Layered double oxides
LEQ	Local equilibrium
MDEA	Methyldiethanolamine
MEA	Monoethanolamine
ODEs	Ordinary differential equations
OECD	Organization for Economic Cooperation and Development
OSR	Oxidative steam reforming
PDEs	Partial differential equations
PROX	Partial oxidation
PSA	Pressure swing adsorption
PSR	Pressure swing regeneration

RDS	Rate-determining step
REQ	Received pressure equalization
SERP	Sorption enhanced reaction process
SE-SMR	Sorption enhanced steam methane reforming
SE-SRE	Sorption enhanced steam reforming of ethanol
SE-WGS	Sorption enhanced water-gas shift
SMR	Steam methane reforming
SOFC	Solid Oxide Fuel Cell
SPECCA	Specific primary energy consumption for CO_2 avoided
SR	Steam reforming
SRE	Steam reforming of ethanol
TSR	Temperature/thermal swing regeneration
WGS	Water-gas shift
ZEG	Zero emission gas

References

Abanades, J.C., 2002. The maximum capture efficiency of CO_2 using a carbonation/calcination cycle of $CaO/CaCO_3$, *Chemical Engineering Journal*, 90, 303–306.

Abanades, J.C., Alvarez, D., 2003. Conversion limits in the reaction of CO_2 with lime, *Energy & Fuels*, 17, 308–315.

Abanades, J.C., Anthony, E.J., Lu, D.Y., Salvador, C., Alvarez, D., 2004. Capture of CO_2 from combustion gases in a fluidized bed of CaO, *AIChE Journal*, 50, 1614–1622.

Abanades, J.C., Grasa, G., Alonso, M., Rodriguez, N., Anthony, E.J., Romeo, L.M., 2007. Cost structure of a postcombustion CO_2 capture system using CaO, *Environmental Science & Technology*, 41, 5523–5527.

Adris, A.M., Pruden, B.B., Lim, C.J., Grace, J.R., 1996. On the reported attempts to radically improve the performance of the steam methane reforming reactor, *Canadian Journal of Chemical Engineering*, 74, 177–186.

Agar, D.W., 2005. The dos and don'ts of adsorptive reactors, in: *Integrated Chemical Processes: Synthesis, Operation, Analysis, and Control*, Sundmacher, K., Kienle, A., Seidel-Morgenstern, A. (Eds.), WILEY, Weinheim, 203–231.

Akande, A., Aboudheir, A., Idem, R., Dalai, A., 2006. Kinetic modeling of hydrogen production by the catalytic reforming of crude ethanol over a co-precipitated $Ni - Al_2O_3$ catalyst in a packed bed tubular reactor, *International Journal of Hydrogen Energy*, 31, 1707–1715.

Akpan, E., Akande, A., Aboudheir, A., Ibrahim, H., Idem, R., 2007. Experimental, kinetic and 2D reactor modeling for simulation of the production of hydrogen by the catalytic reforming of concentrated crude ethanol (CRCCE) over a Ni-based commercial catalyst in a packed-bed tubular reactor, *Chemical Engineering Science*, 62, 3112–3126.

Albrecht, K.O., Wagenbach, K.S., Satrio, J.A., Shanks, B.H., Wheelock, T.D., 2008. Development of a CaO-Based CO_2 sorbent with improved cyclic ctability, *Industrial & Engineering Chemistry Research*, 47, 7841–7848.

Allam, R.J., Chiang, R., Hufton, J.R., Middleton, P., Weist, E.L., White, V., 2005. Development of the sorption enhanced water gas shift process, in: *Carbon dioxide Capture for Storage in Deep Geologic Formations*, Vol. 1, Thomas, D.C., Benson, S.M. (Eds.), Elsevier, Amsterdam, 227–256.

Allen, D.W., Gerhard, E.R., Likins, M.R., 1975. Kinetics of the methane-steam reaction, *Industrial & Engineering Chemistry Research*, 14, 256–259.

Alvarez, D., Abanades, J.C., 2005. Pore-size and shape effects on the recarbonation performance of calcium oxide submitted to repeated calcination/recarbonation cycles, *Energy & Fuels*, 19, 270–278.

Andresen, B., Norheim, A., Strand, J., Ulleberg, Ø., Vik, A., Wærnhus, I., 2014. BioZEG–Pilot plant demonstration of high efficiency carbon negative energy production, *Energy Procedia*, 63, 279–285.

Angeli, S.D., Monteleone, G., Giaconia, A., Lemonidou, A.A., 2014. State-of-the-art catalysts for CH_4 steam reforming at low temperature, *International Journal of Hydrogen Energy*, 39, 1979–1997.

Auerbach, S.M., Carrado, K.A., Dutta, P.K., 2004. *Handbook of Layered Materials*, Marcel Dekker Inc., Hoboken.

Auprêtre, F., Descorme, C., Duprez, D., 2002. Bio-ethanol catalytic steam reforming over supported metal catalysts, *Catalysis Communications*, 3, 263–267.

Balasubramanian, B., Lopez Ortiz, A., Kaytakoglu, S., Harrison, D.P., 1999. Hydrogen from methane in a single-step process, *Chemical Engineering Science*, 54, 3543–3552.

Barelli, L., Bidini, G., Gallorini, F., Servili, S., 2008. Hydrogen production through sorption-enhanced steam methane reforming and membrane technology: A review, *Energy*, 33, 554–570.

Barker, R., 1973. The reversibility of the reaction $CaCO_3 \rightleftarrows CaO + CO_2$, *Journal of Applied Chemistry and Biotechnology*, 23, 733–742.

Barreto, L., Makihira, A., Riahi, K., 2003. The hydrogen economy in the 21st century: A sustainable development scenario, *International Journal of Hydrogen Energy*, 28, 267–284.

Bartels, J.R., Pate, M.B., Olson, N.K., 2010. An economic survey of hydrogen production from conventional and alternative energy sources, *International Journal of Hydrogen Energy*, 35, 8371–8384.

Beebe, T.P., Goodman, D.W., Kay, B.D., Yates, J.T., 1987. Kinetics of the activated dissociative adsorption of methane on the low index planes of nickel single crystal surfaces, *The Journal of Chemical Physics*, 87, 2305–2315.

Bengaard, H.S., Norskov, J.K., Sehested, J., Clausen, B.S., Nielsen, L.P., Molenbroek, A.M., Rostrup-Nielsen, J.R., 2002. Steam reforming and graphite formation on Ni catalysts, *Journal of Catalysis*, 209, 365–384.

Bhatia, S.K., Perlmutter, D.D., 1983. Effect of the product layer on the kinetics of the CO_2-lime reaction, *AIChE Journal*, 29, 79–86.

Bodrov, N.M., Apelbaum, L.O., Temkin, M.I., 1964. Kinetics of the reactions of methane with steam on the surface of nickel, *Kinetics and Catalysis*, 5, 696–705.

Bodrov, N.M., Apelbaum, L.O., Temkin, M.I., 1967. Kinetics of the reaction of methane with water vapour, catalysed nickel on a porous carrier, *Kinetics and Catalysis*, 8, 821–828.

Bodrov, N.M., Apelbaum, L.O., Temkin, M.I., 1968. Kinetics of the reactions of methane with steam on the surface of nickel at $400 - 600°C$, *Kinetics and Catalysis*, 9, 1065–1071.

Boon, J., Cobden, P.D., van Dijk, H.A.J., Annaland, M.V.S., 2015. High-temperature pressure swing adsorption cycle design for sorption-enhanced water-gas shift, *Chemical Engineering Science*, 122, 219–231.

Breen, J.P., Burch, R., Coleman, H.M., 2002. Metal-catalysed steam reforming of ethanol in the production of hydrogen for fuel cell applications, *Applied Catalysis B: Environmental*, 39, 65–74.

Bretado, M.E., Velderrain, V.G., Gutiérrez, D.L., Collins-Martínez, V., Ortiz, A.L., 2005. A new synthesis route to $Li_4 SiO_4$ as CO_2 catalytic/sorbent, *Catalysis Today*, 107, 863–867.

Broda, M., Manovic, V., Imtiaz, Q., Kierzkowska, A.M., Anthony, E.J., Müller, C.R., 2013. High-purity hydrogen via the sorption-enhanced steam methane reforming reaction over a synthetic CaO-based sorbent and a Ni catalyst, *Environmental Science & Technology*, 47, 6007–6014.

Brun-Tsekhovoi, A.R., Zadorin, A.N., Katsobashvili, Y.R., Kourdyumov, S.S., The process of catalytic steam-reforming of hydrocarbons in the presence of carbon dioxide acceptor, Proceedings of the 7th World Hydrogen Energy Conference. Pergamon Press, New York, 885–900.

Bshish, A., Yaakob, Z., Narayanan, B., Ramakrishnan, R., Ebshish, A., 2011. Steam-reforming of ethanol for hydrogen production, *Chemical Papers*, 65, 251–266.

Carrette, L., Friedrich, K.A., Stimming, U., 2000. Fuel cells: Principles, types, fuels, and applications, *Chemphyschem*, 1, 162–193.

Cavallaro, S., Freni, S., 1996. Ethanol steam reforming in a molten carbonate fuel cell. A preliminary kinetic investigation, *International Journal of Hydrogen Energy*, 21, 465–469.

Chanburanasiri, N., Ribeiro, A.M., Rodrigues, A.E., Arpornwichanop, A., Laosiripojana, N., Praserthdam, P., Assabumrungrat, S., 2011. Hydrogen production via sorption enhanced steam methane reforming process using Ni/CaO multifunctional catalyst, *Industrial & Engineering Chemistry Research*, 50, 13662–13671.

Chorkendorff, I., Niemantsverdriet, J.W., 2003. *Concepts of Modern Catalysis and Kinetics Concepts*, Wiley-Vch, Weinheim, Germany.

Ciambelli, P., Palma, V., Ruggiero, A., 2010. Low temperature catalytic steam reforming of ethanol. 2. Preliminary kinetic investigation of Pt/CeO_2 catalysts, *Applied Catalysis B: Environmental*, 96, 190–197.

Cobden, P.D., Elzingaa, G.D., Booneveld, S., Dijkstra, J.W., Jansen, D., van den Brink, R.W., 2009. Sorption-enhanced steam-methane reforming: CaO – $CaCO_3$ capture technology, *Energy Procedia*, 1, 733–739.

Cunha, A.F., 2009. Hydrogen production by catalytic decomposition of methane, Department of Chemical Engineering. Universidade do Porto, PhD Thesis.

Cunha, A.F., Moreira, M.N., Mafalda Ribeiro, A., Ferreira, A.P., Loureiro, J.M., Rodrigues, A.E., 2015. How to overcome the WGS equilibrium using a conventional nickel reformer catalyst, *Energy Technology*, 3, 1205–1216.

Cunha, A.F., Wu, Y.-J., Díaz Alvarado, F.A., Santos, J.C., Vaidya, P.D., Rodrigues, A.E., 2012. Steam reforming of ethanol on a Ni/Al_2O_3 catalyst coupled with a hydrotalcite-like sorbent in a multilayer pattern for CO_2 uptake, *Canadian Journal of Chemical Engineering*, 90, 1514–1526.

Cunha, A.F., Wu, Y.-J., Li, P., Yu, J.-G., Rodrigues, A.E., 2014. Sorption-enhanced steam reforming of ethanol on a novel K–Ni–Cu–hydrotalcite hybrid material, *Industrial & Engineering Chemistry Research*, 53, 3842–3853.

Cunha, A.F., Wu, Y.J., Santos, J.C., Rodrigues, A.E., 2013. Sorption enhanced steam reforming of ethanol on hydrotalcite-like compounds impregnated with active copper, *Chemical Engineering Research and Design*, 91, 581–592.

De Deken, J.C., Devos, E.F., Froment, G.F., 1982. Steam reforming of natural gas: intrinsic kinetics, diffusional influences, and reactor design, in: *Chemical Reaction Engineering-Boston*, Wei, J., Georgakis, C. (Eds.), American Chemical Society, Boston, 181–197.

Derevschikov, V.S., Lysikov, A.I., Okunev, A.G., 2011. Sorption properties of lithium carbonate doped CaO and its performance in sorption enhanced methane reforming, *Chemical Engineering Science*, 66, 3030–3038.

Dewoolkar, K.D., Vaidya, P.D., 2015. Improved hydrogen production by sorption-enhanced steam methane reforming over hydrotalcite- and calcium-based hybrid materials, *Energy & Fuels*, 29, 3870–3878.

Ding, Y., Alpay, E., 2000a. Adsorption-enhanced steam–methane reforming, *Chemical Engineering Science*, 55, 3929–3940.

Ding, Y., Alpay, E., 2000b. Equilibria and kinetics of CO_2 adsorption on hydrotalcite adsorbent, *Chemical Engineering Science*, 55, 3461–3474.

Ding, Y., Alpay, E., 2001. High temperature recovery of CO_2 from flue gases using hydrotalcite adsorbent, *Process Safety and Environmental Protection*, 79, 45–51.

DOE, 2013. Report of the Hydrogen Production Expert Panel: A Subcommittee of the Hydrogen & Fuel Cell Technical Advisory Committee. US Department of Energy, Washington, DC.

Doman, L.E., 2016. International Energy Outlook 2016, World energy demand and economic outlook. U.S. Energy Information Administration, Washington.

Dou, B., Song, Y., Wang, C., Chen, H., Xu, Y., 2014a. Hydrogen production from catalytic steam reforming of biodiesel byproduct glycerol: Issues and challenges, *Renewable and Sustainable Energy Reviews*, 30, 950–960.

Dou, B., Song, Y., Wang, C., Chen, H., Yang, M., Xu, Y., 2014b. Hydrogen production by enhanced-sorption chemical looping steam reforming of glycerol in moving-bed reactors, *Applied Energy*, 130, 342–349.

Dou, B., Wang, C., Chen, H., Song, Y., Xie, B., 2013. Continuous sorption-enhanced steam reforming of glycerol to high-purity hydrogen production, *International Journal of Hydrogen Energy*, 38, 11902–11909.

Dou, B., Wang, C., Song, Y., Chen, H., Jiang, B., Yang, M., Xu, Y., 2016. Solid sorbents for in-situ CO_2 removal during sorption-enhanced steam reforming process: A review, *Renewable and Sustainable Energy Reviews*, 53, 536–546.

Du, H., Ebner, A.D., Ritter, J.A., 2010. Pressure dependence of the nonequilibrium kinetic model that describes the adsorption and desorption behavior of CO_2 in K-promoted hydrotalcite like compound, *Industrial & Engineering Chemistry Research*, 50, 412–418.

Duraiswamy, K., Chellappa, A., Smith, G., Liu, Y., Li, M., 2010. Development of a high-efficiency hydrogen generator for fuel cells for distributed power generation, *International Journal of Hydrogen Energy*, 35, 8962–8969.

Ebner, A.D., Reynolds, S.P., Ritter, J.A., 2006. Understanding the adsorption and desorption behavior of CO_2 on a K-promoted hydrotalcite-like compound (HTlc) through nonequilibrium dynamic isotherms, *Industrial & Engineering Chemistry Research*, 45, 6387–6392.

Ebner, A.D., Reynolds, S.P., Ritter, J.A., 2007. Nonequilibrium kinetic model that describes the reversible adsorption and desorption behavior of CO2 in a K-Promoted Hydrotalcite-like Compound, *Industrial & Engineering Chemistry Research*, 46, 1737–1744.

EIA, US. 2011. International Energy Statistics. US Energy and Information Administration, Washington, DC.

Elnashaie, S.S.E.H., Adris, A.M., Al-Ubaid, A.S., Soliman, M.A., 1990. On the non-monotonic behaviour of methane steam reforming kinetics, *Chemical Engineering Science*, 45, 491–501.

Essaki, K., Kato, M., Nakagawa, K., 2006. CO_2 removal at high temperature using packed bed of lithium silicate pellets, *Nippon Seramikkusu Kyokai Gakujutsu Ronbunshi*, 114, 739–742.

Essaki, K., Muramatsu, T., Kato, M., 2008. Effect of equilibrium-shift in the case of using lithium silicate pellets in ethanol steam reforming, *International Journal of Hydrogen Energy*, 33, 6612–6618.

Feng, H.Z., Lan, P.Q., Wu, S.F., 2012. A study on the stability of a NiO − CaO/Al2O3 complex catalyst by La2O3 modification for hydrogen production, *International Journal of Hydrogen Energy*, 37, 14161–14166.

Ferreira-Aparicio, P., Benito, M.J., Sanz, J.L., 2005. New trends in reforming technologies: From hydrogen industrial plants to multifuel microreformers, *Catalysis Reviews*, 47, 491–588.

Ficicilar, B., Dogu, T., 2006. Breakthrough analysis for CO_2 removal by activated hydrotalcite and soda ash, *Catalysis Today*, 115, 274–278.

Frusteri, F., Freni, S., Spadaro, L., Chiodo, V., Bonura, G., Donato, S., Cavallaro, S., 2004. H_2 production for MC fuel cell by steam reforming of ethanol over

MgO supported Pd, Rh, Ni and Co catalysts, *Catalysis Communications*, 5, 611–615.

Gandía, L.M., Arzamendi, G., Diéguez, P.M., 2013. Chapter 1 — Renewable hydrogen energy: An overview, in: *Renewable Hydrogen Technologies*, Elsevier, Amsterdam, 1–17.

García-Lario, A.L., Grasa, G.S., Murillo, R., 2015. Performance of a combined CaO-based sorbent and catalyst on H_2 production, via sorption enhanced methane steam reforming, *Chemical Engineering Journal*, 264, 697–705.

Gates, S.M., Russell Jr, J.N., Yates Jr, J.T., 1986. Bond activation sequence observed in the chemisorption and surface reaction of ethanol on Ni(111), *Surface Science*, 171, 111–134.

Gazzani, M., Macchi, E., Manzolini, G., 2013a. CO_2 capture in integrated gasification combined cycle with SEWGS — Part A: Thermodynamic performances, *Fuel*, 105, 206–219.

Gazzani, M., Romano, M., Manzolini, G., 2013b. Application of sorption enhanced water gas shift for carbon capture in integrated steelworks, *Energy Procedia*, 37, 7125–7133.

Gazzani, M., Romano, M.C., Manzolini, G., 2015. CO_2 capture in integrated steelworks by commercial-ready technologies and SEWGS process, *International Journal of Greenhouse Gas Control*, 41, 249–267.

Gnanapragasam, N.V., Reddy, B.V., Rosen, M.A., 2014. Hydrogen production using solid fuels, in: *Handbook of Hydrogen Energy*, Sherif, S.A., Goswami, D.Y., Stefanakos, E.K., Steinfeld, A. (Eds.), CRC Press, London, 61–112.

Gorin, E., Retallick, B.W., 1963. Method for the production of hydrogen, US3108857 A, U.S.

Görke, O., Pfeifer, P., Schubert, K., 2009. Kinetic study of ethanol reforming in a microreactor, *Applied Catalysis A: General*, 360, 232–241.

Grande, C.A., 2016. PSA technology for H_2 separation, in: *Hydrogen Science and Engineering: Materials, Processes, Systems and Technology*, Stolten, D., Emonts, B. (Eds.), Wiley, Weinheim, Germany, pp. 489–508.

Gritsch, A., Kolios, G., Nieken, U., Eigenberger, G., 2007. Kompaktreformer für die dezentrale wasserstoffbereitstellung aus erdgas, *Chemie Ingenieur Technik*, 79, 821–830.

Gunduz, S., Dogu, T., 2012. Sorption-enhanced reforming of ethanol over Ni- and Co-incorporated MCM-41 type catalysts, *Industrial & Engineering Chemistry Research*, 51, 8796–8805.

Halabi, M.H., de Croon, M.H.J.M., van der Schaaf, J., Cobden, P.D., Schouten, J.C., 2010. Intrinsic kinetics of low temperature catalytic methane-steam reforming and water-gas shift over $Rh/Ce\alpha Zr_1 - \alpha O_2$ catalyst, *Applied Catalysis A: General*, 389, 80–91.

Halabi, M.H., de Croon, M.H.J.M., van der Schaaf, J., Cobden, P.D., Schouten, J.C., 2012a. High capacity potassium-promoted hydrotalcite for CO_2 capture in H_2 production, *International Journal of Hydrogen Energy*, 37, 4516–4525.

Halabi, M.H., de Croon, M.H.J.M., van der Schaaf, J., Cobden, P.D., Schouten, J.C., 2012b. Kinetic and structural requirements for a CO_2 adsorbent in sorption enhanced catalytic reforming of methane — Part I: Reaction kinetics and sorbent capacity, *Fuel*, 99, 154–164.

Harrison, D.P., 2008. Sorption-enhanced hydrogen production: A review, *Industrial & Engineering Chemistry Research*, 47, 6486–6501.

Harrison, D.P., Peng, Z., 2003. Low-carbon monoxide hydrogen by sorption-enhanced reaction, *International Journal of Chemical Reactor Engineering*, 1, 1–12.

Haryanto, A., Fernando, S., Murali, N., Adhikari, S., 2005. Current status of hydrogen production techniques by steam reforming of ethanol: A review, *Energy & Fuels*, 19, 2098–2106.

He, L., Berntsen, H., Chen, D., 2009. Approaching sustainable H_2 production: Sorption enhanced steam reforming of ethanol, *Journal of Physical Chemistry A*, 114, 3834–3844.

Holladay, J.D., Hu, J., King, D.L., Wang, Y., 2009. An overview of hydrogen production technologies, *Catalysis Today*, 139, 244–260.

Hufton, J.R., Mayorga, S., Sircar, S., 1999a. Sorption-enhanced reaction process for hydrogen production, *AIChE Journal*, 45, 248–256.

Hufton, J.R., Weigel, S.J., Waldron, W.F., Rao, M., Nataraj, S., Sircar, S., Gaffney, T.R., 1999b. Sorption Enhanced Reaction Process (SERP) for Production of Hydrogen. U.S. Department of Energy, Washington, DC.

Hutson, N., Attwood, B., 2008. High temperature adsorption of CO_2 on various hydrotalcite-like compounds, *Adsorption*, 14, 781–789.

Hutson, N.D., Speakman, S.A., Payzant, E.A., 2004. Structural effects on the high temperature adsorption of CO_2 on a synthetic hydrotalcite, *Chemistry of Materials*, 16, 4135–4143.

Ida, J.-I., Lin, Y.S., 2003. Mechanism of high-temperature CO_2 sorption on lithium zirconate, *Environmental Science & Technology*, 37, 1999–2004.

Iulianelli, A., Liguori, S., Wilcox, J., Basile, A., 2016. Advances on methane steam reforming to produce hydrogen through membrane reactors technology: A review, *Catalysis Reviews*, 58, 1–35.

Iwasaki, Y., Suzuki, Y., Kitajima, T., Sakurai, M., Kameyama, H., 2007. Non-equilibrium hydrogen production from ethanol using CO_2 absorption ceramic and precious metal catalysts, *Journal of Chemical Engineering of Japan*, 40, 178–185.

Jang, H.M., Kang, W.R., Lee, K.B., 2013. Sorption-enhanced water gas shift reaction using multi-section column for high-purity hydrogen production, *International Journal of Hydrogen Energy*, 38, 6065–6071.

Jang, H.M., Lee, K.B., Caram, H.S., Sircar, S., 2012. High-purity hydrogen production through sorption enhanced water gas shift reaction using K_2CO_3-promoted hydrotalcite, *Chemical Engineering Science*, 73, 431–438.

Jansen, D., van Selow, E., Cobden, P., Manzolini, G., Macchi, E., Gazzani, M., Blom, R., Henriksen, P.P., Beavis, R., Wright, A., 2013. SEWGS technology is now ready for scale-up!, *Energy Procedia*, 37, 2265–2273.

Jeong, K., Sircar, S., Caram, H.S., 2012. Modeling of heat recovery from a steam-gas mixture in a high-temperature sorption process, *AIChE Journal*, 58, 312–321.

Johnsen, K., Ryu, H.J., Grace, J.R., Lim, C.J., 2006. Sorption-enhanced steam reforming of methane in a fluidized bed reactor with dolomite as CO_2-acceptor, *Chemical Engineering Science*, 61, 1195–1202.

Jones, G., Jakobsen, J.G., Shim, S.S., Kleis, J., Andersson, M.P., Rossmeisl, J., Abild-Pedersen, F., Bligaard, T., Helveg, S., Hinnemann, B., Rostrup-Nielsen, J.R., Chorkendorff, I., Sehested, J., Nørskov, J.K., 2008. First principles calculations and experimental insight into methane steam reforming over transition metal catalysts, *Journal of Catalysis*, 259, 147–160.

Jones, L.W., 1970. Toward a liquid hydrogen fuel economy, in: *Engineering Technical Report*, University of Michigan, Michigan.

Kato, M., Nakagawa, K., Essaki, K., Maezawa, Y., Takeda, S., Kogo, R., Hagiwara, Y., 2005. Novel CO_2 absorbents using lithium-containing oxides, *International Journal of Applied Ceramic Technology*, 2, 467–475.

Kato, M., Yoshikawa, S., Essaki, K., Nakagawa, K., 2003a. Carbon dioxide gas absorbent and carbon dioxide gas separating apparatus. US Patent 20,030,075,050.

Kato, Y., Ando, K., Yoshizawa, Y., 2003b. Study on a regenerative fuel reformer for a zero-emission vehicle system, *Journal of Chemical Engineering of Japan*, 36, 860–866.

Kim, H., Jang, H.D., Choi, M., 2015. Facile synthesis of macroporous Li_4SiO_4 with remarkably enhanced CO_2 adsorption kinetics, *Chemical Engineering Journal*, 280, 132–137.

Koehle, M., Mhadeshwar, A., 2012. Microkinetic modeling and analysis of ethanol partial oxidation and reforming reaction pathways on platinum at short contact times, *Chemical Engineering Science*, 78, 209–225.

Kuramoto, K., Fujimoto, S., Morita, A., Shibano, S., Suzuki, Y., Hatano, H., Shi-Ying, L., Harada, M., Takarada, T., 2003. Repetitive carbonation-calcination reactions of Ca-based sorbents for efficient CO_2 sorption at elevated temperatures and pressures, *Industrial & Engineering Chemistry Research*, 42, 975–981.

Lawrence, P., Grünewald, M., 2006. Multifunctionality at particle level — Studies for adsorptive catalysts, in: *Integrated Reaction and Separation Operations*, Schmidt-Traub, H., Górak, A. (Eds.), Springer, Berlin, pp. 339–359.

Lee, C.H., Lee, K.B., 2014. Application of one-body hybrid solid pellets to sorption-enhanced water gas shift reaction for high-purity hydrogen production, *International Journal of Hydrogen Energy*, 39, 18128–18134.

Lee, C.H., Mun, S., Lee, K.B., 2015. Application of multisection packing concept to sorption-enhanced steam methane reforming reaction for high-purity hydrogen production, *Journal of Power Sources*, 281, 158–163.

Lee, D.K., Baek, I.H., Yoon, W.L., 2004. Modeling and simulation for the methane steam reforming enhanced by in situ CO_2 removal utilizing the CaO carbonation for H_2 production, *Chemical Engineering Science*, 59, 931–942.

Lee, D.K., Hyun Baek, II, Lai Yoon, W., 2006. A simulation study for the hybrid reaction of methane steam reforming and in situ CO_2 removal in a moving bed reactor of a catalyst admixed with a CaO-based CO_2 acceptor for H_2 production, *International Journal of Hydrogen Energy*, 31, 649–657.

Lee, I.-D., Kadlec, R.H., 1988. Effects of adsorbent and catalyst distributions in pressure swing reactors, *AIChE Symposium Series*, 84, 167–176.

Lee, J.M., Min, Y.J., Lee, K.B., Jeon, S.G., Na, J.G., Ryu, H.J., 2010. Enhancement of CO_2 sorption uptake on hydrotalcite by impregnation with K_2CO_3, *Langmuir*, 26, 18788–18797.

Lee, K.B., Beaver, M.G., Caram, H.S., Sircar, S., 2007a. Novel thermal-swing sorption-enhanced reaction process concept for hydrogen production by low-temperature steam−methane reforming, *Industrial & Engineering Chemistry Research*, 46, 5003–5014.

Lee, K.B., Beaver, M.G., Caram, H.S., Sircar, S., 2008a. Production of fuel-cell grade hydrogen by thermal swing sorption enhanced reaction concept, *International Journal of Hydrogen Energy*, 33, 781–790.

Lee, K.B., Beaver, M.G., Caram, H.S., Sircar, S., 2008b. Reversible chemisorbents for carbon dioxide and their potential applications, *Industrial & Engineering Chemistry Research*, 47, 8048–8062.

Lee, K.B., Verdooren, A., Caram, H.S., Sircar, S., 2007b. Chemisorption of carbon dioxide on potassium-carbonate-promoted hydrotalcite, *Journal of Colloid and Interface Science*, 308, 30–39.

LeValley, T.L., Richard, A.R., Fan, M., 2014. The progress in water gas shift and steam reforming hydrogen production technologies — A review, *International Journal of Hydrogen Energy*, 39, 16983–17000.

Li, S., Shi, Y., Cai, N., 2014. Potassium-promoted γ-alumina adsorbent from K_2CO_3 coagulated alumina sol for warm gas carbon dioxide separation, *ACS Sustainable Chemistry & Engineering*, 3, 111–116.

Li, Z.-S., Cai, N.-S., Yang, J.-B., 2006. Continuous production of hydrogen from sorption-enhanced steam methane reforming in two parallel fixed-bed reactors operated in a cyclic manner, *Industrial & Engineering Chemistry Research*, 45, 8788–8793.

Ligthart, D.A.J.M., Van Santen, R.A., Hensen, E.J.M., 2011. Influence of particle size on the activity and stability in steam methane reforming of supported Rh nanoparticles, *Journal of Catalysis*, 280, 206–220.

Llera, I., Mas, V., Bergamini, M.L., Laborde, M., Amadeo, N., 2012. Bio-ethanol steam reforming on Ni based catalyst. Kinetic study, *Chemical Engineering Science*, 71, 356–366.

Llorca, J., de la Piscina, P.R., Dalmon, J.-A., Sales, J., Homs, N., 2003. CO-free hydrogen from steam-reforming of bioethanol over ZnO-supported cobalt catalysts: Effect of the metallic precursor, *Applied Catalysis B: Environmental*, 43, 355–369.

Llorca, J., Homs, N., Sales, J., de la Piscina, P.R., 2002. Efficient production of hydrogen over supported cobalt catalysts from ethanol steam reforming, *Journal of Catalysis*, 209, 306–317.

Llorca, J., Homs, N., Sales, J., Fierro, J.-L.G., Ramírez de la Piscina, P., 2004. Effect of sodium addition on the performance of Co-ZnO-based catalysts for hydrogen production from bioethanol, *Journal of Catalysis*, 222, 470–480.

Lopes, F.V.S., 2010. New Pressure Swing Adsorption Cycles for Hydrogen Purification from Steam Reforming Off-Gases. PhD Thesis. University of Porto, Porto.

Lopes, F.V.S., Grande, C.A., Rodrigues, A.E., 2011. Activated carbon for hydrogen purification by pressure swing adsorption: Multicomponent breakthrough curves and PSA performance, *Chemical Engineering Science*, 66, 303–317.

Lu, Z.P., Loureiro, J.M., Rodrigues, A.E., 1993. Simulation of pressure swing adsorption reactors, CHEMPOR'93, International Chem. Eng. Conf, Porto, Portugal.

Lugo, E.L., Wilhite, B.A., 2016. A theoretical comparison of multifunctional catalyst for sorption-enhanced reforming process, *Chemical Engineering Science*, 150, 1–15.

Lysikov, A., Derevschikov, V., Okunev, A., 2015. Sorption-enhanced reforming of bioethanol in dual fixed bed reactor for continuous hydrogen production, *International Journal of Hydrogen Energy*, 40, 14436–14444.

Lysikov, A.I., Trukhan, S.N., Okunev, A.G., 2008. Sorption enhanced hydrocarbons reforming for fuel cell powered generators, *International Journal of Hydrogen Energy*, 33, 3061–3066.

Maestri, M., Vlachos, D.G., Beretta, A., Groppi, G., Tronconi, E., 2008. Steam and dry reforming of methane on Rh: Microkinetic analysis and hierarchy of kinetic models, *Journal of Catalysis*, 259, 211–222.

Manzolini, G., Jansen, D., Wright, A.D., 2015. Sorption-enhanced fuel conversion, in: *Process Intensification for Sustainable Energy Conversion*, Gallucci, F., Annaland, M.V.S. (Eds.), John Wiley & Sons, Chichester.

Martavaltzi, C.S., Lemonidou, A.A., 2010. Hydrogen production via sorption enhanced reforming of methane: Development of a novel hybrid material — reforming catalyst and CO_2 sorbent, *Chemical Engineering Science*, 65, 4134–4140.

Martunus, Othman, M.R., Fernando, W.J.N., 2011. Elevated temperature carbon dioxide capture via reinforced metal hydrotalcite, *Microporous and Mesoporous Materials*, 138, 110–117.

Mas, V., Dieuzeide, M.L., Jobbágy, M., Baronetti, G., Amadeo, N., Laborde, M., 2008. Ni(II)-Al(III) layered double hydroxide as catalyst precursor for ethanol steam reforming: Activation treatments and kinetic studies, *Catalysis Today*, 133–135, 319–323.

Mathure, P.V., Ganguly, S., Patwardhan, A.V., Saha, R.K., 2007. Steam reforming of ethanol using a commercial nickel-based catalyst, *Industrial & Engineering Chemistry Research*, 46, 8471–8479.

Mattos, L.V., Jacobs, G., Davis, B.H., Noronha, F.B., 2012. Production of hydrogen from ethanol: Review of reaction mechanism and catalyst deactivation, *Chemical Reviews*, 112, 4094–4123.

Mayorga, S.G., Hufton, J.R., Sircar, S., Gaffney, T.R., 1997. Sorption enhanced reaction process for production of hydrogen. Phase 1 final report. U.S. Department of Energy, Washington, DC.

Mayorga, S.G., Weigel, S.J., Gaffney, T.R., Brzozowski, J.R., 2001. Carbon dioxide adsorbents containing magnesium oxide suitable for use at high temperatures. US Patent 6280503.

Meyer, J., Mastin, J., Bjørnebøle, T.-K., Ryberg, T., Eldrup, N., 2011. Techno-economical study of the Zero Emission Gas power concept, *Energy Procedia*, 4, 1949–1956.

Miguel, C.V., Trujillano, R., Rives, V., Vicente, M.A., Ferreira, A.F.P., Rodrigues, A.E., Mendes, A., Madeira, L.M., 2014. High temperature CO_2 sorption with gallium-substituted and promoted hydrotalcites, *Separation and Purification Technology*, 127, 202–211.

Morbidelli, M., Servida, A., Varma, A., 1982. Optimal catalyst activity profiles in pellets. 1. The case of negligible external mass transfer resistance, *Industrial & Engineering Chemistry Fundamentals*, 21, 278–284.

Morgenstern, D.A., Fornango, J.P., 2005. Low-temperature reforming of ethanol over copper-plated raney nickel: a new route to sustainable hydrogen for transportation, *Energy & Fuels*, 19, 1708–1716.

Motay, M.T.D., Marechal, M., 1868. Préparation industrielle de l'hydrogène, *Bulletin Mensuel De La Societe Chimique De Paris*, 9, 334.

Murray, M.L., Hugo Seymour, E., Pimenta, R., 2007. Towards a hydrogen economy in Portugal, *International Journal of Hydrogen Energy*, 32, 3223–3229.

Najmi, B., Bolland, O., Westman, S.F., 2013. Simulation of the cyclic operation of a PSA-based SEWGS process for hydrogen production with CO_2 capture, *Energy Procedia*, 37, 2293–2302.

Nakagawa, K., Kato, M., Ohashi, T., Yoshikawa, S., Essaki, K., 2002. Carbon dioxide gas absorbent. US Patent 20,020,037,810.

Nakagawa, K., Ohashi, T., 1998. A novel method of CO_2 capture from high temperature gases, *Journal of The Electrochemical Society*, 145, 1344–1346.

Ni, M., Leung, M.K.H., Leung, D.Y.C., Sumathy, K., 2007. A review and recent developments in photocatalytic water-splitting using for hydrogen production, *Renewable and Sustainable Energy Reviews*, 11, 401–425.

Nieva, M.A., Villaverde, M.M., Monzón, A., Garetto, T.F., Marchi, A.J., 2014. Steam-methane reforming at low temperature on nickel-based catalysts, *Chemical Engineering Journal*, 235, 158–166.

Ochoa-Fernandez, E., Rønning, M., Grande, T., Chen, D., 2006. Nanocrystalline lithium zirconate with improved kinetics for high-temperature CO_2 capture, *Chemistry of Materials*, 18, 1383–1385.

Ochoa-Fernandez, E., Rusten, H.K., Jakobsen, H.A., Rønning, M., Holmen, A., Chen, D., 2005. Sorption enhanced hydrogen production by steam methane reforming using Li_2ZrO_3 as sorbent: sorption kinetics and reactor simulation, *Catalysis Today*, 106, 41–46.

Oliveira, E.L.G., Grande, C.A., Rodrigues, A.E., 2008. CO_2 Sorption on hydrotalcite and alkali-modified (K and Cs) hydrotalcites at high temperatures, *Separation and Purification Technology*, 62, 137–147.

Oliveira, E.L.G., Grande, C.A., Rodrigues, A.E., 2009. Steam methane reforming in a Ni/Al_2O_3 catalyst: Kinetics and diffusional limitations in extrudates, *Canadian Journal of Chemical Engineering*, 87, 945–956.

Oliveira, E.L.G., Grande, C.A., Rodrigues, A.E., 2010. Methane steam reforming in large pore catalyst, *Chemical Engineering Science*, 65, 1539–1550.

Oliveira, E.L.G., Grande, C.A., Rodrigues, A.E., 2011. Effect of catalyst activity in SMR-SERP for hydrogen production: Commercial vs. large-pore catalyst, *Chemical Engineering Science*, 66, 342–354.

Panagiotopoulou, P., Papadopoulou, C., Matralis, H., Verykios, X., 2014. Production of renewable hydrogen by reformation of biofuels, *Wiley Interdisciplinary Reviews-Energy and Environment*, 3, 231–253.

Poling, B.E., Prausnitz, J.M., O'Connell, J.P., 2001. *The Properties of Gases and Liquids*, McGraw-Hill, New York.

Ram Reddy, M.K., Xu, Z.P., Lu, G.Q., Diniz da Costa, J.C., 2006. Layered double hydroxides for CO_2 capture: structure evolution and regeneration, *Industrial & Engineering Chemistry Research*, 45, 7504–7509.

Ratnasamy, C., Wagner, J.P., 2009. Water gas shift catalysis, *Catalysis Reviews*, 51, 325–440.

Rawadieh, S., Gomes, V.G., 2009. Steam reforming for hydrogen generation with in situ adsorptive separation, *International Journal of Hydrogen Energy*, 34, 343–355.

Reijers, H.T.J., Elzinga, G.D., Cobden, P.D., Haije, W.G., van den Brink, R.W., 2011a. Tandem bed configuration for sorption-enhanced steam reforming of methane, *International Journal of Greenhouse Gas Control*, 5, 531–537.

Reijers, H.T.J., Valster-Schiermeier, S.E.A., Cobden, P.D., van den Brink, R.W., 2005. Hydrotalcite as CO_2 sorbent for sorption-enhanced steam reforming of methane, *Industrial & Engineering Chemistry Research*, 45, 2522–2530.

Reijers, R., van Selow, E., Cobden, P., Boon, J., van den Brink, R., 2011b. SEWGS process cycle optimization, *Energy Procedia*, 4, 1155–1161.

Reßler, S., Elsner, M.P., Dittrich, C., Geisler, S., 2006. Reactive gas adsorption, in: *Integrated Reaction and Separation Operations*, Springer, Heidelberg, 149–190.

Ronning, M., Ochoa-Fernandez, E., Grande, T., Chen, D., 2006. Carbon dioxide gas acceptors, WO2006111343.

Rostrup-Nielsen, J.R., 1984. Catalytic steam reforming, in: *Catalysis: Science and Technology*, Andersen, J.R., Boudart, M. (Eds.), Springer-Verlag, Berlin.

Rostrup-Nielsen, J.R., 2002. Large-scale hydrogen production, *CATTECH*, 6, 150–159.

Rostrup-Nielsen, J.R., 2004. Fuels and energy for the future: The role of catalysis, *Catalysis Reviews*, 46, 247–270.

Rostrup-Nielsen, J.R., 2008. Steam Reforming, in: *Handbook of Heterogeneous Catalysis*, Wiley-VCH Verlag GmbH & Co. KGaA.

Rostrup-Nielsen, J.R., Sehested, J., Nørskov, J.K., 2002. Hydrogen and synthesis gas by steam- and CO_2 reforming, *Advances in Catalysis*, 47, 65–139.

Rout, K.R., Jakobsen, H.A., 2013. A numerical study of pellets having both catalytic- and capture properties for SE-SMR process: Kinetic- and product layer diffusion controlled regimes, *Fuel Processing Technology*, 106, 231–246.

Rusten, H.K., Ochoa-Fernández, E., Chen, D., Jakobsen, H.A., 2007a. Numerical investigation of sorption enhanced steam methane reforming using Li_2ZrO_3 as CO_2-acceptor, *Industrial & Engineering Chemistry Research*, 46, 4435–4443.

Rusten, H.K., Ochoa-Fernández, E., Lindborg, H., Chen, D., Jakobsen, H.A., 2007b. Hydrogen production by sorption-enhanced steam methane reforming using lithium oxides as CO_2-acceptor, *Industrial & Engineering Chemistry Research*, 46, 8729–8737.

Ruthven, D.M., 1984. *Principles of Adsorption and Adsorption Processes*, John Wiley & Sons, New York.

Rydén, M., Ramos, P., 2012. H_2 production with CO_2 capture by sorption enhanced chemical-looping reforming using NiO as oxygen carrier and CaO as CO_2 sorbent, *Fuel Processing Technology*, 96, 27–36.

Sahoo, D.R., Vajpai, S., Patel, S., Pant, K.K., 2007. Kinetic modeling of steam reforming of ethanol for the production of hydrogen over Co/Al_2O_3 catalyst, *Chemical Engineering Journal*, 125, 139–147.

Sakurai, H., Masukawa, H., Kitashima, M., Inoue, K., 2013. Photobiological hydrogen production: Bioenergetics and challenges for its practical application, *Journal of Photochemistry and Photobiology C: Photochemistry*, 17, 1–25.

San Román, M.S., Holgado, M.J., Jaubertie, C., Rives, V., 2008. Synthesis, characterisation and delamination behaviour of lactate-intercalated Mg,Al-hydrotalcite-like compounds, *Solid State Sciences*, 10, 1333–1341.

Satrio, J.A., Shanks, B.H., Wheelock, T.D., 2005. Development of a novel combined catalyst and sorbent for hydrocarbon reforming, *Industrial & Engineering Chemistry Research*, 44, 3901–3911.

Schmidt-Traub, H., Górak, A., 2006. *Integrated Reaction and Separation Operations — Modelling and Experimental Validation*, Springer-Verlag, Berlin.

Seggiani, M., Puccini, M., Vitolo, S., 2011. High-temperature and low concentration CO_2 sorption on Li_4SiO_4 based sorbents: Study of the used silica and doping method effects, *International Journal of Greenhouse Gas Control*, 5, 741–748.

Sehested, J., 2006. Four challenges for nickel steam-reforming catalysts, *Catalysis Today*, 111, 103–110.

Sherif, S.A., Barbir, F., Veziroglu, T.N., 2014. Hydrogen economy, in: *Handbook of Hydrogen Energy*, Sherif, S.A., Goswami, D.Y., Stefanakos, E.K., Steinfeld, A. (Eds.), CRC Press, London, pp. 1–16.

Shokrollahi Yancheshmeh, M., Radfarnia, H.R., Iliuta, M.C., 2016. High temperature CO_2 sorbents and their application for hydrogen production by sorption enhanced steam reforming process, *Chemical Engineering Journal*, 283, 420–444.

Simson, A., Waterman, E., Farrauto, R., Castaldi, M., 2009. Kinetic and process study for ethanol reforming using a Rh/Pt washcoated monolith catalyst, *Applied Catalysis B: Environmental*, 89, 58–64.

Sircar, S., Golden, T.C., 2009. Pressure swing adsorption technology for hydrogen production, in: *Hydrogen and Syngas Production and Purification Technologies*, Liu, K., Song, C., Subramani, V. (Eds.), Wiley, Hoboken, pp. 414–450.

Smith, R., Loganathan, M., Shantha, M.S., 2010. A review of the water gas shift reaction kinetics, *International Journal of Chemical Reactor Engineering*, 8, 1–32.

Soliman, M.A., Adris, A.M., Al-Ubaid, A.S., El-Nashaie, S.S.E.H., 1992. Intrinsic kinetics of nickel/calcium aluminate catalyst for methane steam reforming, *Journal of Chemical Technology & Biotechnology*, 55, 131–138.

Solsvik, J., Jakobsen, H.A., 2011. A numerical study of a two property catalyst/sorbent pellet design for the sorption-enhanced steam–methane reforming process: Modeling complexity and parameter sensitivity study, *Chemical Engineering Journal*, 178, 407–422.

Solsvik, J., Jakobsen, H.A., 2012. A two property catalyst/sorbent pellet design for the sorption-enhanced steam–methane reforming process: mathematical modeling and numerical analysis, *Energy Procedia*, 26, 31–40.

Soria, M.A., Tosti, S., Mendes, A., Madeira, L.M., 2015. Enhancing the low temperature water–gas shift reaction through a hybrid sorption-enhanced membrane reactor for high-purity hydrogen production, *Fuel*, 159, 854–863.

Subramani, V., Sharma, P., Zhang, L., Liu, K., 2009. Catalytic steam reforming technology for the production of hydrogen and syngas, in: *Hydrogen and Syngas Production and Purification Technologies*, Liu, K., Song, C., Subramani, V. (Eds.), Wiley, Hoboken, pp. 14–126.

Sun, J., Luo, D., Xiao, P., Jigang, L., Yu, S., 2008. High yield hydrogen production from low CO selectivity ethanol steam reforming over modified Ni/Y_2O_3 catalysts at low temperature for fuel cell application, *Journal of Power Sources*, 184, 385–391.

Sun, J., Qiu, X.P., Wu, F., Zhu, W.T., 2005. H_2 from steam reforming of ethanol at low temperature over Ni/Y_2O_3, Ni/La_2O_3 and Ni/Al_2O_3 catalysts for fuel-cell application, *International Journal of Hydrogen Energy*, 30, 437–445.

Therdthianwong, A., Sakulkoakiet, T., Therdthianwong, S., 2001. Hydrogen production by catalytic ethanol steam reforming, *ScienceAsia*, 27, 193–198.

Ulibarri, M.A., Pavlovic, I., Barriga, C., Hermosín, M.C., Cornejo, J., 2001. Adsorption of anionic species on hydrotalcite-like compounds: Effect of interlayer anion and crystallinity, *Applied Clay Science*, 18, 17–27.

Vaidya, P.D., Rodrigues, A.E., 2006. Insight into steam reforming of ethanol to produce hydrogen for fuel cells, *Chemical Engineering Journal*, 117, 39–49.

van Selow, E.R., Cobden, P.D., Verbraeken, P.A., Hufton, J.R., van den Brink, R.W., 2009. Carbon capture by sorption-enhanced water-gas shift reaction process using hydrotalcite-based material, *Industrial & Engineering Chemistry Research*, 48, 4184–4193.

Vaporciyan, G.G., Kadlec, R., 1987. Equilibrium-limited periodic separating reactors, *AIChE Journal*, 33, 1334–1343.

Vernon, D.R., Paul, A.E., 2014. Hydrogen enrichment, in: *Handbook of Hydrogen Energy*, Sherif, S.A., Goswami, D.Y., Stefanakos, E.K., Steinfeld, A. (Eds.), CRC Press, London, pp. 903–934.

Voldsund, M., Jordal, K., Anantharaman, R., 2016. Hydrogen production with CO_2 capture, *International Journal of Hydrogen Energy*, 41, 4969–4992.

Wakao, N., Funazkri, T., 1978. Effect of fluid dispersion coefficients on particle-to-fluid mass transfer coefficients in packed beds: correlation of Sherwood numbers, *Chemical Engineering Science*, 33, 1375–1384.

Waldron, W.E., Hufton, J.R., Sircar, S., 2001. Production of hydrogen by cyclic sorption enhanced reaction process, *AIChE Journal*, 47, 1477–1479.

Wang, C., Dou, B., Song, Y., Chen, H., Xu, Y., Xie, B., 2014. High Temperature CO_2 sorption on Li_2ZrO_3 based sorbents, *Industrial & Engineering Chemistry Research*, 53, 12744–12752.

Wang, J.-H., Lee, C., Lin, M., 2009. Mechanism of ethanol reforming: Theoretical foundations, *Journal of Physical Chemistry C*, 113, 6681–6688.

Wang, L., Liu, Z., Li, P., Yu, J., Rodrigues, A.E., 2012a. Experimental and modeling investigation on post-combustion carbon dioxide capture using zeolite 13X-APG by hybrid VTSA process, *Chemical Engineering Journal*, 197, 151–161.

Wang, Q., Luo, J., Zhong, Z., Borgna, A., 2011a. CO_2 capture by solid adsorbents and their applications: current status and new trends, *Energy & Environmental Science*, 4, 42–55.

Wang, Q., Tay, H.H., Zhong, Z., Luo, J., Borgna, A., 2012b. Synthesis of high-temperature CO_2 adsorbents from organo-layered double hydroxides with markedly improved CO_2 capture capacity, *Energy & Environmental Science*, 5, 7526–7530.

Wang, Q., Wu, Z., Tay, H.H., Chen, L., Liu, Y., Chang, J., Zhong, Z., Luo, J., Borgna, A., 2011b. High temperature adsorption of CO_2 on Mg-Al hydrotalcite: Effect of the charge compensating anions and the synthesis pH, *Catalysis Today*, 164, 198–203.

Wang, S., Shen, H., Fan, S., Zhao, Y., Ma, X., Gong, J., 2013. Enhanced CO_2 adsorption capacity and stability using CaO-based adsorbents treated by hydration, *AIChE Journal*, 59, 3586–3593.

Wang, W., Cao, Y., 2011. Hydrogen production via sorption enhanced steam reforming of butanol: Thermodynamic analysis, *International Journal of Hydrogen Energy*, 36, 2887–2895.

Wang, Y.-N., Rodrigues, A.E., 2005. Hydrogen production from steam methane reforming coupled with in situ CO_2 capture: conceptual parametric study, *Fuel*, 84, 1778–1789.

Wei, J., Iglesia, E., 2004. Isotopic and kinetic assessment of the mechanism of reactions of CH_4 with CO_2 or H_2O to form synthesis gas and carbon on nickel catalysts, *Journal of Catalysis*, 224, 370–383.

Wiltowski, T., Mondal, K., Campen, A., Dasgupta, D., Konieczny, A., 2008. Reaction swing approach for hydrogen production from carbonaceous fuels, *International Journal of Hydrogen Energy*, 33, 293–302.

Wright, A.D., White, V., Hufton, J.R., Quinn, R., Cobden, P.D., van Selow, E.R., 2011. CAESAR: Development of a SEWGS model for IGCC, *Energy Procedia*, 4, 1147–1154.

Wu, G., Zhang, C., Li, S., Huang, Z., Yan, S., Wang, S., Ma, X., Gong, J., 2012a. Sorption enhanced steam reforming of ethanol on $Ni-CaO-Al_2O_3$ multifunctional catalysts derived from hydrotalcite-like compounds, *Energy & Environmental Science*, 5, 8942–8949.

Wu, S.-F., Beum, T.H., Yang, J.I., Kim, J.N., 2005. The characteristics of a sorption-enhanced steam-methane reaction for the production of hydrogen using CO_2 sorbent, *Chinese Journal of Chemical Engineering*, 13, 43–47.

Wu, Y.-J., Díaz-Alvarado, F.A., Santos, J.C., Gracia, F., Cunha, A.F., Rodrigues, A.E., 2012b. sorption-enhanced steam reforming of ethanol: thermodynamic comparison of CO_2 sorbents, *Chemical Engineering & Technology*, 35, 847–858.

Wu, Y.-J., Li, P., Yu, J.-G., Cunha, A.F., Rodrigues, A.E., 2013a. K-Promoted hydrotalcites for CO_2 capture in sorption enhanced reactions, *Chemical Engineering & Technology*, 36, 567–574.

Wu, Y.-J., Li, P., Yu, J.-G., Cunha, A.F., Rodrigues, A.E., 2013b. Sorption-enhanced steam reforming of ethanol on NiMgAl multifunctional materials: Experimental and numerical investigation, *Chemical Engineering Journal*, 231, 36–48.

Wu, Y.-J., Li, P., Yu, J.-G., Cunha, A.F., Rodrigues, A.E., 2014a. High-purity hydrogen production by sorption-enhanced steam reforming of ethanol: A cyclic operation simulation study, *Industrial & Engineering Chemistry Research*, 53, 8515–8527.

Wu, Y.-J., Li, P., Yu, J.-G., Cunha, A.F., Rodrigues, A.E., 2014b. Sorption-enhanced steam reforming of ethanol for continuous high-purity hydrogen production: 2D adsorptive reactor dynamics and process design, *Chemical Engineering Science*, 118, 83–93.

Wu, Y.-J., Li, P., Yu, J.-G., Cunha, A.F., Rodrigues, A.E., 2016. Progress on sorption-enhanced reaction process for hydrogen production, *Reviews in Chemical Engineering*, 32, 271–303.

Wu, Y.-J., Li, P., Yu, J.-G., Cunha, A.F., Rodrigues, A.E., 2014c. High-purity hydrogen production by sorption-enhanced steam reforming of ethanol: A cyclic operation simulation study, *Industrial & Engineering Chemistry Research*, 53, 8515–8527.

Wu, Y.-J., Santos, J.C., Li, P., Yu, J.-G., Cunha, A.F., Rodrigues, A.E., 2014d. Simplified kinetic model for steam reforming of ethanol on a Ni/Al_2O_3 catalyst, *Canadian Journal of Chemical Engineering*, 92, 116–130.

Xie, M., Zhou, Z., Qi, Y., Cheng, Z., Yuan, W., 2012. Sorption-enhanced steam methane reforming by in situ CO_2 capture on a $CaO-Ca_9Al_6O_{18}$ sorbent, *Chemical Engineering Journal*, 207–208, 142–150.

Xiong, R., Ida, J., Lin, Y.S., 2003. Kinetics of carbon dioxide sorption on potassium-doped lithium zirconate, *Chemical Engineering Science*, 58, 4377–4385.

Xiu, G.-H., Li, P., Rodrigues, A.E., 2002a. Sorption-enhanced reaction process with reactive regeneration, *Chemical Engineering Science*, 57, 3893–3908.

Xiu, G.-H., Li, P., Rodrigues, A.E., 2003. New generalized strategy for improving sorption-enhanced reaction process, *Chemical Engineering Science*, 58, 3425–3437.

Xiu, G.-H., Soares, J.L., Li, P., Rodrigues, A.E., 2002b. Simulation of five-step one-bed sorption-enhanced reaction process, *AIChE Journal*, 48, 2817–2832.

Xiu, G.H., Li, P., Rodrigues, A.E., 2004. Subsection-controlling strategy for improving sorption-enhanced reaction process, *Chemical Engineering Research and Design*, 82, 192–202.

Xu, J., Froment, G.F., 1989. Methane steam reforming, methanation and water-gas shift: I. Intrinsic kinetics, *AIChE Journal*, 35, 88–96.

Yancheshmeh, M.S., Radfarnia, H.R., Iliuta, M.C., 2016. High temperature CO_2 sorbents and their application for hydrogen production by sorption enhanced steam reforming process, *Chemical Engineering Journal*, 283, 420–444.

Yang, R.T., 2003. *Adsorbents Fundamentals and Applications*, Wiley-Interscience, Hoboken, N.J.

Yong, S.T., Ooi, C.W., Chai, S.-P., Wu, X., 2013. Review of methanol reforming-Cu-based catalysts, surface reaction mechanisms, and reaction schemes, *International Journal of Hydrogen Energy*, 38, 9541–9552.

Yong, Z., Mata, V., Rodrigues, A.E., 2000a. Adsorption of carbon dioxide onto hydrotalcite-like compounds (HTlcs) at high temperatures, *Industrial & Engineering Chemistry Research*, 40, 204–209.

Yong, Z., Mata, V., Rodrigues, A.E., 2000b. Adsorption of carbon dioxide on basic alumina at high temperatures, *Journal of Chemical & Engineering Data*, 45, 1093–1095.

Yong, Z., Mata, V., Rodrigues, A.E., 2002. Adsorption of carbon dioxide at high temperature — a review, *Separation and Purification Technology*, 26, 195–205.

Yong, Z., Rodrigues, A.E., 2002. Hydrotalcite-like compounds as adsorbents for carbon dioxide, *Energy Conversion and Management*, 43, 1865–1876.

Zhang, Q., Han, D., Liu, Y., Ye, Q., Zhu, Z., 2013. Analysis of CO_2 sorption/desorption kinetic behaviors and reaction mechanisms on Li_4SiO_4, *AIChE Journal*, 59, 901–911.

Chapter 3

Membrane Reactors for Water-Gas Shift

In line with the main goal of the book, this chapter addresses another multifunctional reactor configuration designed to overcome thermodynamic limitations related with conversion in reversible reactions: membrane reactors (MRs). According to IUPAC definition, a MR is a device that combines the separation properties of membranes with the typical characteristics of catalytic processes in one unit; this means that the membrane does not only play the role as a separator but also as a part of the reactor itself. As case study, the chapter is focused in an industrially relevant gas-phase reaction, the water-gas shift (WGS), carried out in either MRs or in sorption-enhanced membrane reactors (SEMRs), although studies for the last configuration are still very limited. Therefore, process intensification is reached by combining in a single unit (i) reaction and membranes or (ii) adsorption, reaction and membranes.

Emphasis is given to reaction-related aspects (main types of catalysts employed, mechanisms and kinetics), to the membranes (their nature, properties and permeation mechanism), and to issues related with the process and technology development (reactor configurations, effect of process variables, modelling and simulation). Some examples of applications at pilot scale will also be shown.

3.1. Introduction — The Concept

The water-gas shift (WGS) reaction (Eq. (3.1)) is a key step in fuel processing to generate hydrogen. Particularly, it is the intermediate step used for hydrogen enrichment and CO reduction in the synthesis gas (mixture of carbon monoxide and hydrogen, also called "syngas" or "water gas"). It has several industrial applications but a current challenge is focused on hydrogen-fed fuel cells. It is a reversible, moderately exothermic chemical reaction.

$$CO + H_2O \leftrightarrow CO_2 + H_2 (\Delta H^0_{298K} = -41.09\,kJ/mol) \qquad (3.1)$$

The WGS reaction plays a significant role in several industrial processes, including:

- Tuning the H_2/CO molar ratio in the synthesis gas, the value of which determines its subsequent use in different processes like ammonia manufacture (Haber process), methanol production, Fischer–Tropsch synthesis (FTS) or metals production (by the reduction of the oxide ore, e.g. iron ore) (Ratnasamy and Wagner, 2009);
- Hydrogen production in the petroleum refining and petrochemical industry for a variety of operations (hydrotreating, hydrocracking of petroleum fractions and other hydrogenations);
- Hydrogen production as fuel for power generation and transportation (Smith *et al.*, 2010);
- FTS for e.g. coal-to-liquid (CTL) processes with iron-based (Bukur *et al.*, 2016) and other catalysts (Laan and Beenackers, 1999);
- Hydrogen production coupled with steam reforming of hydrocarbons (natural gas, petroleum gas, naphtha, gasoline, coals and various types of biomass), particularly of methane (Levalley *et al.*, 2014).

To circumvent the thermodynamic limitations of the reversible WGS reaction, further detailed in the next section, several strategies can be adopted. One possibility is through the use of membrane reactors (MRs), which remove one reaction product (e.g. H_2) from the reaction medium by using permselective membranes. As illustrated in Fig. 3.1, the idea is to use a hydrogen-selective membrane as a way to extract it from the reaction medium and thus shift the reversible reaction towards the products side. Moreover, a high pressure CO_2 stream is produced in the retentate side (along with high purity

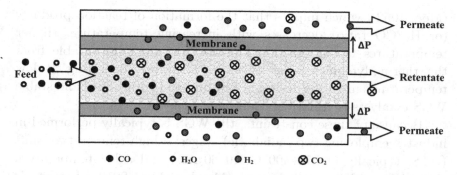

Figure 3.1 Sketch of a WGS MR using an H_2-selective membrane. Adapted from Smart *et al.* (2010). Copyright (2010), The Royal Society of Chemistry.

H_2 in the permeate), which can facilitate the economics of power generation with carbon sequestration (because compression energy required is reduced). The concept has been proposed as long as in the 1960s (Rothfleisch, 1964); many publications refer instead to the pioneering work of Uemiya *et al.* (1991) in the 1990s, and others to previous works, from Gryaznov in the 1960s, or even in the 19th century. Since then different types of membranes have been employed (cf. Section 3.5), in different MR configurations (cf. Section 3.6); membranes selective to the other reaction product, CO_2 (cf. Section 3.5.4), have also been used. Alternatively, the shift of the WGS equilibrium can be reached by removing CO_2 using a selective sorbent — so-called sorption enhanced; this topic has also been the focus of extensive research (as detailed in Chapter 2), and more recently through the integration of both an H_2-selective membrane and a CO_2 sorbent in the same multifunctional device (cf. Section 3.6.4).

3.2. Thermodynamic Aspects

There are several parameters influencing the WGS in what concerns equilibrium. As shown above (Eq. (3.1)), the WGS reaction is reversible and moderately exothermic, yielding 41.1 kJ/mol of carbon monoxide converted. The equilibrium constant (K_p) for the WGS reaction decreases with the temperature, as expressed by Eq. (3.2)

(Moe, 1962), which implies that the formation of reaction products (or H_2/CO ratio) decreases with increasing temperature. Higher temperatures favour reaction kinetics, but are unfavourable from the thermodynamics point of view. Therefore, operating at low temperatures might be desirable, but finding highly active and stable WGS catalysts is still a big challenge.

Because of these constraints, the WGS is typically performed in industry employing two (adiabatic) stages: a high-temperature shift (HTS, typically at 370–400°C, 10–60 atm) and a low-temperature shift (LTS, ~200°C, 10–40 atm) (Mendes *et al.*, 2010a; Lee *et al.*, 2013). The units are connected via an interstage cooler, also usually with intermediate acid gas removal and desulphurization, which is required to eliminate poisons to the second-stage catalyst. HTS is characterized by fast kinetics (thus decreasing catalyst bed volume), but the final CO conversion is limited by equilibrium. In contrast, LTS undergoes slow kinetics, but the thermodynamic limitation is much less severe than that of HTS. CO levels at the exit of the HTS reactor are around 3–5 wt.% while values around 0.3 wt.% can be achieved at an exit temperature of 200 °C in the LTS reactor (Ratnasamy and Wagner, 2009). By combining HTS and LTS in series and matching both properly, adjustment of the final gas composition (H_2/CO ratio) becomes feasible, which ultimately determines the industrial use of the syngas stream.

The reformate gas composition also influences the CO equilibrium conversion ($X_{CO,eq}$), which can be determined from the equilibrium constant and gas composition using Eq. (3.3),

$$K_y = K_p = \frac{y_{CO_2,eq}y_{H_2,eq}}{y_{CO,eq}y_{H_2O,eq}} = \exp\left(\frac{4557.8}{T} - 4.33\right) \qquad (3.2)$$

$$K_p = \frac{(y_{CO_2,in} + y_{CO,in}X_{CO,eq})(y_{H_2,in} + y_{CO,in}X_{CO,eq})}{[y_{CO,in}(1 - X_{CO,eq})](y_{H_2O,in} - y_{CO,in}X_{CO,eq})} \qquad (3.3)$$

where $y_{i,eq}$ is the mole fraction of species i at equilibrium at absolute temperature T and $y_{i,in}$ is the mole fraction of that component in the initial/feed reactant mixture. The CO, H_2 and CO_2 contents in the reformate stream will change depending on the process and

Table 3.1 Typical dry exit gas (molar compositions) of (a) steam reforming, (b) auto-thermal reforming and (c) coal gasification processes for H_2 production. Reprinted from Mendes *et al.* (2010a) with permission from John Wiley and Sons.

Component	Feedstock		
	CH_4^a	CH_4^b	$Coal^c$
CO	0.08	0.128	0.570
CO_2	0.12	0.080	0.049
H_2	0.73	0.578	0.381
CH_4	0.04	0.004	—
N_2	0.03	0.211	—

[a] 830–850°C, 25–50 atm, Ni-based catalyst, S/C = 3.0.
[b] 830–850°C, 25–50 atm, Ni-based catalyst, O_2/C = 0.30, S/C = 3.0.
[c] 1100–1400°C, 75–85 atm, O_2/coal (kg/kg) = 0.899, H_2O/coal (kg/kg) = 0.318.

reformate operation (cf. Table 3.1), thereby affecting the maximum (equilibrium) attainable CO conversion.

The dependence of the equilibrium constant from temperature given in Eq. (3.2) has been proposed by Moe (1962) and is a simple empirical equation that reasonably represents such dependence. Other equations and thermodynamic data can be found elsewhere (e.g. Smith *et al.*, 2010).

Steam-to-carbon (S/C) monoxide molar ratio is another factor that determines performance of WGS reactors, and of course $X_{CO,eq}$. First, the equilibrium CO conversion increases with S/C, increasing the final H_2/CO ratio. On the other hand, steam is a mild oxidant that slows the reduction of the component (metal) oxides in a WGS catalyst, which largely prevents the excessive reduction and activity loss of the catalyst during the reaction (Lee *et al.*, 2013). Moreover, steam also slows methanation, which is an undesirable side reaction of WGS between CO and H_2 because it decreases the hydrogen concentration and causes a loss of surface area because of its exothermicity (Lee *et al.*, 2013). For these reasons, an appropriate amount of extra steam is usually added to the gas stream before the

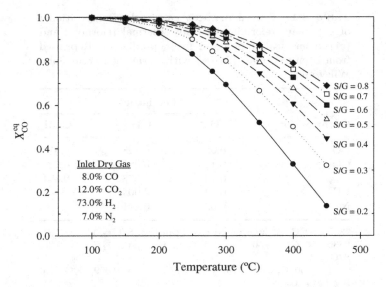

Figure 3.2 CO equilibrium conversion of a typical reformate stream from a methane steam reforming process at various S/G ratios. Reprinted from Mendes *et al.* (2010a) with permission from John Wiley and Sons.

WGS reactor. However, excess steam is costly, and thus needs to be controlled. Moreover, the amount of water added must be balanced taking also into account the desired CO composition at the end of the process, the catalyst capacity for H_2O activation and process operating conditions (Mendes *et al.*, 2010a). Figure 3.2 illustrates the effect of both temperature and water concentration (molar steam-to-dry gas (S/G) ratio) on the CO equilibrium conversion for a typical dry reformate gas, excluding any residual hydrocarbons.

Total pressure has no effect on the WGS equilibrium, because there is no change in the volume (number of moles) from reactants to products. However, it should positively affect the CO conversion in a conventional reactor because it increases the reaction rate. In MRs, total pressure also affects permeation rate, thus performance in WGS processes, which will be discussed later.

3.3. Catalysts

There is a huge amount of studies available where different catalysts, with different compositions, employing a variety of supports, and prepared by diverse methodologies, have been employed in the WGS reaction. So, it was decided just to give an overview of the main types, providing the interested reader with a few reviews and other references on this thematic, as the focus of this book is more centred on the process/reactor rather than on catalysis issues.

In general, and as classified by Ratnasamy and Wagner (2009), there are five main types of WGS catalysts, although the first two and the third are the main classes used in industry as shift catalysts:

(1) Promoted iron oxide (or iron oxide–chromium oxide, also called ferrochrome) catalysts, usually employed at moderately high temperatures (ca. 350–450°C) — so-called HTS catalysts;
(2) Copper–zinc oxide (and copper-based) catalysts, usually employed at relatively low temperatures (ca. 190–250°C) — so-called LTS catalysts;
(3) Catalysts based on cobalt and molybdenum sulphides as the active ingredients — so-called sour gas shift catalysts as they are sulphur-tolerant (thus can be used in sulphur-containing "sour gas" streams);
(4) Modified copper-zinc catalysts (usually with iron oxide) to operate at temperatures above LTS catalysts, which are usually employed at intermediate temperatures (ca. 275–350°C) — so-called medium-temperature shift (MTS) catalysts;
(5) Precious metal-based catalysts (mainly platinum and gold), particularly for use in fuel cell applications.

Table 3.2 presents some reviews, which by themselves contain numerous references, with important information about preparation methods, effect of several promoters and supports, characterization data by different techniques (e.g. TPR, BET, TEM, XRD, XPS, TPD, etc.), catalytic performances (and activation energies) under diverse conditions, stability along time-on-stream, resistance to

Table 3.2 Main types of WGS catalysts used, typical formulations and some review articles where detailed information is available.

Catalyst type	Typical (base) formulations	References
(1) HTS	Fe_2O_3–Cr_2O_3 + MgO; Fe_3O_4–Cr_2O_3; CuO–Fe_3O_4–Cr_2O_3	Ratnasamy and Wagner (2009), Smith *et al.* (2010), Newsome (1980), Mendes *et al.* (2010a), Zhu and Wachs (2016) and Gradisher *et al.* (2015)
(2) LTS	Cu–ZnO–Al_2O_3; CuO–ZnO–Cr_2O_3	Ratnasamy and Wagner (2009), Smith *et al.* (2010), Newsome (1980), Mendes *et al.* (2010a) and Gradisher *et al.* (2015)
(3) Sulphur-tolerant	Ni or Co with Mo; Mo-magnesia; Co–Mo–Ni + alkali; sulfided Co–Mo–Cs	Ratnasamy and Wagner (2009), Newsome (1980), Mendes *et al.* (2010a), Ghenciu (2002) and Hulteberg (2012)
(4) MTS	CuO–ZnO–Fe_2O_3	Smith *et al.* (2010)
(5) Precious metal-based	Au–Fe_2O_3/Au–CeO_2; Au–TiO_2/Ru–ZrO_2; Rh–CeO_2/Pt–CeO_2; Pt–ZrO_2/Pt–TiO_2; Pt–Fe_2O_3/Pd–CeO_2; Ru-based/Rh-based; Pd-based	Ratnasamy and Wagner (2009), Smith *et al.* (2010), Mendes *et al.* (2010a), Ghenciu (2002), Andreeva (2002), Nguyen-Phan *et al.* (2015), Gorte and Zhao (2005) and Gradisher *et al.* (2015)

sintering, poisoning, etc., mechanistic and kinetic studies for each type of the above-mentioned (and other) WGS catalyst categories.

3.4. Mechanisms and Kinetic Models

Knowledge of a reaction mechanism is important, among other factors, to allow establishing possible reaction kinetic equations which, in turn, are crucial for reactor modelling and reactor design, including MRs.

Although the WGS reaction is apparently simple, involving only four species, and in spite of the huge amount of studies reported in the literature, there is no consensus regarding the reaction mechanism for each type of catalysts, which is also determined by the conditions employed in each work. Therefore, many kinetic equations and some mechanistic studies can be found, each one being very specific and limited to the range of conditions analyzed and catalyst formulation employed.

In this section, an overview will be given for some mechanisms and kinetics equations, being suggested some references (particularly reviews/book chapters) on this matter that contain a huge amount of information for the interested reader (e.g. Smith *et al.*, 2010; Mendes *et al.*, 2010a; Ratnasamy and Wagner, 2009; Newsome, 1980; Platon and Wang, 2010).

Very often, WGS rate equations are written as follows (Mendes *et al.*, 2010a):

$$r = R_f(1 - \beta) \tag{3.4}$$

$$R_f = R_{fo} \exp\left(-\frac{E_a}{RT}\right) \tag{3.5}$$

$$\beta = \frac{p_{CO_2} p_{H_2}}{K_p p_{CO} p_{H_2O}} \tag{3.6}$$

where r is the reaction rate, R_f is the forward reaction rate, β is the approach to equilibrium, E_a is the apparent activation energy, R is the ideal gas constant, T is the absolute temperature, K_p is the WGS reaction equilibrium constant and p_i is the partial pressure for species i. Another way of presenting the kinetics is in a power law formulation:

$$r = k p_{CO}^a p_{H_2O}^b p_{CO_2}^c p_{H_2}^d (1 - \beta) \tag{3.7}$$

where k is the forward reaction rate constant and a–d are the forward reaction orders. Some values of kinetic parameters are reported in Table 3.3 for a few relevant catalysts. Smith *et al.* (2010) also report a long list of kinetic parameters corresponding to power law type kinetics for high and low-temperature WGS catalysts, with particular

Table 3.3 Comparison of activation energies and reaction orders for the forward WGS reaction. Reprinted from Mendes et al. (2010a) with permission from John Wiley and Sons. For more info, please see Mendes et al. (2010a) and references therein.

Catalyst	Operating conditions*	E_a (kJ mol^{-1})	Reaction orders** CO	H$_2$O	CO$_2$	H$_2$
40 wt.% Cu–ZnO–Al$_2$O$_3$	1 bar, 190°C	79	0.8 (5–25%)	0.8 (10–46%)	−0.9 (5–30%)	−0.9 (25–60%)
8% Cu–Al$_2$O$_3$	1 bar, 200°C	62	0.9 (5–25%)	0.8 (10–46%)	−0.7 (5–30%)	−0.8 (25–60%)
8% Cu–CeO$_2$	1 bar, 240°C	56	0.9 (5–25%)	0.4 (10–46%)	−0.6 (5–30%)	−0.6 (25–60%)
10 atom% Cu–Ce (30 atom% La)O$_x$	1 bar, 450°C	70.4	0.8 (1–10%)	0.2 (11–50%)	−0.3 (5–35%)	−0.3 (5–40%)
Fe$_2$O$_3$–Cr$_2$O$_3$	1 bar, 450°C	118	1 (20–75%)	0 (10–40%)	—	—
Pd–CeO$_2$	1 atm, 200–240°C	38	0	0.5	−0.5	−1
1% Pt–Al$_2$O$_3$	1 atm, 285°C	68	0.06 (5–25%)	1.0 (10–46%)	−0.09 (5–30%)	−0.44 (25–60%)
1% Pt–Al$_2$O$_3$	1 atm, 315°C	84	0.1 (5–25%)	1.1 (10–46%)	−0.07 (5–30%)	−0.44 (25–60%)
1% Pt–CeO$_2$	1 atm, 200°C	75	−0.03 (5–25%)	0.44 (10–46%)	−0.09 (5–30%)	−0.38 (25–60%)
1.4% Pt, 8.3% CeO$_2$–Al$_2$O$_3$	1 bar, 200–260°C	86	0.13	0.49	−0.12	−0.45
2% Pt–CeO$_2$-ZrO$_2$	1.3 bar, 210–240°C	71	0.07	0.67	−0.16	−0.57
2% Pt–1% Re/CeO$_2$–ZrO$_2$	1.3 bar, 210–240°C	71	−0.05	0.85	−0.05	−0.32
2 wt.% Au–CeZrO$_4$	170°C	—	0.7	0.6	−0.3	−0.9
4.5 wt.% Au–CeO2	1 bar, 180°C	—	1	1	−0.5	−0.7
2.6 wt.% Au–CeO2***	1 bar, 180°C	40	0.5 (0.2–2 kPa)	0.5 (0.7–10 kPa)	−0.5 (1.2–3.4 kPa)	−0.5 (3.2–75 kPa)

*Temperature and total pressure at which the reaction order measurements were carried out.

**Values between brackets represent the ranges of concentrations for each species in the feed, or their partial pressures.

***Partial orders for water and carbon monoxide were obtained without reaction products in the feed.

emphasis being given to noble metal catalysts. It should be noted that specific conditions vary from study to study, in some cases using idealized reformate feeds, thus a comparison should be made with care. Although such equations are merely empirical, they are very useful for reactor design. Moreover, the differences in reaction orders may be an indication of the existence of different reaction mechanisms and/or distinct active sites e.g. for H_2O activation (Mendes *et al.*, 2010a). However, thorough analysis is required, because different mechanisms can lead to the same overall kinetic equation.

Two main reaction mechanisms have been proposed in the literature for the WGS reaction (Mendes *et al.*, 2010a; Smith *et al.*, 2010; Ratnasamy and Wagner, 2009), from which kinetic models (based on the awareness regarding elementary steps involved in the reaction) can be formulated. The regenerative (or redox) mechanism (Eqs. (3.8)–(3.9)), firstly proposed by Temkin (1979), considers the oxidation of the reduced vacant site (*) in the catalyst surface by water molecules to produce hydrogen. To complete the redox cycle, a carbon monoxide molecule promotes the reduction of an oxidized site (O), yielding carbon dioxide:

$$H_2O + (^*) \leftrightarrow H_2 + (O) \qquad (3.8)$$

$$CO + (O) \leftrightarrow CO_2 + (^*) \qquad (3.9)$$

This mechanism implies that the catalyst undergoes changes in oxidation state during the course of the reaction. It is essentially the same as the well-known Mars–Van Krevelen mechanism, with the difference that the oxygen used to oxidize the catalyst comes from the water rather than from oxygen. According to Ratnasamy and Wagner (2009), there is some consensus on the redox mechanism prevailing over the iron–chromia catalysts at high temperatures. Gorte and Zhao (2005) also support the existence of a redox mechanism over ceria-supported precious metals. Newsome (1980) reports kinetic equations based on this redox mechanism over different catalysts, wherein different rate-limiting steps have been assumed. For instance, Eq. (3.10) by assuming that step (3.9) was the rate-controlling one (over an iron-based catalyst), or Eq. (3.11) by assuming a heterogeneous surface with no rate-controlling steps, with a low concentration of oxygen vacancies on the catalyst surface,

$$r = kp_{CO}\left(\frac{p_{H_2O}}{p_{H_2}}\right)^{1/2} - k_- p_{CO_2}\left(\frac{p_{H_2}}{p_{H_2O}}\right)^{1/2} \tag{3.10}$$

$$r = k\frac{p_{H_2O}p_{CO} - \dfrac{p_{CO_2}p_{H_2}}{K_p}}{Ap_{H_2O} + p_{CO_2}} \tag{3.11}$$

where k and k_- denote the forward and backward rate constants, respectively, and A is a constant. As reported by Newsome (1980), a good evidence of this mechanism is the similarity between experimentally measured rates of oxidation and reduction of a given catalyst (e.g. iron-based shift catalyst) with H_2O and CO and those of CO conversion in the WGS reaction.

The other most common mechanism is the associative one, of the Langmuir–Hinshelwood (LH) type (Smith *et al.*, 2010; Mendes *et al.*, 2010a). Such mechanism, illustrated through Eqs. (3.12)–(3.17), involves adsorption–desorption steps. Firstly, the dissociative adsorption of water molecules that form reactive hydroxyl groups occurs; such groups, when combined with adsorbed CO produce a surface intermediate that decomposes into the adsorbed reaction products; last steps correspond to the desorption to the gas phase of both CO_2 and H_2.

$$H_2O + 2(*) \leftrightarrow H^* + OH^* \tag{3.12}$$

$$CO + (*) \leftrightarrow CO^* \tag{3.13}$$

$$OH^* + CO^* \leftrightarrow HCOO^* + (*) \tag{3.14}$$

$$HCOO^* + (*) \leftrightarrow CO_2^* + H^* \tag{3.15}$$

$$CO_2^* \leftrightarrow CO_2 + (*) \tag{3.16}$$

$$2H^* \leftrightarrow H_2 + 2(*) \tag{3.17}$$

In situ techniques can be used to prove the formation of the above-mentioned intermediates (formate and/or carbonate). Moreover, and again, depending on the considered rate-controlling step different rate equations can be derived, which should be matched with experimentally collected rate data. For instance, Ayastuy *et al.* (2005) proposed the following kinetic equation for a commercial Cu-based catalyst in the low-temperature range (considering the formation of a formate intermediate and atomically adsorbed hydrogen as rate-controlling):

$$r = \frac{k \left(p_{CO} p_{H_2O} - \frac{p_{CO_2} p_{H_2}}{K_p} \right)}{\left(1 + K_{CO} p_{CO} + K_{H_2O} p_{H_2O} + K_{H_2}^{0.5} p_{H_2}^{0.5} + K_{CO_2} p_{CO_2} p_{H_2}^{0.5} \right)^2}$$

(3.18)

where k is the rate constant for the forward reaction, and K_i is the equilibrium adsorption constant for species i (again, K_p stands for the reaction equilibrium constant).

Depending on the catalyst formulation/composition, nature of the support, metal dispersion and size, and even operating conditions, different mechanisms have been observed. For more information, it is suggested the reading of some reviews on the topic (e.g. Newsome, 1980; Smith *et al.*, 2010; Mendes *et al.*, 2010a; Laan and Beenackers, 1999). Particularly, Smith *et al.* (2010) provide an extensive list of kinetic expressions, either for high or low temperature WGS catalysts.

3.5. Membrane Types

Membranes have been widely used in several applications, including gas separation/purification. In the perspective of gas-phase (and other) reaction processes, the membrane can be positioned downstream the reactor, or integrated with such unit, in the latter case providing a more compact unit with synergistic benefits, in line with the process intensification strategy. In particular, for reversible equilibrium-limited reactions (which include the WGS), the continuous removal of a product species from the reaction medium shifts the reaction towards its formation, thereby providing a higher conversion. The combination of low cost and long-lasting (hydrogen- or carbon dioxide-selective) membranes and the WGS reaction has been the focus of attention. Depending on the type of membrane, either H_2 or CO_2 can be selectively removed, having both strategic advantages and disadvantages, as detailed along the following subsections.

One possible classification for membranes is based in their nature. Having such criteria in mind, membranes can be roughly classified into organic, inorganic, and hybrid (organic/inorganic). For the particular case of inorganic membranes, two subtypes can be considered, which will be analyzed below in the perspective of hydrogen separation and integration in a MR: (i) dense metal and (ii) (micro) porous membranes (Lu *et al.*, 2007; Mendes *et al.*, 2010a). For carbon dioxide selective extraction and for integration in a WGS MR, mostly organic (polymeric) membranes will be considered (cf. Section 3.5.4).

As discussed elsewhere (Lu *et al.*, 2007), the proper selection of the membrane type to be coupled with a MR depends on several issues, including productivity, separation selectivity, membrane life time, mechanical and chemical integrity at the operating conditions employed and, particularly, the cost. Selectivity and permeation are among the most critical. In fact, the use of more selective membranes leads to the requirement of lower driving forces, namely pressure ratio, lowering operating costs. In addition, the use of membranes with higher permeances lead to less requirements in terms of membrane area, this way decreasing capital costs.

Inorganic membranes offer several advantages over organic ones, mostly due to their stability and both chemical and mechanical resistance in processes operating at temperatures above $100°C$, which is the case of the WGS. However, organic membranes might also be competitive, as described later on, and particularly in terms of price (Lu *et al.*, 2007).

3.5.1. Dense metal membranes

Dense metal membranes have been particularly interesting for hydrogen purification purposes due to their permselective characteristics, with very high selectivities (typically in the range of 10^3). In particular, there are a few types of dense metal hydrogen permselective membranes commercially available, namely those based on pure metal (Pd) or metal alloys (e.g. Pd-alloy).

Palladium membranes are quite selective towards hydrogen. However, well-known drawbacks like embrittlement, deposition of carbonaceous impurities and poisoning (e.g. by CO and H_2S) are important disadvantages for integrating them into hydrogen-related technologies, and particularly WGS MRs. The embrittlement phenomenon occurs due to exposure of the Pd membrane to H_2 at low temperatures, producing pinholes on the material that will detrimentally affect its permselectivity. This can however be overcome by using Pd-alloy membranes, obtained by alloying Pd with Ni, Cu, and, in particular, Ag; such alloying elements also increase hydrogen permeance.

Pd–Ag membranes show better mechanical properties as compared to pure Pd membranes. Moreover, although Pd–Ag membranes show lower H_2 diffusivity, they exhibit higher H_2 solubility and inherently higher permeability with respect to pure Pd membranes. The silver load has to be optimized in this alloy, being common to use an Ag content of ca. 23–25 wt.% because H_2 permeability shows a maximum at such compositions.

Several types of Pd-based membranes are available, including relatively thick dense metal membranes (ca. $50–100 \, \mu m$), usually in a self-supporting configuration. In particular, Pd and Pd–Ag thin wall tubes have been produced by a cold rolling procedure, followed by wrapping and diffusion welding (Bettinali *et al.*, 2002). Although they present some relevant advantages (namely nearly infinite H_2 selectivity, good chemical and physical stability), they are costly, and have low permeance. For these reasons, efforts have been directed towards obtaining thin layers ($<20 \, \mu m$) of the palladium or palladium alloy deposited onto a porous ceramic (e.g. alumina) or metal (e.g. sintered stainless steel) substrate (Lu *et al.*, 2007), which is porous and not selective for the permeation of hydrogen. The permselective layer ranges in thickness from about $2–20 \, \mu m$; thinner membranes are difficult to achieve without introducing pinholes and other adverse defects into the permselective layer, although they present several advantages like higher hydrogen fluxes and better MR economy.

There are mainly three techniques for coating metallic thin films onto porous metallic or ceramic supports: electroless plating, chemical vapor deposition (CVD) and physical sputtering. For more details regarding each of these techniques (among others), and pros and cons of each membrane type, it is recommended the reading of some literature in this matter (e.g. Lu *et al.*, 2007; Basile *et al.*, 2008; Ockwig and Nenoff, 2007, and references therein). It is worth mentioning that there is a wide variation in the permeability values reported in the literature depending on the substrate, manufacture conditions employed, etc., although in general they follow the order: electroless deposition > CVD deposition > sputtering method.

Different groups around the world have been trying to improve the performance of metallic membranes, employing these and other techniques. Goals are to maximize H_2 selectivity and flux, while decreasing the amount of metals (to decrease cost) that ensure high mechanical and chemical stability. Some of such groups include AIST in Japan, the Colorado School of Mines (CSM) in US, the Dalian Institute of Chemical Physics (DICP) in China, SINTEF in Norway, the Southwest Research Institute (SwRI) and the Worcester Polytechnic Institute (WPI), both in the US, the Energy research Centre of the Netherlands (ECN), and several companies such as CRI/Criterion (a company owned by Shell), Eltron Research, Inc., Green Hydrotec, Hy9, Media and Process Technology, Inc., Membrane Reactor Technologies, Pall Corporation or REB Research and Consulting (Gallucci *et al.*, 2013).

As mentioned above, one critical parameter when comparing different materials in the search for the best one is the permeance. As an example, in Fig. 3.3 are reported hydrogen flux data for supported membranes prepared by CVD (Uemiya *et al.*, 2001). It is seen that the hydrogen flux for the Pd-based membrane is higher than for those prepared from other metals for the whole temperature range. Moreover, the permeability for the Pd supported membrane is in the order of $1 \times 10^{-7}\,\mathrm{mol\,m^{-1}\,s^{-1}\,kPa^{-0.5}}$ at 750°C

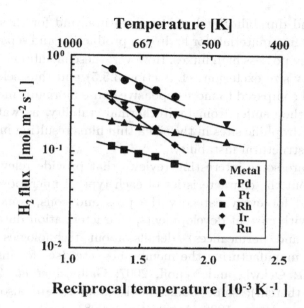

Figure 3.3 Hydrogen fluxes over various supported Pt-group membranes prepared by CVD. Transmembrane pressure 196 kPa. Reprinted from Uemiya *et al.* (2001) with permission from Elsevier.

(Uemiya *et al.*, 2001), which is typically considerably lower than those prepared by electroless deposition (Lu *et al.*, 2007).

3.5.2. Microporous membranes

Microporous membranes are another type of material generally employed for H_2 production and purification, in this case having pore diameters smaller than 2 nm. In terms of structure, they can be either crystalline (made of zeolites or metal-organic frameworks) or amorphous (e.g. silica, carbon, etc.) (Gallucci *et al.*, 2013). However, only some of them (namely the inorganic ones, and particularly the silica-based) are promising in WGS reaction, usually at higher temperatures. For instance, carbon molecular sieves are not considered to be feasible candidates because of the oxidative nature of their surface (Lu *et al.*, 2007).

Microporous ceramic membranes, namely silica and functionalized silica, are normally prepared by sol–gel methods. They have high stability and durability at high temperature, and for these reasons are encouraging materials for hydrogen production and separation. In general, they possess high fluxes. However, these membranes separate hydrogen by size exclusion (cf. Section 3.5.5) and thus selectivity is still limited compared to more expensive dense inorganic membranes. Moreover, they suffer from hydrothermal stability, as water reacts with the hydrophilic sites in the silica thin films resulting in chemical and micro-structural instability.

There are some interesting reviews that provide relevant information about the characteristics of each type of microporous membranes, and for every category the pros and cons, some relevant examples with current developments, characterization data (namely selectivity and permeances), details about technologies used for processing/manufacturing the membranes, etc. (see for instance Lu *et al.*, 2007; Ockwig and Nenoff, 2007; Gallucci *et al.*, 2013, and references therein); some books/book chapters are also of great relevance (e.g. Hsieh, 1996; Ayral *et al.*, 2008).

3.5.3. Others

Other types of membranes that are worth mentioning are proton conducting membranes (dense ceramic membranes and composite ceramic-metal (cermet) membranes). Dense ceramic membranes (perovskite-type and non-perovskite-type — e.g. doped rare earth metal oxides and fluorite-structured metal oxides) can recover H_2 at very high purity due to a proton transport mechanism, but they have to operate at temperatures as high as $900°C$, not interesting for WGS. Gallucci *et al.* (2013) present some of the most relevant current developments on cermet membranes for hydrogen separation.

3.5.4. Membranes for H_2 vs. CO_2 removal

As described earlier, shift of the WGS reaction towards the product side can be reached with either H_2- or CO_2-selective membranes. The latter are particularly interesting because (1) a H_2-rich product is recovered at high pressure (feed gas pressure), thus avoiding the need for an additional compressor; (2) high cost of Pd-based membranes is avoided; (3) air can be used as sweep gas and (4) CO_2 at high concentration can be obtained on the permeate side for sequestration. Carbon dioxide permselective membranes for incorporation in a MR are still a challenge, because of temperature limitations, and the difficulty of retaining small molecular size molecules like H_2 while permeating larger ones as CO_2 (Scholes *et al.*, 2010). Polymeric materials with CO_2 selectivity can be used, particularly facilitated transport membranes, which include a facilitator species (e.g. amine) that undergoes a reversible complexation reaction with CO_2 within the membrane, thus enhancing the solubility. In this case, the molecular transport mechanism involves both diffusion and reaction. Carbon dioxide reacts on the feed side with the carriers loaded in the membrane. The complexation product moves across the membrane due to driving force (concentration gradient), being released in the low-pressure permeate side. Figure 3.4 illustrates such transport mechanism. This way, transport of CO_2 is enhanced due to the complexation reaction, while other gases (non-polar) only permeate by the physical solution-diffusion, which is limited by their low solubility on the highly polar sites in the membrane (Zou *et al.*, 2007).

To date, the best CO_2-selective membrane for WGS MRs are those that incorporate facilitated transport, because they provide both a high CO_2 flux as well as high selectivity, to limit H_2 loss in the permeate stream (Scholes *et al.*, 2010). Ho and co-workers have tested (Zou *et al.*, 2007) and modelled (Huang *et al.*, 2005; Zou *et al.*, 2007) the WGS MRs using CO_2-selective facilitated transport membranes. Particularly, their 1D non-isothermal model allowed to better understand the effect that important variables (as membrane CO_2/H_2 selectivity, membrane permeability towards CO_2, and sweep-to-feed molar flow rate) have in the process response (CO and H_2 concentration, H_2 recovery, membrane permeation area required) (Huang *et al.*, 2005). More recently, the modelling was extended to spiral-wound modules, once such configuration is easier of fabrication and catalyst housing (Ramasubramanian *et al.*, 2013).

Experimentally, Ho and co-workers have incorporated fixed and mobile carriers in cross-linked poly(vinyl alcohol), obtaining

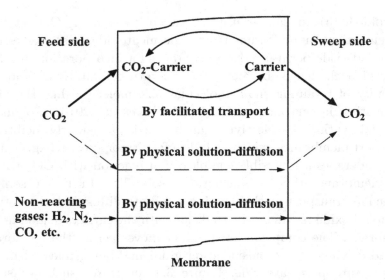

Figure 3.4 Schematic of CO_2 transport mechanism in facilitated transport membranes. Reprinted with permission from Zou *et al.* (2007). Copyright (2007), American Chemical Society.

membranes with good selectivities and high CO_2 permeabilities (Zou *et al.*, 2007). When tested in a MR assembly, exit CO concentrations below 30 ppm (dry basis) have been reached, as shown in Fig. 3.5. Moreover, higher H_2O contents allowed reaching smaller dry CO concentrations on the retentate side because this way both the WGS reaction equilibrium is shifted forward and the CO_2 transfer in the membrane is promoted. Under certain conditions, CO concentrations of less than 10 ppm were reached, thus meeting the purity requirement of hydrogen for Proton Exchange Membrane (PEM) fuel cells (of course, this depends upon the operating conditions and design of the fuel cell). Moreover, the experimental retentate CO concentrations were much lower than the equilibrium CO values (without CO_2 removal) — cf. Fig. 3.5, clearly highlighting the advantage of *in situ* CO_2 removal.

For the reasons already described, most WGS MR applications have been focused on H_2-selective rather than on CO_2-selective membranes. According to Boutikos and Nikolakis, the latter could improve CO conversion only if the CO_2 content of the feed is higher

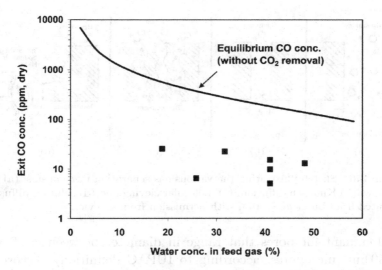

Figure 3.5 Retentate CO concentration vs. feed water concentration in the circular MR. Feed gas 1% CO, 17% CO_2, 45% H_2, and 37% N_2; $T = 150°C$; $P_f = 2.2$ atm, $P_s = 1.0$ atm; feed/sweep flow rates = 1/1 (dry basis); membrane thickness = 41 μm. Reprinted with permission from Zou *et al.* (2007). Copyright (2007), American Chemical Society.

than that of H_2 (Boutikos and Nikolakis, 2010). According to their predictions, H_2 recovery was higher in reactors using an H_2-selective membrane, and small recoveries were estimated when a CO_2-selective membrane was used.

3.5.5. Hydrogen permeation mechanism

Mechanisms of molecular transport through membranes are diverse, as nicely described by Lu *et al.* (2007) and shown in Fig. 3.6. In short, such mechanisms are of the following types:

(a) Viscous flow, which does not provide any separation among the upstream and downstream sides of the membrane (cf. Fig. 3.6(a));

(b) Knudsen flow, which occurs when the mean free path of the molecules is relatively long compared to the pore size, so the molecules collide frequently with the pore wall, rather than between the gas molecules themselves; Knudsen diffusion is

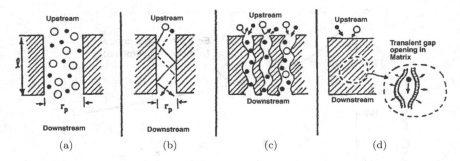

Figure 3.6 Sketch illustrating the various gas separation mechanisms: (a) viscous flow, (b) Knudsen diffusion, (c) molecular sieving and (d) solution-diffusion. Reprinted from Lu *et al.* (2007), with permission from Elsevier.

dominant for pores that range in diameter between ca. 2 and 50 nm (mesopores, according to IUPAC definition). According to the kinetic theory of gases, Knudsen diffusivity (D_K, $m^2\,s^{-1}$) is given by:

$$D_K = \frac{4}{3}r_p\sqrt{\frac{2}{\pi}\frac{RT}{M_i}} \qquad (3.19)$$

where r_p stands for the pore radius (m), R is the gas constant (Pa $m^3\,mol^{-1}\,K^{-1}$), T is the temperature (K) and M_i is the species i molecular weight (kg mol^{-1}). Therefore, at constant temperature, separation between two species A and B is based on the inverse square root ratio of their molecular weights, as given by the separation factor $\alpha_{A/B}$:

$$\alpha_{A/B} = \sqrt{\frac{M_B}{M_A}} \qquad (3.20)$$

(c) Molecular sieving (also called activated diffusion or size exclusion), which prevails in very small pores, i.e. when the molecules dimensions are similar to that of the pores–micropores (cf. Fig. 3.6(c)). In this case, separation is due to diffusion rates of gas species, which are commonly higher for smaller molecules, if adsorption in pores walls is neglected.

Table 3.4 Various characteristics of porous and dense membranes. Reprinted from Basile *et al.* (2008) with permission from Elsevier.

Membrane type	d_{pore} (nm)	Diffusion mechanism	$\alpha_{H_2/other gas}$	H_2 permeability	Reactant loss
Macroporous	>50	Poiseuille (viscous flow)	1	Very high	High
Mesoporous	2–50	Knudsen	$\alpha_{H_2/N_2} = 3.74$	High	Average
Microporous	<2	Activated process	High	Average	Low
Dense Pd	—	Fick	Infinite	Very low	—

(d) Solution-diffusion, in which case the transport is governed by both solubility and diffusivity. This is dominant in non-porous polymeric membranes (cf. Fig. 3.6(d)) and dense metallic.

Among the inorganic H_2 permselective membrane types mentioned above, i.e. (i) microporous ceramic or molecular sieves, (ii) dense phase metal or metal alloys, and (iii) dense ceramic perovskites, the former follows the molecular sieving mechanism (activated process), while the 2nd and 3rd the solution-diffusion (Lu *et al.*, 2007). Knudsen diffusion might be also important, depending on the membrane pore size. Table 3.4 provides an overview of the main membranes types according to pores dimensions, prevailing diffusion mechanism and other characteristics. Of course in a composite membrane system, the transport mechanisms change from activated transport for the microporous top layer to Knudsen diffusion and Poiseuille permeation for the support (mesoporous and macroporous materials). Hence, the transport resistance of the support has to be taken into account, the overall resistance being derived from analogous resistance circuits (although it is generally observed that the top layer limits the diffusion, i.e. rate-determining). Emphasis is given herein below to the solution-diffusion (or Fick-based) due to the wide use of dense metallic membranes.

The solution-diffusion mechanism means that permeability, L (mol/m $Pa^{0.5}$ s), of hydrogen through a metal is a function of diffusivity, D $(\mathrm{m}^2/\mathrm{s})$ — seen as the kinetic term, and solubility, S $(\mathrm{mol}/\mathrm{m}^3 \ \mathrm{Pa}^{0.5})$ — thermodynamic term (Edlund, 2010; Paglieri and Way, 2002):

$$L = D \cdot S \qquad (3.21)$$

Although the transport mechanism of hydrogen through metallic membranes is solution-diffusion, the process is in this case a bit more complex than in polymeric films. The process involves several stages, namely:

(i) sorption of H_2 molecules on the metallic surface;
(ii) dissociation of the sorbed hydrogen molecules;
(iii) diffusion of the hydrogen atoms through the metal lattice;
(iv) reassociation (or recombination) of the hydrogen atoms on the permeate side;
(v) desorption of hydrogen molecules.

The diffusion of hydrogen atoms through the metal film (stage iii) is generally the rate-limiting step, particularly in thick (i.e. dense, defect-free) palladium-based membranes, and at temperatures above ca. 150°C.

The H_2 mass transfer through dense membranes can be described by the Fick's law, which for a single dimension yields (Paglieri and Way, 2002):

$$J_{H_2} = -D_H \frac{dC_H}{dz} \qquad (3.22)$$

being J_{H_2} the hydrogen flux through the membrane, D_H the effective diffusion coefficient of atomic hydrogen, and C_H the atomic hydrogen concentration at the spatial position z along the membrane thickness. This concentration is in turn related to the hydrogen partial pressure by Sieverts' law (equilibrium conditions) (Sieverts and Krumbhaar, 1910):

$$C_H = S p_{H_2}^{0.5} \qquad (3.23)$$

where S is the solubility of hydrogen in the lattice. If the sorbed hydrogen is in equilibrium with gas phase at both membrane surfaces, combining Eqs. (3.22) and (3.23) and the definition of permeability (Eq. (3.21)), and integrating for the membrane thickness, δ, one obtains:

$$J_{H_2} = \frac{L_{H_2}}{\delta} \left(\sqrt{p_{H_2,ret}} - \sqrt{p_{H_2,perm}} \right) \qquad (3.24)$$

where the subscripts ret and perm refer to both sides of the membrane (retentate and permeate, respectively). Very often, the expression is written as:

$$J_{H_2} = \frac{L_{H_2}}{\delta} (p_{H_2,ret}^n - p_{H_2,perm}^n) \qquad (3.25)$$

where n stands for the power dependency of the hydrogen partial pressure (usually in the range from 0.5 to 1), a parameter that should be obtained experimentally. Its value is often used as an indicator for the rate-controlling step of the permeation through a metal composite membrane. If the diffusion of atomic hydrogen through the dense metal layer is rate-limiting, then n equals 0.5 (Sieverts's law) — Eq. (3.24) (Dittmeyer *et al.*, 2001). This is the case shown, as an example, in Fig. 3.7 for a 50-μm thick PdAg membrane (Mendes *et al.*, 2010b). In the case of polymeric membranes where selective transport of a gas is by a solution-diffusion process, the exponent n is always unity (Lu *et al.*, 2007). This is also the case of exceptionally thin dense metallic membranes (Edlund, 2010).

The dependence of permeation from temperature can be described by an Arrhenius-type equation:

$$L_{H_2} = L_{H_2,0} e^{-\frac{E_a}{RT}} \qquad (3.26)$$

Typical permeation activation energies (E_a, which is equal to the sum of the diffusion energy and the heat of dissolution), and pre-exponential factors ($L_{H_2,0}$) for H_2 permeation through dense Pd–Ag membranes are reported in Table 3.5. The reader should however be alert that hydrogen flux (and permeation parameters) will always depend on both membrane material nature and thickness of the selective layer, apart from operating conditions; the same applies to selectivity.

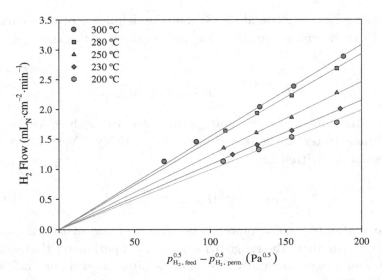

Figure 3.7 Hydrogen flux through a 50-μm thick Pd–Ag membrane as a function of the driving force. Reprinted from Mendes *et al.* (2010b) with permission from Elsevier.

Table 3.5 Typical apparent activation energies and pre-exponential factors for hydrogen permeation through dense Pd–Ag membranes. Reprinted from Mendes *et al.* (2010b) with permission from Elsevier. For more info, please see Mendes *et al.* (2010b) and references therein.

E_a (kJ mol^{-1})	$L_{H_2,0}$ (mol m m^{-2}s^{-1} Pa$^{-0.5}$)
10.72	5.44×10^{-8}
11.24	6.64×10^{-8}
17.60	2.06×10^{-7}
18.45	3.23×10^{-7}
18.56	3.85×10^{-7}

However, in WGS reactors (as well as in other processes, e.g. steam reformers), other species are present in the reaction medium which can interfere with H_2 permeation. This is the case, for instance, of either CO or CO_2 (Peters *et al.*, 2008). The presence of such surface adsorbed chemical species may lead to deviations to the Sieverts' law.

In this regard, Barbieri *et al.* (2008) proposed a new equation where the Sieverts' law and the Langmuir isotherm are coupled to describe the decrease on H_2 permeation in the presence of CO as follows:

$$J_{H_2}^{SL} = \left(1 - \alpha \frac{K_{CO}p_{CO}}{1 + K_{CO}p_{CO}}\right) J_{H_2}^{Siev} \qquad (3.27)$$

where $J_{H_2}^{SL}$ is the permeated hydrogen molar flux in the presence of e.g. CO, $J_{H_2}^{Siev}$ is the hydrogen flux assessed accordingly to the Sieverts law (Eq. (3.24)), p_{CO} is the carbon monoxide partial pressure and K_{CO} is the corresponding adsorption equilibrium constant. Finally, α is a parameter that depends only on the temperature and accounts for additional effects of the adsorbed gas, for instance in the dissociation of H_2 molecules. Figure 3.8 illustrates the effect of CO content in the feed in the decline of the hydrogen flux (normalized by the flux reached under identical conditions, with the same driving force, but without CO) at different temperatures for

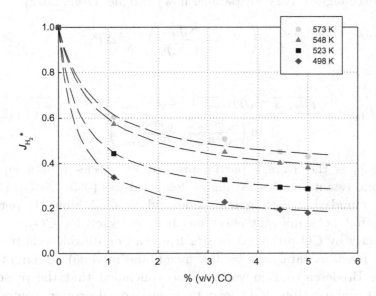

Figure 3.8 Normalized H_2 flux ($J_{H_2}^*$) as a function of the CO concentration in the feed (% (v/v) CO) for different temperatures. The fitted function (dashed lines) based on SL model (Eq. (3.27)) is also shown. Reprinted from Miguel *et al.* (2012), with permission from Elsevier.

a 50-μm thick Pd–Ag (25 wt.% of silver) tubular membrane (Miguel *et al.*, 2012). Under the conditions employed, it was observed that CO significantly reduces the amount of permeated H_2, even at low concentrations — for a CO content of only 5%, hydrogen permeation flux decreases 57% at 573 K, being this effect even more remarkable at lower temperatures because adsorption of CO is favoured. A good agreement between experimental data and the Sieverts–Langmuir (SL) model at all conditions tested was also observed (see the fitting lines).

Miguel *et al.* (2012) have also extended the SL equation to CO_2. More importantly, and because H_2 is removed along the membrane tube length, the local driving force for permeability changes in the axial direction, as well as the species i (can be either CO or CO_2) partial pressure. For that reason, a new SL equation was proposed, wherein the driving force in the Sieverts' term has been replaced by the logarithmic mean driving force, $\Delta P_{\ln}$ (calculated based on the heat exchanger theory for parallel flow) (Miguel *et al.*, 2012)

$$J_{H_2}^{SL} = \left(1 - \alpha \frac{K_i \overline{p_i}}{1 + K_i \overline{p_i}}\right) \frac{L_{H_2}}{\delta} \Delta P_{\ln} \qquad (3.28)$$

with

$$\Delta P_{\ln} = \frac{\left(\sqrt{p_{H_2,\text{feed}}} - \sqrt{p_{H_2,\text{perm}}}\right) - \left(\sqrt{p_{H_2,\text{ret}}} - \sqrt{p_{H_2,\text{perm}}}\right)}{\ln\left(\frac{\sqrt{p_{H_2,\text{feed}}} - \sqrt{p_{H_2,\text{perm}}}}{\sqrt{p_{H_2,\text{ret}}} - \sqrt{p_{H_2,\text{perm}}}}\right)} \qquad (3.29)$$

where $\overline{p_i}$ is the average partial pressure of species i between the feed and retentate sides. At higher temperatures (350–420°C), Boon *et al.* found that carbon monoxide and steam inhibit H_2 permeation, but no significant effect has been detected for CO_2, except indirectly by CO produced *in situ* from carbon dioxide (via reverse WGS reaction catalyzed by the membrane module) (Boon *et al.*, 2015). Bredesen and co-workers also concluded that the presence of CO and possibly H_2O lead to a surface adsorption controlled hydrogen flux (Fig. 3.9(a)), while the presence of CO_2 also reduces the flux, which is again considered to be due to CO formation (Peters *et al.*, 2008). However, they state that when using the thin

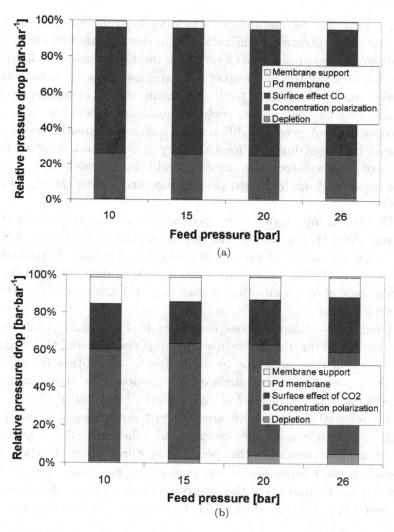

Figure 3.9 Hydrogen pressure drop due to depletion, concentration polarization, surface effects, transport in the palladium membrane and porous support, compared to the total hydrogen partial pressure drop as a function of the applied feed pressure for (a) CO–H_2 mixture and (b) CO_2–H_2 mixture; $T = 400°$C. Reprinted from Peters *et al.* (2008) with permission from Elsevier.

($\sim$2 μm) Pd–23%Ag/stainless steel composite membrane prepared at SINTEF, concentration polarization is the dominating flux inhibiting parameter in the presence of CO_2 only or inert dilutants (Fig. 3.9(b)). This is a result of the build-up of a hydrogen-depleted concentration polarization layer adjacent to the membrane due to insufficient mass transport in the gas phase; such phenomenon lowers the effective hydrogen partial pressure difference which consequently results in reduced hydrogen flux. As for CO, they also conclude that surface effects of WGS-related gases are responsible for the major part of the total pressure drop/hydrogen permeation flux decline (Peters *et al.*, 2008).

Of course, apart of these species others present in industrial streams affect H_2 permeance and metallic membranes, namely H_2S — a common contaminant present in coal-derived syngas. Particularly, ppm levels of H_2S in the feed mixture can rapidly deactivate Pd-based membranes; alloys with other metals (e.g. Pd–Cu) exhibit however higher tolerance.

Extensions to approaches described in Eqs. (3.24)–(3.28) can also be found in the literature, namely considering the surface mass transfer resistance due to the hydrogen adsorption–desorption reactions over the metal surfaces, as proposed by Tosti and co-workers (cf. e.g. Vadrucci *et al.*, 2013; Pérez *et al.*, 2015).

As referred above, there are different mechanisms of gas permeation through solid membranes and inherently the applicable equations that describe the permeance differ. Such mechanisms depend on the relative size of the permeating gas molecules and the diameter of the pores. For the interested reader, other references are suggested (e.g. Oyama *et al.*, 2011; Ockwig and Nenoff, 2007).

3.6. Reactor Configurations

There are several features associated with MRs, and particularly for the reversible WGS reaction, including the fact of not being limited by the chemical equilibrium as they accelerate the forward reaction by selective separation of one product (in this case H_2 or CO_2) using permselective membranes. Other advantages include, for this case study, the feasibility of producing pure/ultra-pure hydrogen, the possibility of CO_2 sequestration, as well as the reduction in the number of process items. In fact, (WGS) MRs are multiphase reactors integrating the catalytic reaction and the separation through the membrane in a single unit, in the perspective of process intensification, also yielding additional advantages in terms of process efficiency and reduced reactor (or catalyst) volume, often operating under less severe conditions as compared with conventional reactors. Several MR configurations have been implemented, as nicely reviewed by Gallucci *et al.* (2013) for hydrogen production. Such configurations will be detailed below, with a few examples regarding application in the WGS reaction, including more recent hybrid reactors with simultaneous H_2 removal through a membrane and CO_2 capture in a selective sorbent (the so-called SEMRs).

3.6.1. Packed bed membrane reactors

The packed bed membrane reactor (PBMR) is the more widely used reactor configuration among MRs, including for WGS. The most used PBMR geometry is the tubular one where the catalyst may be packed either in the membrane tube (Fig. 3.10(a)) or in the shell side (Fig. 3.10(b)), being the permeate collected in the other side of the membrane (the figure shows the particular case of using hydrogen permselective membranes), usually making use of a sweep gas flowing either in co- or counter-current mode (Gallucci *et al.*, 2013).

Among the PBMRs, it is worth mentioning the use of a multi-tube arrangement, which allows increasing productivity. Tosti and co-workers have used such a configuration, employing a multi-tube Pd–Ag MR for pure hydrogen production, using permeator tubes of wall thickness 50–60 μm, diameter 10 mm and length 250 mm (Tosti *et al.*, 2010). Their module makes use of up to 19 Pd–Ag membrane tubes in a finger-like configuration inside the shell module, as illustrated in Fig. 3.11(a), which have been prepared from commercial Pd–Ag foils by cold-rolling and then the tubes have been produced by joining the metal foils via diffusion welding. This way, one end of the permeator tubes is closed and the feed stream, in this case

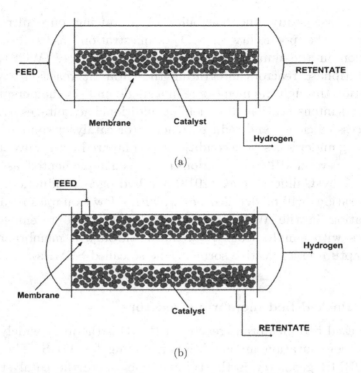

Figure 3.10   PBMR with catalyst in tube (a) and in shell (b) configurations. Reprinted from Gallucci *et al.* (2013) with permission from Elsevier.

coming from an ethanol reformer, is introduced into the membrane lumen through a stainless steel tube. The feed stream passes along the catalyst bed while hydrogen permeates the membrane and is recovered in the shell side by sweeping with an inert gas (nitrogen) or by vacuum pumping. Because the Pd–Ag tubes are fixed to the module at only one end, the free elongation/contraction that is a result of the thermal and hydrogenation cycles is permitted and a good durability is ensured (Tosti *et al.*, 2010). Figure 3.11(b)) shows a view of the open multi-tube MR. However, the authors report that, as a consequence of the interaction between the WGS catalyst and the Pd–Ag membrane, possible corrosion phenomena have produced the formation of holes and microcracks on the permeator tubes, and hence the loss of their selectivity.

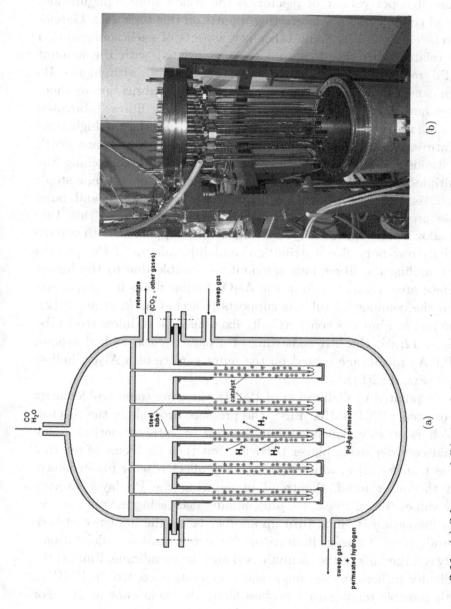

Figure 3.11 (a) Scheme and (b) view of the multi-tube MR (open). Reprinted from Tosti *et al.* (2010) with permission from Elsevier.

Another interesting system, which allows increasing the membrane area per volume of reactor, is the hollow fibre configuration. Li and co-workers have interesting reports on this topic (e.g. García-García *et al.*, 2012). Some other key aspects of such configuration are related with the decreased costs associated with the amount of Pd required (ultra-thin Pd-based membranes, with higher H_2 permeances, are aimed), and membrane support fabrication method. They have focused on asymmetric Al_2O_3 hollow fibres, fabricated by a phase inversion technique, followed by sintering at high temperature. Among the advantages claimed, the following are worth mentioning: (i) low-cost fabrication technique (by combining the multiple fabrication steps of commercial supports in a single step); (ii) it is a simple process, which enables reproducible and large scale production; (iii) it allows the direct deposition of thin Pd–Pd-alloy layers (thickness $\leq 5\,\mu m$) due to the support smooth surface and narrow pore size distribution; and iv) scale-up of the process by bundling the fibres into a module is feasible due to the higher surface area–volume ratio of the Al_2O_3 hollow fibres in comparison with the commercial tubular supports (García-García *et al.*, 2012). The hollow fibre was concentrically fixed inside a stainless steel tube (Fig. 3.12(a)), the MR consisting of a packed catalyst bed around a Pd–Ag membrane coated on the outer surface of a Al_2O_3 hollow fibre (Fig. 3.12(b)).

As pointed by Gallucci *et al.* (2013), there are some problematics associated with PBMRs. First, the pressure drop along the reactor, which is strictly dependent on the size of catalyst particle. One could consider using bigger particles, but then problems of internal mass transfer arise, which detrimentally affect reactor performance. On the other hand, the trend in reducing the Pd layer to very low values (down to ca. 1–$2\,\mu m$), aiming decreasing membrane cost and increasing H_2 flux, turn up another issue: the influence of bed-to-wall mass transfer limitations (or concentration polarization), between the bulk of the catalytic bed and the membrane. Finally, the difficulty in heat management and temperature control in PMBRs, with possible temperature profiles along the membrane length. For these reasons, research efforts have been driven towards new reactor configurations, shortly addressed in the following sections.

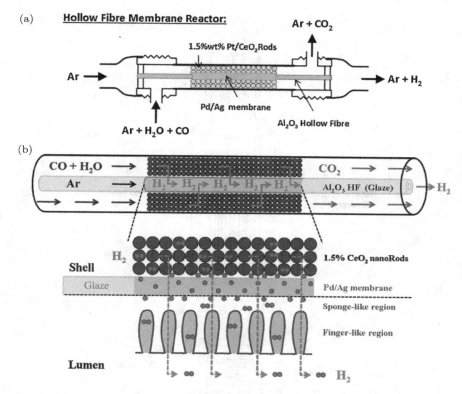

Figure 3.12 (a) Scheme of the hollow fibre fixation inside a stainless steel tube and (b) schematic representation of the hollow fibre MR. Reprinted from García-García *et al.* (2012) with permission from Elsevier.

3.6.2. Fluidized-bed membrane reactors

Although almost no reports are found focused exclusively on WGS employing fluidized-bed MRs, some interesting examples have been reported integrating steam reforming of e.g. methane and WGS in such reactor configuration. This reactor arrangement has several important advantages, as mentioned above, including excellent mass and heat transfer characteristics, and also ease of inserting membranes. This way, processes like autothermal reforming of methane are possible to be operated in such reactors with oxygen (or air) addition without hotspot formation.

In the papers of van Sint Annaland and co-workers, some nice examples can be found, for instance, the use of a multifunctional

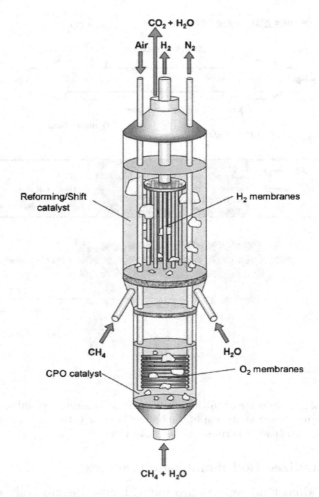

Figure 3.13 Schematic of the fluidized-bed MR concept for pure hydrogen production through autothermal reforming of methane and WGS. Reprinted from Gallucci *et al.* (2013) with permission from Elsevier.

fluidized-bed reactor that integrates permselective membranes for (i) hydrogen removal (to shift methane steam reforming and WGS equilibriums — top section of the reactor) and (ii) oxygen addition (to supply the required energy via partial oxidation of part of the methane feed and enable pure CO_2 capture — bottom section of the reactor) (e.g. Patil *et al.*, 2007). The two sections of the

proposed novel reactor concept, shown in Fig. 3.13, are required due
to the quite distinct operating temperatures of the permselective
H_2 membranes ($<$973 K) and the permselective O_2 membranes
($>$1173 K), i.e. Pd-based and dense perovskite membranes, respectively. Ten dead-end Pd-based membranes have been used (procured
from REB Research, Ferndale, US; dimensions 3.2 mm diameter and
20 cm length), which consist of a metal tube reinforced with Inconel
with a dense 4–5 μm Pd layer deposited on both sides (Patil *et al.*,
2007). Figure 3.14 shows two pictures of the reactor and membrane
assembly.

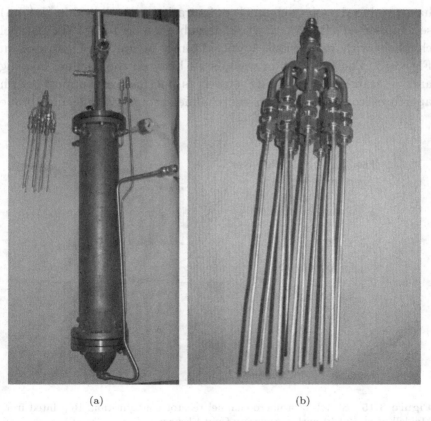

(a)   (b)

Figure 3.14 Picture of the (a) fluidized-bed MR and (b) membrane assembly.
Reprinted from Patil *et al.* (2007) with permission from Elsevier.

3.6.3. Micro-membrane reactors

Micro-structured MRs have been the focus of recent attention as they provide even better heat management, reduce gas-phase diffusion limitations (decreased mass transfer resistances — concentration polarization) and increase the membrane area to reactor volume ratio. However, their use and particularly in the WGS reaction is still limited, in some cases being only characterized in terms of membrane permeance, selectivity and durability (without reaction). The design of micro-membrane systems by the teams of both the Norwegian University of Science and Technology and SINTEF, in Norway, is worth mentioning (e.g. Mejdell *et al.*, 2009; Peters *et al.*, 2015). They have produced a micro-channel configuration with ultra-thin ($1.4\,\mu m$) self-supported (Pd–Ag 23 wt.%) membranes on top of six parallel channels with dimensions $1\,mm \times 1\,mm \times 13\,mm$, as illustrated in Fig. 3.15. The Pd–Ag film was placed between the channel housing and a stainless steel plate (used as mechanical support) with apertures corresponding to the channel geometry. The membrane

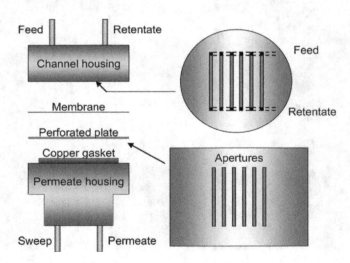

Figure 3.15 Sketch of a micro-channel reactor configuration. Reprinted from Mejdell *et al.* (2009) with permission from Elsevier.

was found to withstand differential pressures up to 4.7 bar, which allow hydrogen separation from gas mixtures without the use of a sweep gas. Possible avoidance of sweep gas is advantageous since H_2 dilution or its subsequent separation from the sweep gas can be omitted. However, they report that after ~7 days a small nitrogen leakage evolved, resulting in the H_2/N_2 separation factor to decrease from ~5700 to ~390 at ~3 bar differential pressure (Mejdell *et al.*, 2009). More recently, they have concluded that for pressures above ca. 5 bar, the application of micro-channel supported modules is not feasible, suggesting alternatively the use of a continuous porous stainless steel support that has been introduced to allow for a sufficient stabilization of the thin Pd–Ag 23 wt.% films (Peters *et al.*, 2015). The porous stainless steel supported micro-channel module has shown a very good stability up to 15 bar during the complete operation of 1100 h (ca. 46 days) at 723 K, with a minimum value for the H_2–N_2 permselectivity of 39,000. Even so, pore formation in the feed side film surface has been observed, which still requires further investigation.

Interesting configurations of micro-MRs applied to the WGS reaction are those using hollow fibres and reported by Li and co-workers from the Imperial College, London, UK. One example relates to the use of asymmetric Al_2O_3 hollow fibres that were employed as a substrate for both coating of the Pd membrane (6 μm thick, deposited by electroless plating on the outer layer) and impregnation of a 30% CuO–CeO_2 catalyst (into the inner finger-like structure) (Rahman *et al.*, 2011). The authors have firstly performed some catalytic tests using the Al_2O_3 hollow fibres without the Pd membrane in two distinct configurations: (i) open-both-end configuration (which enabled the reactants to flow through the lumen of the hollow fibres) and (ii) dead-end configuration (which forced the reactant to permeate through the finger-like structure and sponge-like structure) — cf. Fig. 3.16. Because both configurations gave nearly the same CO conversion in the WGS reaction, they concluded that the open-both-end configuration enabled the reactants to reach the 30% CuO–CeO_2 catalyst that was placed on the inner surface

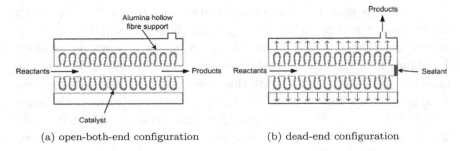

(a) open-both-end configuration   (b) dead-end configuration

Figure 3.16 Schematic diagram of operational modes for the catalytic hollow fibre micro-reactor. Reprinted from Rahman *et al.* (2011) with permission from Elsevier.

of the hollow fibres. The finger-like region in the Al_2O_3 hollow fibre was made up by a set of hundreds of conical micro-channels, which were perpendicularly distributed around the fibre lumen in which the reactants could be efficiently mixed.

SEM images of the Al_2O_3 hollow fibre put into evidence its asymmetric pore structure (Figs. 3.17(a) and 3.17(b)). The narrow pore size distribution of the sponge-like structures enabled the deposition of a thin, defect-free, Pd membrane on the outer surface of the Al_2O_3 hollow fibre, whereas the open pore structure of the finger-like region enabled the impregnation of the 30% $CuO–CeO_2$ catalyst in the inner surface, with a uniform dispersion of both Cu and Ce (Fig. 3.17(c)). Finally, when tested in the WGS reaction, the micro-MR provided a CO conversion that was above (17% higher) than the corresponding thermodynamic equilibrium conversion while high purity H_2 has been produced in the shell side (Rahman *et al.*, 2011).

3.6.4. Sorption enhanced membrane reactors (SEMRs)

The combination, in the same device, of reaction–separation strategies described along this book to shift reversible reactions has been increasing in number and complexity along the years. Particularly, the combination of the sorption enhanced concept in a MR (thus removing simultaneously two reaction products, e.g. carbon dioxide with a CO_2-selective sorbent and hydrogen with an H_2-permselective

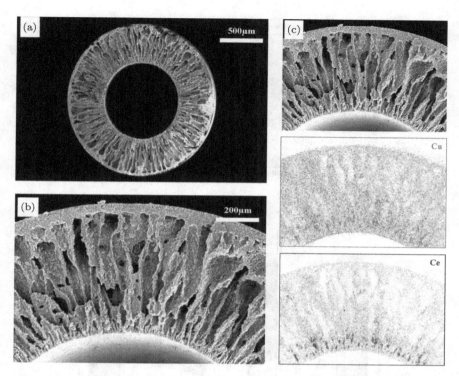

Figure 3.17 SEM pictures of the (a) and (b) Al$_2$O$_3$ hollow fibre at different magnifications before Pd deposition and catalyst impregnation and (c) Cu and Ce distribution (EDS images) of the fibres after catalyst impregnation. Reprinted from Rahman *et al.* (2011) with permission from Elsevier.

membrane) is a new reactor configuration. It has been tested/studied theoretically for steam reforming of methane (e.g. Andrés *et al.*, 2011; Anderson *et al.*, 2015; Wu *et al.*, 2015), reforming of glycerol (Leal *et al.*, 2016; Silva *et al.*, 2015) and methanol synthesis (Bayat *et al.*, 2014), the latter with water sorption. Studies dedicated to the WGS reaction are also limited and quite recent.

The hybrid SEMR (HSEMR, designation adopted by Soria *et al.* (2015)) carries out simultaneously the WGS reaction while removing hydrogen and carbon dioxide from the reaction zone, with an H$_2$-permselective membrane and a CO$_2$ sorbent, respectively (Fig. 3.18). For a continuous implementation, this requires the cyclic

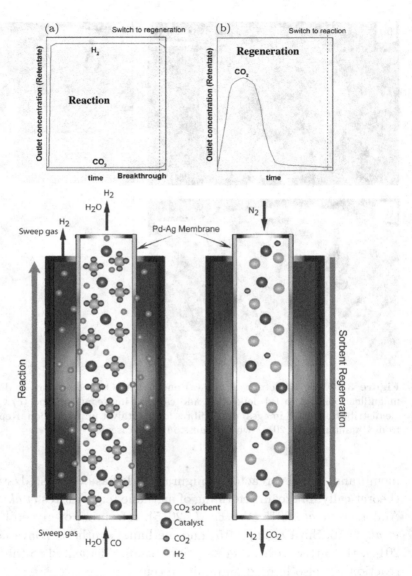

Figure 3.18   Schematic view of the conceived HSEMR based on two parallel reactors configuration for continuous operation and corresponding outlet concentrations histories in the retentate stream during (a) reaction and (b) regeneration stages. Reprinted from Soria *et al.* (2015) with permission from Elsevier.

operation of at least two parallel reactors (if regeneration is not longer than the production stage), producing (Fig. 3.18(a)) and regenerating (Fig. 3.18(b)) out of phase. Once the CO_2 sorbent gets saturated (and the gas starts breaking through the column), the reaction is ended in a given column and the regeneration stage is initiated.

Soria *et al.* (2015) have tested the HSEMR in the LT-WGS reaction integrating a Cu–ZnO–Al_2O_3 catalyst, a K_2CO_3-promoted hydrotalcite (as CO_2 sorbent — MG30-K) and a self-supported Pd–Ag membrane. It is worth noting that the presence of water vapour enhanced the sorption capacity of the hydrotalcite, which is beneficial for WGS as it is present as reactant species. It was shown that when both CO_2 and H_2 are removed from the reaction zone, the hydrogen production is enhanced compared to either a traditional (fixed bed) or a sorption enhanced (only CO_2 is removed) reactor, allowing overcoming equilibrium limitations and to obtain a pure H_2 stream. Particularly, at 5.5 bar and 250°C, in the pre-breakthrough zone complete CO conversion is reached (Fig. 3.19), allowing to obtain two hydrogen streams feasible to be fed to a fuel cell (with 100% H_2 recovery). Moreover, in the pre-breakthrough

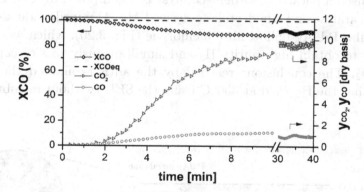

Figure 3.19 CO conversion and CO and CO_2 composition (vol.%) history during WGS reaction in a HSEMR at 250°C and 5.5 bar; catalyst–sorbent weight ratio = 1/5; $W_{cat}/Q_{CO} = 3.7 \times 10^{-4}\,\mathrm{g\,h\,mL_N^{-1}}$; feed composition $(CO/H_2O/N_2) =$ 10/15/75 (vol.%). Reprinted from Soria *et al.* (2015) with permission from Elsevier.

zone, the CO_2 is sorbed by the hydrotalcite and the CO is not produced (the equilibrium is completely shifted), which is favourable for the Pd–Ag membrane since CO_2 and mainly CO in gas phase can poison the membrane, affecting the H_2 permeability, as detailed before (cf. Section 3.5.5). As in the tests carried out with catalyst and sorbent only — sorption enhanced reactor (SER) — in the HSEMR is also observed that the CO concentration reaches a steady-state value when CO_2 concentration stabilizes, i.e. when the hydrotalcite becomes saturated (from this point on it works as a MR).

The preliminary results obtained by Maroño *et al.* (2014) with a SEMR are also encouraging. They argue that the advantages of using such a reactor configuration seem to be positive because complete conversion of CO is obtained up to CO_2 breakthrough, while a pure stream of H_2 is being produced in the permeate side. However, they recommend that, in order to avoid the occurrence of the reverse WGS reaction on the membrane surface, additional excess of steam is used (they employed a molar S/C ratio of 3, at 633–653 K, 7.5 bar).

García-García *et al.* (2014) also published their preliminary data. They used a packed adsorbent-catalyst bed composed by 10% CuO–CeO_2 catalyst and a hydrotalcite-derived Mg–Al mixed oxide around a tubular Pd–Ag hollow fibre membrane (Fig. 3.20), which was conceived to obtain high purity H_2 and simultaneously *in situ* capture of CO_2. The conclusions reached by the authors allowed them to infer that the H_2 yield at 350°C using the SER (without membrane)

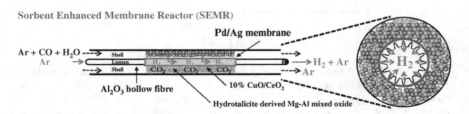

Figure 3.20   A schematic representation of the SEMR used by García-García *et al.* (employing asymmetric Al_2O_3 hollow fibres coated with Pd–Ag). Reprinted from García-García *et al.* (2014) with permission from Elsevier.

was 80%, which is 33% higher than that obtained in the traditional fixed-bed reactor (FBR) and 18% higher than the corresponding thermodynamic equilibrium. However, due to the high $CO–H_2O$ ratio $(R > 1)$ employed in such a work, undesirable side reactions such as C deposition became important at temperatures higher than 300–400°C. A similar behaviour was observed using the SEMR, however in this case a high purity CO_x free H_2 stream was obtained. The performance of their Pd–Ag membrane in the SEMR during the WGS reaction is summarized in Fig. 3.21. H_2 permeation across the Pd–Ag membrane started at 200°C and increased with the reaction temperature. The H_2 recovery (i.e. percentage of H_2 permeated through the membrane) followed the same trend, suggesting that H_2 permeation is not limited by boundary layer or concentration polarization effects (García-García *et al.*, 2014).

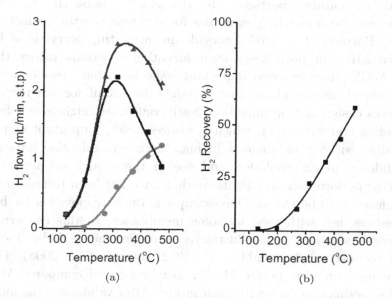

Figure 3.21 (a) H_2 production as a function of the temperature during the WGS reaction using a SEMR: total (▲), retained (■), and permeated (●). (b) H_2 recovery as a function of the temperature during the WGS reaction. Volumetric flow rates are measured at standard conditions ($P = 105\,\text{Pa}$, $T = 273\,\text{K}$). Reprinted from García-García *et al.* (2014) with permission from Elsevier.

3.7. Modelling and Simulation

Studies reporting modelling/simulation of WGS MRs are numerous. Of course, variations depend on several factors like the catalyst considered (and inherently the kinetic equations adopted — cf. Section 3.4), the membrane employed (and inherently the permeation–selectivity equations and data considered — cf. Section 3.5), the reactor flow pattern (with advective and eventually axial dispersion terms, for instance), etc. Moreover, the hypothesis considered will also determine the model complexity, and these include assuming isothermicity or not along the reactor (so that energy equation might be or not ignored, apart from the mass balance), take or not into account total pressure variation (being required the momentum balance — e.g. Ergun equation), use 1D or 2D models (the latter being particularly important for highly permeable, thin membranes), consider or not mass/heat transfer resistances in the film and inside the catalyst particles, etc. Hsieh in his book (Hsieh, 1996) reports some interesting examples for different reactor geometries, while Barbieri *et al.* (2008) provide an interesting overview of Pd-based MRs, in both cases even for other reactions rather than the WGS. However, one important issue is to have confidence in the model developed, so that it might be useful for reactor and process design and optimization. Such confidence might result from experimental validation, which is always a very important step, if possible without parameters fitting. Moreover, this also improves confidence in extrapolation and use of the model for predicting reactor performance in conditions that have not been tested, or are not feasible to be tested. One example is the extrapolation for high pressures, not withstand by some membranes. In this concern, it is worth mentioning, as illustrative example, the work by Tsotsis and co-workers (Abdollahi *et al.*, 2012; Hwang *et al.*, 2008). They have used an ultra-permeable Pd membrane and compared WGS MR experimental data with their model. After validation, the model has been used to study the design aspects of the process. In short, and using whenever feasible authors' nomenclature, their isothermal co-current MR model considers firstly an equation describing H_2

transport through the Pd membrane (cf. Eq. (3.25)):

$$J_{H_2} = U_{H_2} \left(\left(p_{H_2}^F\right)^n - \left(p_{H_2}^P\right)^n \right) \qquad (3.30)$$

where J_{H_2} is the H_2 molar flux (mol/m^2 h), $p_{H_2}^F$ and $p_{H_2}^P$ represent its partial pressure (bar) in the feed and permeate side, respectively, n is the pressure exponent (determined independently from pure hydrogen permeation experiments to be 0.96), and U_{H_2} is the H_2 permeance (mol/m^2 h bar^n), also obtained from single-gas permeation experiments. A similar equation was used for other species (i) that may permeate as well, namely through small cracks and defects:

$$J_i = U_i \left(p_i^F - p_i^P \right) \qquad (3.31)$$

and the associated parameters have also been determined independently.

Hydrogen mass balances in the feed and permeate sides at steady state have been described by the following equations, respectively, where the left-side term refers to the molar flow rate variation along the MR tube (or reactor volume) — with convective contribution only in the overall flux, and the terms in the right refer to permeation and reaction rate (the latter only in the feed/retentate side equation):

$$\frac{dn_{H_2}^F}{dV} = -\alpha_m U_{H_2} \left(\left(p_{H_2}^F\right)^n - \left(p_{H_2}^P\right)^n \right) + (1 - \varepsilon_v) \beta_c \rho_c r^F \qquad (3.32)$$

$$\frac{dn_{H_2}^P}{dV} = \alpha_m U_{H_2} \left(\left(p_{H_2}^F\right)^n - \left(p_{H_2}^P\right)^n \right) \qquad (3.33)$$

where n_{H_2} is the H_2 molar flow rate (mol/h), F and P superscripts stand again for the feed and permeate side, respectively, and V is the reactor volume variable (m^3). Because permeation of H_2 occurs through the membrane surface only, the geometric factor α_m is required, which denotes the surface area of the membrane per unit reactor volume (m^2/m^3). On the other hand, reaction occurs only on the solid phase, and that is why the parameters associated to the reaction rate term are required: ε_v bed porosity in the feed side (thus $(1 - \varepsilon_v)$ denotes the fraction of volume occupied by the solid; β_c — fraction of the solid volume occupied by the catalyst; ρ_c — catalyst density (g/m^3); r^F — WGS reaction rate (mol/g h).

Similarly, steady-state mass balances for components other than H$_2$ in the feed and permeate sides are described by the following equations (assuming again a 1D pseudo-homogeneous model and isothermal operation, being negligible mass and heat transfer resistances):

$$\frac{dn_i^F}{dV} = -\alpha_m U_i(p_i^F - p_i^P) + \nu_i(1 - \varepsilon_v)\beta_c\rho_c r^F \tag{3.34}$$

$$\frac{dn_i^P}{dV} = \alpha_m U_i(p_i^F - p_i^P) \tag{3.35}$$

being that ν_i represents the stoichiometric coefficient for component i, negative for reactants and positive for products. Again, the term $(p_i^F - p_i^P)$ represents the driving force for permeation through the membrane surface, which has the associated negative sign in Eq. (3.34) because it represents a depletion in the feed/retentate side, while it is positive in Eq. (3.35) as it represents a gain in the permeate (co-current) chamber.

The pressure drop in the packed bed has been calculated by the authors (Abdollahi *et al.*, 2012) using the Ergun equation:

$$\frac{dP^F}{dV} = -10^{-6}\frac{f(G^F)^2}{A^F g_c d_p \rho^F} \tag{3.36}$$

$$f = \frac{(1 - \varepsilon_v)}{\varepsilon_v^3}\left(1.75 + \frac{150(1 - \varepsilon_v)}{N_{\text{Re}}^F}\right) \tag{3.37}$$

$$N_{\text{Re}}^F = \frac{d_p G^F}{\mu^F} \tag{3.38}$$

where P^F is the feed side total pressure (bar), G^F is the superficial mass flow velocity in the feed side or massic flux (g/m^2 h), A^F is the cross-sectional area available to flow for the reactor feed side (m^2), g_c is the gravity conversion factor, d_p is the particle diameter in the feed side (m), ρ^F is the average fluid density (g/m^3), and μ^F is the fluid viscosity (g/m h).

The above first-order differential equations require one boundary condition each:

$$at\ V = 0: n_i^F = n_{i0}^F, \quad n_i^P = n_{i0}^P, \quad P^F = P_0^F, \quad P^P = P_0^P \quad (3.39)$$

wherein the subscript "0" refers to inlet conditions in either feed (F) or permeate (P) side; the system of ODEs has been solved numerically.

As stated before, gas permeation experiments provided membrane-related parameters (based on the experimental observations that mixed gas permeances do not differ substantially from their single gas counterparts). In the same way, independent kinetic studies should provide the reaction rate equation and kinetic parameters. In this particular example, a commercial Cu–Zn–Al$_2$O$_3$ LT-WGS catalyst has been used and a reaction rate expression consistent with a Hougen–Watson type surface mechanism developed (Abdollahi *et al.*, 2012). The authors validated their model both in the MR configuration (which provided satisfactory agreement with experimental data, in terms of CO conversions and H$_2$ recoveries), and also as a packed-bed reactor (PBR), as shown in the parity plot of Fig. 3.22.

The MR model was then very useful for predicting performance in different conditions, namely higher pressures (due to the range of pressures that could be employed experimentally), and for scale-up (due to limitations in terms of the membrane area), thus allowing simulating the MR in conditions closer to industrial ones. In particular, it is noteworthy the effect of the total pressure, which was increased up to 20 bar in the simulations. It was found that increasing reactor pressure has a significant effect on both CO conversion (Fig. 3.23(a)) and H$_2$ recovery (Fig. 3.23(b)) because this way the driving force (H$_2$ partial pressure difference across the membrane) is increased. However, hydrogen purity is detrimentally affected at high reactor pressures, because the driving force for other species permeation is also increased; this is particularly noticeable at high contact times (Fig. 3.23(c)). The authors conclude that, even so, in the whole range of conditions simulated the hydrogen purity is >99.9%, indicating that the use of Pd-MRs for industrially relevant

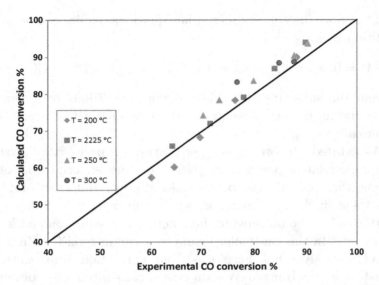

Figure 3.22   Measured vs. fitted CO conversion data using a Hougen–Watson rate expression. Adapted from Abdollahi *et al.* (2012).

conditions is indeed promising, particularly for fuel cell applications (Abdollahi *et al.*, 2012).

The use of these simulation tools has some other advantages apart from those already identified, like allowing studying the effect of parameters that experimentally might be difficult to isolate and analyze, as is the case of the heat transfer coefficients (in this perspective see e.g. Adrover *et al.* (2009)), or the hydrogen permeances — corresponding to different membranes. Moreover, they also allow better comprehending what is occurring inside and along the reactor (because model outputs include concentration, temperature and other properties profiles along the reactor coordinates — axial for 1D models and also radial for 2D models). In this perspective, and as an example, one could mention the work reported by Mendes *et al.* (2011) where a PBMR in a "finger-like" membrane configuration has been experimentally tested and modelled, considering also axially dispersed plug flow pattern (apart from the hypothesis of the previous example). Adopting the same nomenclature and formulation as the authors, the partial mass

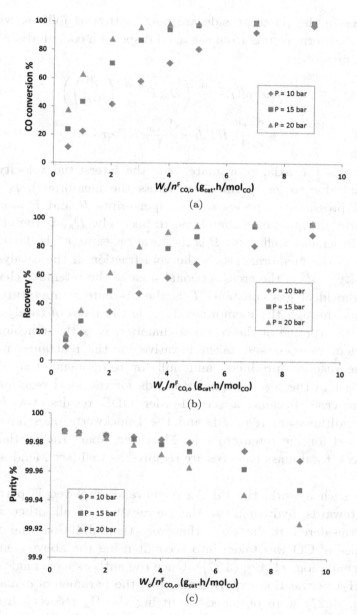

Figure 3.23 Simulated effect of pressure on (a) CO conversion, (b) H₂ recovery and (c) H₂ purity, at $T = 300\,°C$ and SR (steam-to-sweep gas ratio) = 0.1. Adapted from Abdollahi *et al.* (2012).

balances in the retentate side are now written as follows, wherein the second-term results from the axial dispersion contribution to the overall mass flux:

$$\frac{d}{dz}(u^R p_i^R) - \frac{d}{dz}\left(D_{ax}P^R \frac{d}{dz}\left(\frac{p_i^R}{P^R}\right)\right)$$

$$+ \frac{2\pi r^m}{\varepsilon_b A^R}RTJ_i - \frac{W_{cat}}{\varepsilon_b V^R}RT\nu_i r = 0 \qquad (3.40)$$

where z is the axial coordinate, u is the interstitial velocity (not constant due to gas permeation across the membrane), p_i is the partial pressure of species i, the superscripts R and P stand for retentate and permeate chambers, respectively, D_{ax} is the effective axial dispersion coefficient, P is the total pressure, r^m is the internal radius of the membrane, ε_b is the void fraction of the catalyst bed (porosity), A^R is the cross-sectional area of the retentate chamber, R is the ideal gas constant, T is the absolute temperature, J is the flux through the membrane, W_{cat} is the mass of catalyst bed, V^R is the volume of the reaction chamber, ν_i is the stoichiometric coefficient of species i, taken negative for the reactants, positive for the reaction products, and null for components that do not take part in the reaction, and r stands for the local reaction rate. In this case, because a second-order ODE results, two boundary conditions are required, and the Danckwerts ones have been employed for the retentate side. Moreover, apart the partial mass balances, total mass balances are required as well (see Mendes *et al.*, 2011).

In such a work, the Pd–Ag membrane employed is permeable only towards hydrogen, so the permeation of all other species was considered to be null. However, the inhibition due to the presence of CO was taken into account using the above-mentioned SL formulation (cf. Eq. (3.27)). Once the authors have made single-gas (H_2) permeation experiments only, the parameters α and K_{CO} in Eq. (3.27) were obtained by fitting the H_2 recovery from the model to the experimental results, obtained with either sweep gas or vacuum in the permeated side. Very good agreement has been obtained with only two fitting parameters, as illustrated in Fig. 3.24,

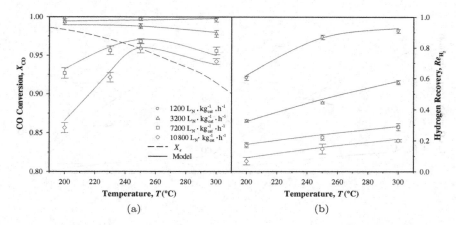

Figure 3.24 Effect of the reaction temperature and feed gas space velocity on the CO conversion (a) and H_2 recovery (b) as a function of the reaction temperature in the MR, operating in counter-current mode. $Q_{\text{sweep}} = 1.0\,L_N\,\text{min}^{-1}$, $P^F = 2.0\,\text{bar}$, and $P^P = 1.0\,\text{bar}$. Error bars are based on t-student distribution and 95% confidence limit (with three replicates). Reprinted from Mendes *et al.* (2011), with permission from Elsevier.

either in terms of CO conversion or H_2 recovery (the particular effect of each operating condition, temperature and flow rate in this case, will be detailed later on along Section 3.8).

As mentioned above, one of the advantages of the simulation model is to better understand what happens inside and along the reactor. This can be perceived from the same work, and particularly from Fig. 3.25, which provides the simulated axial composition profiles (in the retentate side) for each species for the two operating modes of the MR, vacuum (a) and sweep gas (b) (Mendes *et al.*, 2011). It is worth mentioning the good agreement with the experimental compositions obtained at the reactor outlet (cf. circle symbols in each plot). Such profiles allowed comprehending that there are two distinct regions inside the reactor. In the first one (ca. 10% of the reactor length), CO concentration decreases sharply, while H_2 fraction slightly increases (it is worth mention that a simulated reformate feed has been used). At such high reaction temperature (300°C), reaction kinetics is favoured decreasing rapidly the CO concentration. Although H_2 permeation is also favoured at

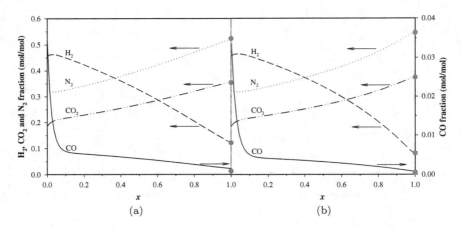

Figure 3.25 Simulated profiles on the reaction side as a function of the dimensionless reactor length: vacuum mode, $P^F = 2.0\,\text{bar}$ and $P^P = 30\,\text{mbar}$ (a) and sweep gas mode, $P^F = 2.0\,\text{bar}$, $P^P = 1.0\,\text{bar}$ and $Q_{\text{sweep}} = 1.0\,L_N\,\text{min}^{-1}$, (b). Other experimental conditions: $T = 300°\text{C}$, GHSV $= 1200\,L_N\,\text{kg}_{\text{cat}}^{-1}\,\text{h}^{-1}$. Symbols represent the experimental molar fraction (dry basis) of each species measured at the exit of the reactor. Reprinted from Mendes *et al.* (2011) with permission from Elsevier.

such temperature, it does not follow the rate of production from the WGS reaction, and thus its concentration increases, though slowly. In the second region, the shift of the reaction towards the reaction products depends only on the H_2 removal due to permeation. Consequently, CO concentration decreases slowly. So, in summary, the MR can be divided into a region of chemical control (first part) and a region of permeation control (second part), which account respectively for 10% and 90% of the reactor length, under such conditions. Such knowledge might be very useful for process/reactor optimization.

The assumption of isothermal operation in a reactor, and also in a MR, depends on the extent of the generated reaction heat compared to the heat loss/gain through the reactor walls. So, in some cases it is required to consider energy balances as well, due to the exothermicity of the WGS reaction. One example is provided in the work by Brunetti *et al.* (2007), where a non-isothermal mathematical

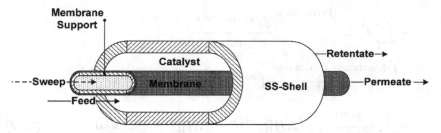

Figure 3.26 MR scheme used in the simulations by Brunetti *et al.* Reprinted from Brunetti *et al.* (2007) with permission from Elsevier.

model was formulated to describe the behaviour of a PBMR involving a dense Pd–Ag commercial permselective membrane (considered to have an infinite selectivity towards H_2 compared with other species). The MR configuration is shown in Fig. 3.26, which consists of two concentric tubes: the outer tube is the shell; the inner tube is the Pd-alloy self-supported membrane. The catalyst has been packed in the shell side, because such configuration is more efficient e.g. for the heat exchange.

Plug flow in the retentate and permeate streams has been assumed in the model. The diffusion limitations in the catalyst pellets were considered by means of the effectiveness factor, which is associated with the reaction term. So, comparing with previous models, the main differences are related with the need of solving also energy balances, which have been written as follows (using again the same nomenclature and formulation as the authors) (Brunetti *et al.*, 2007):

$$-\sum_{i=1}^{n} \bar{F}_i C_{p_i} \bigg|^{\text{Reaction}} \frac{dT^{\text{Reaction}}}{d\bar{z}}$$

$$+ Da\,\overline{r_{\text{CO}}}\left(-\Delta H_{\text{Reaction}}\right) + \frac{U^{\text{Shell}} A^{\text{Shell}}}{F_{\text{CO}}^{\text{Feed}}} \left(T^{\text{Furnace}} - T^{\text{Reaction}}\right)$$

$$- \frac{U^{\text{Membrane}} A^{\text{Membrane}}}{F_{\text{CO}}^{\text{Feed}}} \left(T^{\text{Reaction}} - T^{\text{Permeation}}\right) = 0 \qquad (3.41)$$

$$-\sum_{i=1}^{n} \bar{F}_i C_{p_i} \bigg|^{\text{Permeation}} \frac{dT^{\text{Permeation}}}{d\bar{z}}$$

$$+\frac{U^{\text{Membrane}} A^{\text{Membrane}}}{F_{\text{CO}}^{\text{Feed}}} \left(T^{\text{Reaction}} - T^{\text{Permeation}}\right)$$

$$+\frac{A^{\text{Membrane}}}{F_{\text{CO}}^{\text{Feed}}} J_{\text{H}_2} \left(\Delta H_{\text{H}_2}^{\text{Reaction}} - \Delta H_{\text{H}_2}^{\text{Permeation}}\right) = 0 \qquad (3.42)$$

In these equations, $\bar{F}$ is a dimensionless feed flow rate (normalized by the CO feed flow rate, $F_{\text{CO}}^{\text{Feed}}$), T is the temperature (°C), $\bar{z}$ is the dimensionless reactor length, Da is the Damköhler's number, a dimensionless group relating characteristic space time per characteristic reaction time (calculated both using the inlet operating conditions, $\text{Da} = r_{\text{CO},z=0} V_{\text{Reaction}}/F_{\text{CO}}^{\text{Feed}}$), $\overline{r_{\text{CO}}}$ is a dimensionless reaction rate (normalized by the one calculated at feed conditions, $r_{\text{CO},z=0}$), ΔH is the heat of reaction (J/mol), U is the overall heat transfer coefficient (W/m^2 K), A is the membrane or heat exchange surface (m^2), and J is the permeation flux (mol/m^2 s, calculated using Sievert's law). Subscripts and superscripts are clearly identifiable.

The energy balance for the reaction side (Eq. (3.41)) contains the following terms (from the left to the right): convective flux of energy, heat produced by chemical reaction, heat exchanged with the furnace and heat exchanged with the permeation side. As can be inferred from the inspection of the MR configuration (Fig. 3.26), the heat produced by reaction and the exchange with the furnace is not present in the energy balance on permeate side (Eq. (3.42)). In addition, a further term related to energy associated with the H$_2$ permeating flux is also considered, although authors state its contribution is small ($\sim$5%) (Brunetti *et al.*, 2007).

After model validation, some simulations were carried out to assess the effect of several parameters, namely feed pressure, Da number, feed composition, etc. Particularly, Fig. 3.27 shows the effect of feed pressure on the CO conversion and temperature profiles, and compares them with a traditional reactor (TR). As reference, the traditional reactor equilibrium conversion (TREC) is also reported

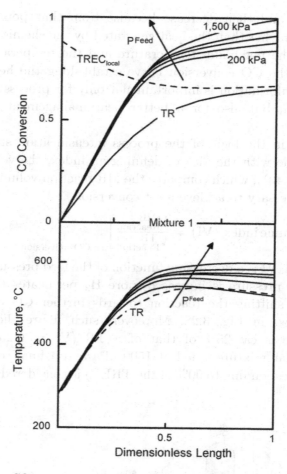

Figure 3.27 CO conversion and temperature profiles as a function of the dimensionless reactor length at different feed pressures in the range 200–1500 kPa. Furnace temperature = 280 °C; Da = 1. Reprinted from Brunetti *et al.* (2007) with permission from Elsevier.

in the figure, which was calculated at the temperature of the referred abscissa, thus using the temperature profile calculated for the TR and reported in the same figure (lower section). As the feed pressure increases, higher CO conversions are obtained in the MR, owing to the positive effect of the pressure on the H_2 permeation. It is also seen that temperature profiles increase on the first part of the reactor and then decrease because from that point the heat exchange is higher than the heat generated by the chemical reaction. Therefore, the maximum temperature and the temperature profile depend on the CO conversion curve establishing the heat production. Such information is important not only for process design and optimization, but also for a better heat management and safety issues.

Moving in the logic of the process intensification strategy, the authors came with the idea of defining an index, the volume index (VI) (Eq. (3.43)), which compares the MR reaction volume with that of a TR necessary to achieve a set conversion.

$$\text{Volume index (VI)} = \left. \frac{V^{\text{MR}}_{\text{Reaction}}}{V^{\text{TR}}_{\text{Reaction}}} \right]_{\text{set CO conversion}} \tag{3.43}$$

As expected, VI is a decreasing function of the feed pressure (because as the feed pressure is increased more H_2 permeates through the membrane, shifting the reaction towards further CO conversion), as illustrated in Fig. 3.28. Moreover, such figure shows that a MR volume is ca. 25% of that of a TR (VI = 0.25) when an equimolecular mixture is fed at 1500 kPa and a final conversion of $\sim$80% (corresponding to 90% of the TREC) is considered. Therefore,

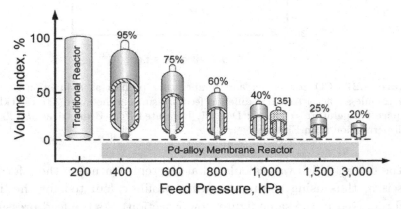

Figure 3.28 VI as a function of feed pressure for a mixture containing only the reactants (CO and H_2O, in a ratio 1/1). Furnace temperature = 280°C. Set CO conversion 90% of the TREC. Reprinted from Brunetti *et al.* (2007) with permission from Elsevier.

in such conditions, the catalyst amount is drastically reduced with clear gain also in terms of plant size reduction.

As mentioned above, one can find a lot of published studies dealing with modelling/simulation of WGS MRs. Increased complexity compared with the previous examples can be added, but that might also require added effort in terms of computation. A "cost-benefit" analysis should be done, and particularly consider if such added complexity is really required or not. Some examples can be found for instance dealing with the use of bi-dimensional models (which allow predicting reactor performances in terms of both axial and radial coordinates) (Basile *et al.*, 2003), which are particularly relevant for thin, highly permeable membranes (where concentration polarization phenomena are important). Combination of a 2D and non-isothermal hypothesis in the same model, along with internal and external mass transfer resistance, has also been reported recently (Sanz *et al.*, 2015); use of CFD tools is of relevance nowadays, and a few works dedicated to modelling of WGS MRs can be found.

3.8. Parametric Study: Effect of Operating Variables

There are more variables affecting the performance of a MR as compared with a conventional FBR, because apart from the reaction kinetics, flow pattern, etc. one must also consider the effect of the membrane itself. So, factors like transmembrane pressure, sweep gas flow rate, etc. must be also taken into account. Given the huge number of reports found in the open scientific literature about this topic, in the following sections just a few examples will be presented to illustrate which are the more relevant parameters that have to be considered in a MR operation/optimization, again focused on the WGS reaction.

3.8.1. Effect of temperature

Temperature has a very important effect in a WGS MR. On one hand, it affects equilibrium conversion. On the other hand, it affects reaction kinetics. Finally, its effect on the membrane permeation

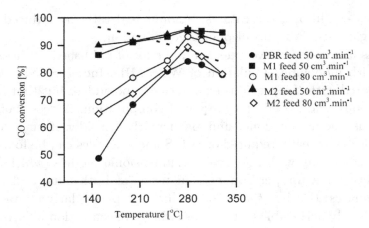

Figure 3.29 Temperature dependence of CO conversion for a PBMR and a PBR (H_2O/CO molar ratio $= 1.0$, no sweep gas). Reprinted from Giessler *et al.* (2003) with permission from Elsevier.

towards hydrogen (and possibly other species) should be considered as well.

Figure 3.29 shows, as an example, data obtained by Diniz da Costa and co-workers using a molecular sieve silica (MSS) membrane PBR using a Cu–ZnO–Al$_2$O$_3$ catalyst in the LT-WGS reaction (Giessler *et al.*, 2003). They used both hydrophilic membranes (M2, which underwent pore widening during the reaction caused by water vapour) and hydrophobic membranes (M1, which indicated no such behaviour and also showed increased H$_2$ permeation with temperature, a characteristic of activated transport — cf. Section 3.5.5). For all cases, maximum CO conversions were observed at 280 °C: 84% in the PBR and 93% in the MRs (for a feed flow rate of $50\,\text{cm}^3\,\text{min}^{-1}$). It is also noticed a drop in CO conversion at higher temperatures (seen also in the PBR, due to reaction exothermicity), which in the case of the hydrophobic membrane (M1) is not very pronounced. A more noticeable drop in CO conversion was obtained for the hydrophilic membrane (M2), especially at a higher feed flow rate of $80\,\text{cm}^3\,\text{min}^{-1}$. This was ascribed to the membrane structural degradation (Giessler *et al.*, 2003). In conclusion, the equilibrium conversion was even surpassed with the MR, namely at 280°C, the

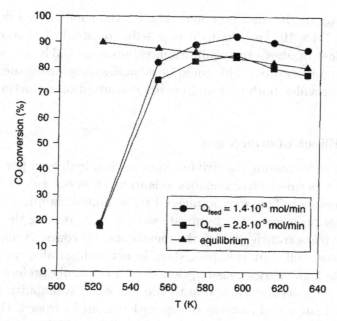

Figure 3.30 Effect of temperature on conversion of carbon monoxide at two different total feed flows for a composite palladium membrane. Experimental conditions: H_2O/CO molar ratio $= 0.96$; $P_{lumen} = 1.1$ bar. Reprinted from Basile *et al.* (1996) with permission from Elsevier.

reason being due to the withdrawal of H_2 (reaction product), which shifts the chemical equilibrium in the forward direction, resulting in higher CO conversions.

Another example is provided in Fig. 3.30, for a composite palladium membrane that was obtained by coating an ultra-thin double-layer palladium film on the inner surface of the support of a commercial tubular ceramic membrane. Basile *et al.* (1996) came also to the conclusion that while the thermodynamic equilibrium conversion of CO decreases with temperature, for their catalytic MR there is a maximum CO conversion around 600 K. This behaviour results from the trade-off between the reaction kinetics (which increases with temperature) and thermodynamic considerations for the WGS reaction. Of course, in this analysis, the promoted hydrogen permeation with temperature should be considered as well (kinetic term). It must be noted that for the smaller feed flow rate ($Q_{\text{feed}} = 1.4 \times 10^{-3}$ mol/min) it is possible to obtain CO conversion values close or above the equilibrium values until 553 K, i.e. working at lower temperatures. This could be advantageous for the membrane life and provides both cost and energy consumption reductions.

3.8.2. Effect of sweep gas

One way of increasing the driving force for e.g. hydrogen permeation through a permselective membrane is to use a sweep gas in the permeate chamber. This way permeated H_2 is diluted/swept, decreasing its partial pressure in the permeate side, thus increasing the driving force for mass transfer across the membrane. Of course, vacuum can be used as well, but this procedure is not industrially encouraged due to its high energy consumption. Steam is usually preferred to be used in the sweep stream, due to the low cost, availability in most industrial sites, and easiness of separation from hydrogen. However, at lab scale, inert species like nitrogen are often used.

Figure 3.31 shows the CO conversion and H_2 recovery as a function of the contact time ($W_c/n_{CO,o}^F$, mass of undiluted catalyst (g) over the feed molar flow rate of CO (mol/h) at different sweep ratios, namely SR = 0 (no sweep), SR = 0.1 and SR = 0.3 (SR stands for the ratio between inlet steam molar flow rate in the permeate side and the inlet molar feed flow rate in the feed side); the calculated equilibrium conversion (X_e) is also included (dash line) (Abdollahi *et al.*, 2012). It is noticed that although the sweep ratio (SR) has almost no effect upon CO conversion, which is already above X_e and close to completion, it improves hydrogen recovery (up to almost 90%). As mentioned above, increasing the SR (as it occurs with increasing the feed pressure — cf. Section 3.8.4) increases the hydrogen partial pressure difference across the membrane, thus providing a higher driving force for H_2 to diffuse through the membrane.

Figure 3.32 shows the effect of the sweep gas (nitrogen) flow rate in the CO conversion reached by Giessler *et al.* (2003) using the two types of molecular sieve silica (MSS) membranes mentioned

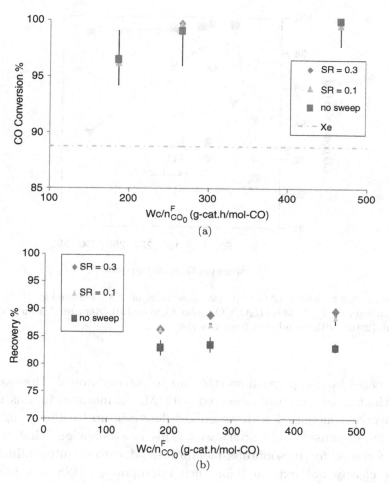

Figure 3.31 Effect of sweep ratio (SR) on (a) CO conversion and (b) H_2 recovery, $T = 300°C$ and $P = 4.46\,bar$; membrane used is the one shown in Fig. 3.35. Reprinted from Abdollahi *et al.* (2012) with permission from Elsevier.

in Section 3.8.1, Fig. 3.29. It is noticed that CO conversion was in general above the equilibrium conversion, but was generally higher for the hydrophobic (M1) membrane than for the hydrophilic one (M2). It is particularly noteworthy the increase in CO conversion for M1 upon the addition of sweep gas to a flow rate of only $40\,cm^3\,min^{-1}$, remaining then almost unaffected with further

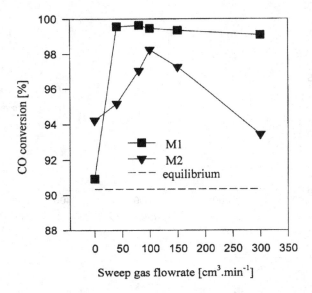

Figure 3.32 Effect of sweep gas flow rate on CO conversion (feed flow $50\,\mathrm{cm^3\,min^{-1}}$, $T = 523\,\mathrm{K}$, H_2O/CO molar ratio $= 1.0$). Reprinted from Giessler *et al.* (2003) with permission from Elsevier.

increases in sweep gas flow rates up to $300\,\mathrm{cm^3\,min^{-1}}$. However, a distinct behaviour was observed with M2 membrane. In this case, conversion increased with sweep gas flow rate up to $100\,\mathrm{cm^3\,min^{-1}}$, declining considerably afterwards. This is a behaviour that would be foreseen for mesoporous membranes (due to counter-diffusion), but clearly not expected for such microporous MSS membranes. Such behaviour was ascribed to the degradation in the hydrophilic membrane micro-structure that occurred in presence of water vapour, which underwent pore widening (Giessler *et al.*, 2003).

3.8.3. Effect of the H_2O/CO molar ratio

In a WGS MR, the use of excess steam can have several effects, being particularly important the fact that it can help preventing carbon deposition. It also shifts the reaction in the forward direction, but excess of water vapour can also have negative effects (among them economics, due to energy required in excess water vapourization). So, it has to be properly optimized.

Ma and co-workers have performed a very interesting study aiming the scale-up of the WGS process in a MR to obtain high hydrogen production rates through the use of large surface area, $\sim 0.02\,m^2$, composite Pd membranes (Catalano *et al.*, 2013). They used thin, $\delta < 10\,\mu m$, defect-free composite membranes that were prepared by the electroless plating method on porous stainless steel tubular supports. The MR was tested in a wide range of conditions, namely total flow rates up to $1.5\,Nm^3\,h^{-1}$, 20 bar maximum pressure, and temperatures from 693 K to 713 K. Figure 3.33 reports the effect of the steam-to-carbon (S/C) ratio in the feed, which was changed between 2.5 and 3.5, in the CO conversion and hydrogen recovery. It is noticed that CO conversion slightly increased with S/C, as a consequence of the higher CO conversion at equilibrium in the presence of high steam concentration (CO conversion at

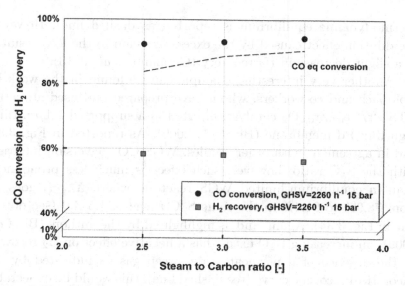

Figure 3.33 Carbon monoxide conversion and hydrogen recovery as a function of the steam-to-carbon ratio (S/C). Experiments performed at 713 K with a GHSV = $2260\,h^{-1}$ ($0.6\,Nm^3\,h^{-1}$ of feed flow rate) at 15 bar of retentate pressure. Reprinted with permission from Catalano *et al.* (2013). Copyright (2013), American Chemical Society.

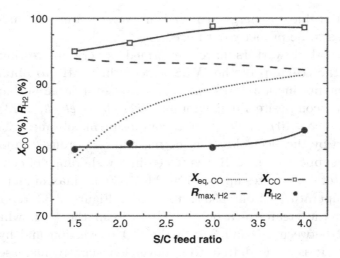

Figure 3.34 Influence of steam-to-carbon ratio on CO conversion and H_2 recovery in a MR at $P_{total} = 1.2\,MPa$, GHSV = $4050\,l\,kg^{-1}\,h^{-1}$ and $T = 623\,K$. Reprinted from Bi *et al.* (2009) with permission from Elsevier.

thermodynamic equilibrium is reported as dashed line). However, the dilution effect caused by the excess of steam in the feed resulted in a slight decrease in H_2 recovery (Catalano *et al.*, 2013).

Another very interesting example can be found in the work by Goldbach and co-workers, whom have prepared and used an active HTS $Pt/Ce_{0.6}Zr_{0.4}O_2$ catalyst, inserted in a supported $1.4\,\mu m$ thick high flux Pd membrane (Bi *et al.*, 2009). As reported in Fig. 3.34, and in agreement with other studies, WGS CO conversion increases with the S/C ratio; however, such effect is much less pronounced than in equilibrium-limited WGS reactors, where $X_{eq,CO}$ goes up from 78.9% to 91.4% in the same S/C range of 1.5–4.0 (see dotted line in Fig. 3.34). Again, and as highlighted by the authors (Bi *et al.*, 2009), an increasing H_2O excess has a negative effect on H_2 recovery in the absence of a permeate side sweep gas as indicated by the theoretical recovery curve (see dashed line); this would be expectable because steam dilutes H_2 in the product mixture thus lowering its partial pressure, and inherently the driving force for H_2 transport through the membrane. They claim that H_2O can also block the Pd membrane surface sites and thus impede the dissociative adsorption of H_2. However, it was experimentally observed a slight increase in H_2 recovery, which was not the result of an increased driving force due to the larger CO conversion (in fact, average H_2 partial pressure difference was indicated to be $\Delta P_{H_2} = 295\,\text{kPa}$ at S/C = 1.5 vs. $\Delta P_{H_2} = 282\,\text{kPa}$ at S/C = 4.0) but probably owed to the further reduction of the CO level. The authors attribute therefore such behaviour to an inhibition of the Pd membrane surface by CO, which seems to play a role at concentrations <1% and at 623 K, therefore affecting membrane performance; this aspect has been previously described in Section 3.5.5.

As mentioned before, from the practical point of view the selection of the optimum S/C ratio should however take into account also economical aspects, namely due to heat requirements to vapourize water at high S/C ratios.

3.8.4. Effect of total pressure

In the case of the WGS reaction, equilibrium is not affected by total pressure variation, as there is no variation in the total number of moles. However, total pressure, or transmembrane pressure, is a critical parameter affecting performance (conversion and recovery) of MRs mostly because it affects permeation through the membrane (for permselective membranes).

Tsotsis and co-workers (Abdollahi *et al.*, 2012) have used an ultra-thin, high-performance supported Pd membrane for pure hydrogen production using as feed a simulated reformate, which had the following composition (in molar ratios): $H_2{:}CO{:}CO_2{:}CH_4{:}H_2O = 5.22{:}1{:}0.48{:}0.1{:}2.8$. Among other factors, the feed pressure effect has been analyzed, keeping the permeate side under atmospheric pressure conditions and using steam as sweep gas. Their membrane had a thickness, in the Pd layer, of ca. 3 μm (Fig. 3.35). The membrane performance in the WGS MR (using a Cu–Zn–Al$_2$O$_3$ catalyst) is shown in Fig. 3.36 for two different operating pressures (namely 3.08 and 4.46 bar), in terms of both CO conversion and H_2 recovery as a function of $W_c/n_{CO,o}^F$. Clearly, the removal of hydrogen from the reaction mixture has a substantial beneficial effect on

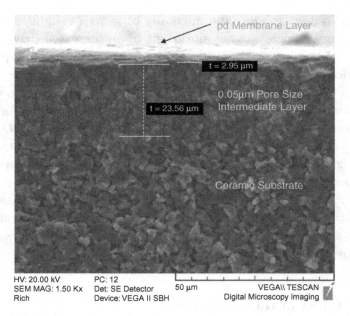

Figure 3.35 SEM cross-section of the Pd layer deposited on the alumina support using the electroless plating method. Reprinted from Abdollahi *et al.* (2012) with permission from Elsevier.

reactor conversion, which is above the equilibrium conversion based on the feed composition. Increasing the reactor pressure improves conversion and particularly hydrogen recovery, with almost complete CO conversion and hydrogen recovery in excess of 90% being attained at the highest contact time (Abdollahi *et al.*, 2012). This means that the product hydrogen purity is >99.9% with CO concentration <75 ppm, which may be acceptable for some PEM fuel cells.

A dense metallic permeator tube, which was assembled by an annealing and diffusion welding technique from a commercial flat sheet membrane of Pd–Ag, was used by Mendes *et al.* (2010b). A "finger-like" configuration of the self-supported membrane was adopted, and the performance of the MR assessed in a wide range of conditions, making use of a commercial $CuO–ZnO–Al_2O_3$ catalyst and a simulated reformate feed. Hydrogen permeation through the membrane can be improved by increasing total feed pressure or by decreasing the permeate chamber pressure by applying vacuum to

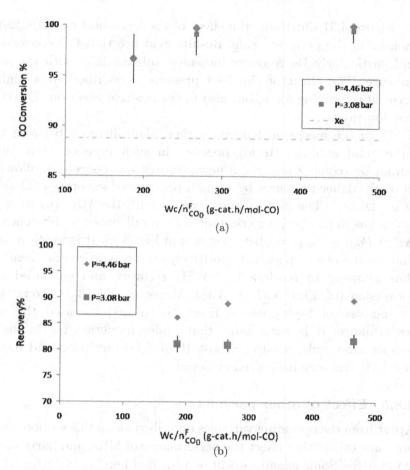

Figure 3.36 Effect of pressure on (a) CO conversion and (b) H_2 recovery, $T = 300°C$ and SR (ratio of inlet steam molar flow rate in the permeate side to the inlet molar feed flow rate in the feed side) = 0.3. Reprinted from Abdollahi *et al.* (2012) with permission from Elsevier.

the permeate side (or alternatively by using a sweep gas, as described in Section 3.8.2). In such a work, the effect of the feed pressure in the MR performance was studied for different operation modes: (i) vacuum mode — the permeate chamber was at ∼30 mbar, and (ii) sweep gas mode, using nitrogen in the permeate side in a counter-current manner (Mendes *et al.*, 2010b).

Figure 3.37 illustrates the effect of the feed total pressure using vacuum in the permeate side. Results evidence that CO conversion and particularly H_2 recovery increased substantially with the feed pressure. The effect of the feed pressure is ascribed to a higher permeation driving force, but also to the positive effect on the WGS reaction rate.

From the mechanical point of view, the difference between the lumen and external (total) pressure in such type of membranes cannot overcome 2 bar. So, a new strategy was conceived, allowing to reach higher pressures by using a pressurized sweep gas (Mendes *et al.*, 2010b). The performance reached with the MR operating on sweep gas mode, keeping constant the overall pressure difference at $\Delta P = P_{feed} - P_{perm} = 1$ bar, is shown in Fig. 3.38. It is worth noting that the use of sweep gas acts positively on the hydrogen permeation, thus allowing to reach a higher H_2 recovery and enhanced CO conversion (cf. Figs. 3.37 vs. 3.38). Moreover, Fig. 3.38 shows that an increase of feed pressure from 2.0 to 4.0 bar raised the MR performance; it is remarkable that under moderate conditions of temperature and pressure, almost 100% CO conversion and nearly 100% H_2 recovery have been achieved.

3.8.5. Effect of other parameters

Apart from the operating variables described along this section, there are some others that affect the performance of MRs, and particularly WGS MRs. Some examples will be provided herein.

One of such factors is the time factor (W/F). Its effect is easy to anticipate, as increasing catalyst mass (W) or decreasing feed flow rate of carbon monoxide (F) leads to improved contact time between reactants and the catalyst, improving reaction conversion. This is illustrated in Fig. 3.39, in the work by Uemiya *et al.* already in the 1990s, using a palladium membrane (thickness of 20 μm, which was by that time claimed to be the least thickness reported with 100% hydrogen selectivity) supported on a porous glass cylinder (Uemiya *et al.*, 1991). As shown, the conversion of CO increased with W/F, reaching values beyond equilibrium, as a result of selective and rapid

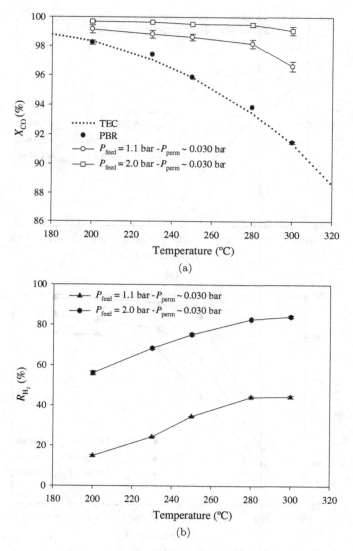

Figure 3.37 (a) CO conversion and (b) H_2 recovery as a function of the feed pressure for the WGS reaction at different temperatures in a Pd–Ag MR operating on vacuum mode (GHSV = 1200 L_N kg_{cat}^{-1} h^{-1}, feed composition: 4.70% CO, 34.78% H_2O, 28.70% H_2, 10.16% CO_2 balanced in N_2). TEC stands for thermodynamic equilibrium conversion. Reprinted from Mendes *et al.* (2010b) with permission from Elsevier.

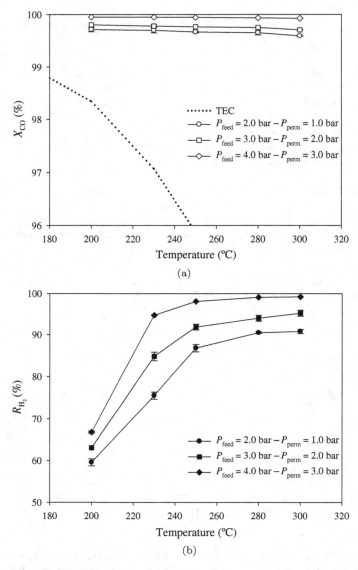

(a)

(b)

Figure 3.38 CO conversion (a) and H₂ recovery (b) as a function of the feed pressure for the WGS reaction at different temperatures in a Pd–Ag MR operating on sweep gas mode (N₂ sweep gas flow rate = 1000 mL_N min⁻¹, GHSV = 1200 L_N kg$_{cat}^{-1}$ h⁻¹, feed composition: 4.70% CO, 34.78% H₂O, 28.70% H₂, 10.16% CO₂ balanced in N₂). TEC stands for thermodynamic equilibrium conversion. Reprinted from Mendes *et al.* (2010b) with permission from Elsevier.

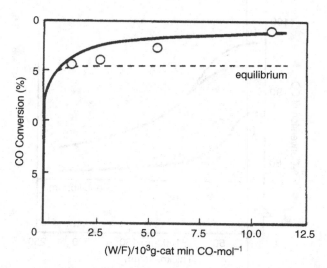

Figure 3.39 Effect of time factor (W/F) on conversion of carbon monoxide. O — experimental results; solid curve represents calculated results. Experimental conditions: temperature — 673 K; H_2O/CO molar ratio — 1; flow rate of sweep argon $-400 \, cm^3$ (STP) min^{-1}. Reprinted with permission from Uemiya *et al.* (1991). Copyright (1991), American Chemical Society.

extraction of produced hydrogen. Other works (e.g. Basile *et al.*, 2001; Giessler *et al.*, 2003) report the effect of feed flow rate, which is equivalent to changing W/F. Again, the conversion in general increases as the CO feed flow rate decreases, because the residence times are increased; this way the reagents have more time to react inside the reactor and hydrogen has more time to permeate through the membrane.

After appropriate validation of their MR model, also shown in the previous figure, these authors were able to analyze the effect of other parameters, namely of the membrane thickness (Uemiya *et al.*, 1991). Results shown in Fig. 3.40 evidence that the level of CO conversion increased with decreasing thickness of palladium film, as a consequence of improved rate of H_2 permeation flux (cf. e.g. Eq. (3.24)).

Apart from the membrane thickness, its properties and ultimately its nature/composition are for sure of paramount importance. This can be perceived from the following example, wherein

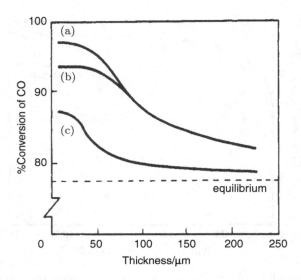

Figure 3.40 Conversion of carbon monoxide as a function of palladium thickness. Feed rate of CO: (a) 25, (b) 25 and (c) $100 \, cm^3$ (STP) min^{-1}. Flow rate of sweep argon: (a) 3200, (b) 400 and (c) $400 \, cm^3$ (STP) min^{-1}. Reprinted with permission from Uemiya *et al.* (1991). Copyright (1991), American Chemical Society.

Criscuoli *et al.* (2000) have compared the performance of MRs using a mesoporous ceramic membrane with another one employing a palladium (70–75 μm thick) membrane. The mesoporous ceramic membrane was a commercial asymmetric tubular (furnished by SCT, France) alumina membrane — inner layer was 4 μm thick γ-alumina with an average pore diameter of 4 nm, while other layers in α-alumina had larger pores dimensions (0.2–12 μm) and were thicker. Results obtained for a given feed mixture are shown in Fig. 3.41. As expected, by using membranes it is possible to reach higher conversions than in a traditional FBR; however, the equilibrium value was only overcome with the Pd-based MR. It was concluded that the better performance of the Pd membrane with respect to the ceramic one was due not only to the hydrogen selectivity (which is nearly infinite in the Pd-based) but also to the hydrogen permeation rate (Criscuoli *et al.*, 2000).

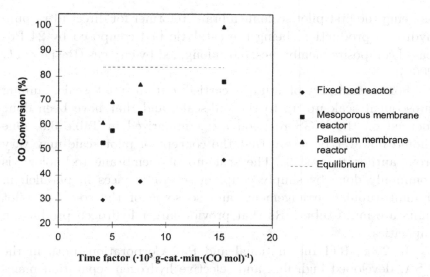

Figure 3.41 Effect of the time factor on the CO conversion for three reaction systems and reaction mixture 1 (CO 32%; CO_2 12%; H_2 4%; N_2 52% on dry basis). $T = 595\,\mathrm{K}$; $H_2O/CO = 1.1$; FBR: $P = 1\,\mathrm{atm}$; mesoporous MR: $P_{\mathrm{lumen}} = P_{\mathrm{shell}} = 1\,\mathrm{atm}$; no sweep gas; palladium MR: $P_{\mathrm{lumen}} = P_{\mathrm{shell}} = 1\,\mathrm{atm}$; sweep gas flow rate $= 43.6\,\mathrm{mL/min}$. Reprinted from Criscuoli *et al.* (2000) with permission from Elsevier.

3.9. Practical Applications and Pilot Scale

To the best of the authors' knowledge, no industrial applications of WGS MRs exist. The best known examples of relatively large MR technology are focused on natural gas steam reforming. Such works have been carried out by Tokyo Gas Company (in Japan) and Membrane Reactor Technology (in Vancouver, British Columbia, Canada) that developed hydrogen production demonstration plants with target scales of about 50–100 $\mathrm{Nm^3/h}$ of hydrogen (Edlund, 2010), although nominal capacity for the TGC unit is 40 $\mathrm{Nm^3/h}$ (Shirasaki and Yasuda, 2013). Tokyo Gas Company uses a packed bed reactor design while Membrane Reactor Technology uses a fluidized-bed approach, in both cases incorporating palladium-alloy membranes (Edlund, 2010). For the TGC case, it has been reported as being the first pilot scale membrane reformer for direct ultra-pure hydrogen production, being the catalytic bed composed by 24 Pd-based composite membranes, 0.6 m long, fed by city gas (Basile *et al.*, 2008).

Some examples of studies carried out at pilot scale, aiming subsequent scale-up up to the full-scale, and that have been done focused on the WGS reaction are summarized in Table 3.6. One should however be aware that the concept of pilot scale may vary from author to author. The scale-up of membrane technology is commonly done by simply coupling several devices in parallel, in a multi-tubular arrangement, and so some of the so-called pilot units are multi-tube MRs that provide larger hydrogen production capacities.

In 2008, RTI International and Pall Corporation, both in the USA, developed high flux and selective hydrogen separation membranes by depositing thin Pd-alloy films on tubular ceramic/porous stainless steel composite substrates (Damle *et al.*, 2008). Such membranes were then tested in a WGS MR configuration in a pilot scale module holding three membrane tubes. They were able to reach more than 80% of maximum possible hydrogen recovery while operating at 648 K and 823 K and 6.9 and 10.3 bar pressure. This and other examples are illustrated in Table 3.6.

To put forward research in this area, several programs and projects have been running. One example is the DEMCAMER (http://www.demcamer.org/), a large scale collaborative project focused on new materials and/or membranes for catalytic reactors. Among project objectives is worth mentioning the goal to improve performance, cost effectiveness and sustainability of some selected chemical processes, including WGS. The DEMCAMER plan involves several activities related to the whole product chain, including development of materials/components (membranes, supports, seals, catalysts, etc.) through integration/validation at lab-scale, until development/validation of semi-industrial pilot scale MR prototypes.

On the other hand, Compact Multifuel-Energy To Hydrogen converter (CoMETHy) collaborative project was focused on developing

Table 3.6 Examples of pilot scale studies on WGS MRs.

Membrane/module characteristics	Process characteristics	Maximum H_2 production	References
Module with 3 Pd membranes of approx. 5.6 μm thick and 44 cm long in parallel configuration; total membrane surface area of 580.6 cm².	$T = 673$ K, $P = 20$–35 bar(a), $P_{perm} = 15$ bar(a)	ca. 32 L min^{-1} (1.92 m³ h^{-1})	Li et al. (2011)
Module holding three 12" (active length) membrane tubes — total membrane area of 300 cm²; membranes based on thin (ca. 4 μm) Pd-alloy films on tubular ceramic/porous stainless steel composite substrates.	Pressures of 6.9–10.3 bar; temperatures of 648–823 K	14 L min^{-1} (0.84 m³ h^{-1})	Damle et al. (2008)
~0.02 m² composite Pd and Pd-based membranes, ca. 25 cm long, $\delta < 10$ μm prepared by the electroless plating method on porous stainless steel tubular supports.	Total feed flow rate up to 1.5 Nm³ h^{-1}, total pressure up to 20 bar, $T = 693$–713 K	5.6 Nm³ day^{-1} (0.23 Nm³ h^{-1})	Catalano et al. (2013)
Flat plate multilayer Pd–Cu membrane module after HTS and LTS reactors (not MR), total area = 5 × 16.6 cm².	$T = 473$–673 K, $P = 10$–20 bar	ca. 4 L min^{-1} in the permeate side (0.24 m³ h^{-1})	Lee et al. (2012)
Multi-tube MR with 19 Pd–Ag tubes having $\delta = 50$–60 mm, $d = 10$ mm and length = 250 mm.	$P = 1$–2.5 bar, $T = 623$–673 K	ca. 3.5 NL min^{-1} (0.21 Nm³ h^{-1})	Tosti et al. (2010)

an innovative, compact and modular steam reformer to convert reformable fuels (methane, ethanol, etc.) to pure hydrogen, adaptable to several heat sources (solar, biomass, fossil, etc.), depending on the locally available energy mix. By operating the process at lower temperatures, it is possible to combine steam reforming and WGS reactions in a single stage, with simultaneous hydrogen purification through permselective membranes. Among CoMETHy objectives, it is worth mentioning the proof-of-concept at the $2\,Nm^3/h$ hydrogen production scale.

Interestingly, the CACHET project, leaded by BP, concluded that among the seven technologies studied, the integrated membrane WGS process was the one with the lowest CO_2 capture cost (Beavis, 2011). Using dense metal membranes ensures that carbon dioxide does not pass through the membrane and comes out of the reactor in the retentate stream at high pressure (thus ready for compression and storage) whereas hydrogen emerges in the permeate, ready for electric power generation. Membranes obtained by magnetron sputtering have been developed at SINTEF, scaled up to 50 cm long, and long-term stability tests pointed for membrane lifetime of >1 year in WGS conditions (Beavis, 2011). Membranes prepared at the Dalian Institute of Chemistry and Physics, DICP, by the electroless plating method, also ca. 50 cm long, have been tested at ECN (Li *et al.*, 2011).

In the follow-up of such a project, a new one CACHET 2 was started in 2010 with the goal of developing a scaled-down membrane module, able to generate a hydrogen flux of ca. $12\,Nm^3/m^2 \cdot h$ (van Berkel *et al.*, 2013). Moreover, the demonstration of such scaled-down membrane module, which includes 3×1 m long Pd membranes (with a diameter of 14 mm, membrane area of $0.12\,m^2$) equipped with robust sealing (cf. Fig. 3.42), was one of the prime goals of the project. It should be also demonstrated long-term stability over a testing period of 1000 h under realistic conditions. An enhanced full scale membrane module was also envisaged in CACHET 2 project, in which the diameter of the membranes is increased to 1 inch and the number of membranes to 19, bringing the total membrane area to a value of $9\,m^2$ (van Berkel *et al.*, 2013).

Figure 3.42 Drawing and picture of the down-scaled membrane module from CACHET 2 project with 1 m long Pd membranes. Reprinted from van Berkel *et al.* (2013) with permission from Elsevier.

Among several factors, the cost of Pd-based membranes is certainly a hurdle for their commercialization and large scale implementation. Therefore, they are most likely to be commercially viable for small scale applications, as concluded by Damle *et al.* (2008) in their techno-economical analysis. They highlight as an additional benefit of the MR approach the increased hydrogen yield per unit of fuel feedstock used and reduced energy and operational cost due to simpler MR operation compared to a PSA system (the competing technology for the MR is the conventional WGS reactor followed by PSA for hydrogen purification).

3.10. Conclusions

The WGS is an important reaction in several industrial processes, and its use is growing due to the shift towards the hydrogen economy. Therefore, numerous catalytic formulations have been developed, conceived to operate under specified conditions (namely of temperature), and for which mechanistic and kinetic studies proposed have been shortly addressed along this chapter. The reaction has been widely studied, particularly using conventional reactors, but more recently using MRs as well. Using such reactor configuration, the increase of the CO conversion above the equilibrium value appears to be possible as hydrogen (or carbon dioxide) is selectively removed through the membrane, which is itself integral part of the reactor. This brings enormous advantages, in line with the trends of process intensification. Moreover, by properly tuning the operating conditions, eventually guided with robust models, maximum performance can be reached with a well-designed WGS MR, either in terms of CO conversion or H_2 recovery.

Different reactor configurations have been presented, from the packed to the fluidized-bed MRs, which should be carefully chosen taking into account process specifications. Micro-structured MRs are also worth mentioning, given their huge potential. More recently, the incorporation of a CO_2 sorbent with the H_2 permselective membrane has also been demonstrated, in a so-called SEMR configuration. Although preliminary results are promising, more studies are needed to allow for a better understanding and design of such multifunctional catalytic reactors, as it is still a very immature technology.

As detailed along this chapter, the development of catalytic MRs is a multistep task, requiring efforts in numerous areas. The growth of this technology needs a parallel development of more active and stable catalysts, high H_2 permeable and selective membranes, fast capture/release kinetics with good mechanical strength of CO_2 sorbents (for SEMR), along with advanced control strategy and engineering reactor design. However, and as pointed by Armor already some years ago (Armor, 1998), adequate membrane materials do not yet exist with a large surface area on a commercial scale to meet the demands of process operations, which is still valid for the WGS reaction. The membranes should be highly permselective, crack free, uniform, stable, preferably very thin, not susceptible to poisoning or fouling, and as cheap as possible. However, big progress has been done in the last years, so that some pilot scale units and demonstration projects have emerged, pointing for the potential industrial breakthrough of such technology in the short term. Yet, it appears that the limiting stage is still centered in the materials science, particularly at the membrane level, although results reached so far are quite encouraging with state-of-art membranes of each type herein described.

Nomenclature

A	Cross-sectional area (or membrane or heat exchange surface — Eqs. (3.41)–(3.42))	m^2
C_H	Atomic hydrogen concentration	mol/m^3
D	Diffusivity	m^2/s
Da	Damköhler's number	
D_{ax}	Effective axial dispersion coefficient	m^2/s
D_H	Effective diffusion coefficient of atomic hydrogen	m^2/s
D_K	Knudsen diffusivity	m^2/s
d_p	Particle diameter	m
d_{pore}	Pore diameter	nm
E_a	Activation energy (or apparent activation energy)	J/mol
$\overline{F}$	Dimensionless feed flow rate	
F_i	Flow rate of species i	mol/s or mol/h
G	Superficial mass flow velocity or massic flux	$g/(m^2h)$
g_c	Gravity conversion factor	
ΔH	Heat of reaction	J/mol
J_i	Molar flux of species i (e.g. that permeates through a membrane)	$mol/(m^2s)$
J_i^*	Normalized flux of species i	
k	Forward reaction rate constant	
k_-	Backward reaction rate constant	
K_i	Equilibrium adsorption constant for species i	Pa^{-1} or bar^{-1} (or others)
K_y, K_p	Reaction equilibrium constant, in terms of molar fractions or partial pressures	
L, L_i	Permeability or permeability towards species i	$mol/(m\,Pa^{0.5}s)$
$L_{i,o}$	Pre-exponential factor related with the permeability for species i	$mol/(m\,Pa^{0.5}s)$
M_i	Molecular weight of species i	kg/mol
n	Power dependency of the hydrogen partial pressure in the permeation flux	
n_i	Molar flow rate of species i	mol/h
P	Total pressure	Pa or bar

p_i	Partial pressure of species i	Pa or bar
$\overline{p}_i$	Average partial pressure of species i between the feed and retentate sides	Pa or bar
ΔP_{ln}	Logarithmic mean driving-force, in terms of partial pressures	Pa or bar
r	Reaction rate	$\text{mol}/(\text{g}_{\text{cat}}.\text{s})$ or $\text{mol}/(\text{m}^3_{\text{cat}}.\text{s})$
R	Ideal gas constant	
$\overline{r_i}$	Dimensionless reaction rate for species i (normalized by the one calculated at feed conditions)	
r^m	Internal radius of the membrane	m
r_p	Pore radius	m
R_i	Recovery of species i	
Re	Reynolds number	
R_f	Forward reaction rate	$\text{mol}/(\text{g}_{\text{cat}}.\text{s})$ or $\text{mol}/(\text{m}^3_{\text{cat}}.\text{s})$
S	Solubility (e.g. of hydrogen in the lattice of a membrane)	$\text{mol}/(\text{m}^3\,\text{Pa}^{0.5})$
T	Temperature	K or °C
u	Interstitial velocity	m/s
U	Overall heat transfer coefficient	$\text{W}/(\text{m}^2\text{K})$
U_i	Permeance for species i	$\text{mol}/(\text{m}^2\,\text{h}\,\text{bar}^n)$
V	Reactor volume	m^3
W, W_c	Mass of catalyst or mass of catalyst bed	g
x	Dimensionless reactor length	
X_i	Conversion of species i	
y_i	Molar fraction of species i in the gas phase	
z	spatial position or axial coordinate	m
$\overline{z}$	Dimensionless reactor length	

Greek symbols

α	Parameter in Eq. (3.27) that accounts for effects of the adsorbed gas in hydrogen permeation	
$\alpha_{A/B}$	Separation factor between species A and B	
α_m	Geometric factor, denoting the surface area of a membrane per unit reactor volume	m^2/m^3
β	Approach to equilibrium	
β_c	Fraction of the solid volume occupied by the catalyst	
$\varepsilon_b,\,\varepsilon_v$	Void fraction of the catalyst bed or bed porosity	
δ	Membrane thickness	m (or μm)
μ	Fluid viscosity	$\text{g}/(\text{ms})$

ν_i	Stoichiometric coefficient for component i	
ρ	Fluid density	g/m^3
ρ_c	Catalyst density	g/m^3

Subscripts

Cat	Catalyst
e, eq	Equilibrium
Feed	Feed
i	Species i
in	Initial/feed
perm	Permeate
ret	Retentate
0	Inlet conditions

Superscripts

$a-d$	Reaction orders (for power laws)
F	Feed
P	Permeate
R	Retentate
Siev	Sieverts
SL	Sieverts–Langmuir

Acronyms

BET	Brunauer, Emmett and Teller (specific surface area)
CFD	Computational fluid dynamics
CTL	Coal-to-liquid
CVD	Chemical vapor deposition
FBR	Fixed-bed reactor
FTS	Fischer–Tropsch synthesis
GHSV	Gas hourly space velocity
HSEMR	Hybrid sorption enhanced membrane reactor
HTS	High-temperature shift
IUPAC	International Union of Pure and Applied Chemistry
LH	Langmuir–Hinshelwood
LT	Low-temperature
LTS	Low-temperature shift
MR	Membrane reactor
MSS	Molecular sieve silica
MTS	Medium temperature shift
PBMR	Packed-bed membrane reactor
PBR	Packed-bed reactor
PEM	Proton exchange membrane
PSA	Pressure swing adsorption
S/C	Steam-to-carbon (or steam-to-carbon monoxide) ratio

S/C   Steam-to-carbon ratio
S/G   Steam-to-dry gas ratio
SEM   Scanning electron microscopy
SEMR   Sorption enhanced membrane reactor
SER   Sorption enhanced reactor
SL   Sieverts–Langmuir
SR   Steam-to-sweep gas ratio or sweep ratio
STP   Standard temperature and pressure (conditions)
TEC   Thermodynamic equilibrium conversion
TEM   Transmission electron microscopy
TGC   Tokyo Gas Company
TPD   Temperature-programmed desorption
TPR   Temperature-programmed reduction
TR   Traditional reactor
TREC   Traditional reactor equilibrium conversion
VI   Volume index
WGS   Water-gas shift
XPS   X-ray photoelectron spectroscopy
XRD   X-ray diffraction

References

Abdollahi, M., Yu, J., Liu, P.K.T., Ciora, R., Sahimi, M., Tsotsis, T.T., 2012. Ultra-pure hydrogen production from reformate mixtures using a palladium membrane reactor system, *Journal of Membrane Science*, 390–391, 32–42.

Adrover, M.E., López, E., Borio, D.O., Pedernera, M.N., 2009. Theoretical study of a membrane reactor for the water gas shift reaction under nonisothermal conditions, *AIChE Journal*, 55, 3206–3213.

Anderson, D.M., Nasr, M.H., Yun, T.M., Kottke, P.A., Fedorov, A.G., 2015. Sorption-enhanced variable-volume batch-membrane steam methane reforming at low temperature: Experimental demonstration and kinetic modeling, *Industrial and Engineering Chemistry Research*, 54, 8422–8436.

Andreeva, D., 2002. Low temperature water gas shift over gold catalysts, *Gold Bulletin*, 35, 82–88.

Andrés, M.-B., Boyd, T., Grace, J.R., Lim, C.J., Gulamhusein, A., Wan, B., Kurokawa, H., Shirasaki, Y., 2011. *In-situ* CO_2 capture in a pilot-scale fluidized-bed membrane reformer for ultra-pure hydrogen production, *International Journal of Hydrogen Energy*, 36, 4038–4055.

Armor, J.N., 1998. Applications of catalytic inorganic membrane reactors to refinery products, *Journal of Membrane Science*, 147, 217–233.

Ayastuy, J.L., Gutierrez-Ortiz, M.A., Gonzalez-Marcos, J.A., Aranzabal, A., Gonzalez-Velasco, J.R., 2005. Kinetics of the low-temperature WGS reaction over a $CuO/ZnO/Al_2O_3$ catalyst, *Industrial & Engineering Chemistry Research*, 44, 41–50.

Ayral, A., Julbe, A., Rouessac, V., Roualdes, S., Durand, J., 2008. Microporous silica membranes: Basic principles and recent advances, in *Inorganic Membranes: Synthesis, Characterization and Applications*, R. Mallada, M. Menendez (Eds.), Elsevier, Amsterdam, The Netherlands, Chapter 2, 33–79.

Barbieri, G., Scura, F., Brunetti, A., 2008. Mathematical Modeling of Pd-Alloy Membrane Reactors, in *Inorganic Membranes: Synthesis, Characterization and Applications*, R. Mallada, M. Menendez (Eds.), Elsevier, Amsterdam, The Netherlands, Chapter 9, 325–400.

Barbieri, G., Scura, F., Lentini, F., De Luca, G., Drioli, E., 2008. A novel model equation for the permeation of hydrogen in mixture with carbon monoxide through Pd–Ag membranes, *Separation & Purification Technology*, 61, 217–224.

Basile, A., Criscuoli, A., Santella, F., Drioli, E., 1996. Membrane reactor for water gas shift reaction, *Gas Separation & Purification*, 10, 243–254.

Basile, A., Gallucci, F., Tosti, S., 2008. Synthesis, characterization and applications of Palladium membranes, *Membrane Science and Technology*, 13, 255–323.

Basile, A., Chiappetta, G., Tosti, S., Violante, V., 2001. Experimental and simulation of both Pd and Pd/Ag for a water gas shift membrane reactor, *Separation and Purification Technology*, 25, 549–571.

Basile, A., Paturzo, L., Gallucci, F., 2003. Co-current and counter-current modes for water gas shift membrane reactor, *Catalysis Today*, 82, 275–281.

Bayat, M., Dehghani, Z., Rahimpour, M.R., 2014. Membrane/sorption-enhanced methanol synthesis process: Dynamic simulation and optimization, *Journal of Industrial and Engineering Chemistry*, 20, 3256–3269.

Beavis, R., 2011. The EU FP6 CACHET project — Final results, *Energy Procedia*, 4, 1074–1081.

Bettinali, L., Lecci, D., Marini, F., Tosti, S., Violante, V., 2002. Method of diffusion bonding thin foils made of metal alloys selectively permeable to hydrogen, particularly providing membrane devices, and apparatus for carrying out the same, European Patent EP 1184125.

Bi, Y., Xu, H., Li, W., Goldbach, A., 2009. Water-gas shift reaction in a Pd membrane reactor over $Pt/Ce_{0.6}Zr_{0.4}O_2$ catalyst, *International Journal of Hydrogen Energy*, 34, 2965–2971.

Boon, J., Pieterse, J.A.Z., van Berkel, F.P.F., van Delft, Y.C., van Sint Annaland, M., 2015. Hydrogen permeation through palladium membranes and inhibition by carbon monoxide, carbon dioxide, and steam, *Journal of Membrane Science*, 496, 344–358.

Boutikos, P., Nikolakis, V., 2010. A simulation study of the effect of operating and design parameters on the performance of a water gas shift membrane reactor, *Journal of Membrane Science*, 350, 378–386.

Brunetti, A., Caravella, A., Barbieri, G., Drioli, E., 2007. Simulation study of water gas shift reaction in a membrane reactor, *Journal of Membrane Science*, 306, 329–340.

Bukur, D.B., Todic, B., Elbashir, N., 2016. Role of water-gas-shift reaction in Fischer–Tropsch synthesis on iron catalysts: A review, *Catalysis Today*, 275, 66–75.

Catalano, J., Guazzone, F., Mardilovich, I.P., Kazantzis, N.K., Ma, Y.H., 2013. Hydrogen production in a large scale water-gas shift Pd-based catalytic membrane reactor, *Industrial & Engineering Chemistry Research*, 52, 1042–1055.

Criscuoli, A., Basile, A., Drioli, E., 2000. An analysis of the performance of membrane reactors for the water-gas shift reaction using gas feed mixtures, *Catalysis Today*, 56, 53–64.

Damle, A., Richardson, C., Powers, T., Love, C., Acquaviva, J., 2008. Demonstration of a pilot–scale membrane reactor process for hydrogen production, *ECS Transactions*, 12, 499–510.

Dittmeyer, R., Höllein, V., Daub, K., 2001. Membrane reactors for hydrogenation and dehydrogenation processes based on supported palladium, *Journal of Molecular Catalysis A: Chemical*, 173, 135–184.

Edlund, D., 2010. Hydrogen membrane technologies and application in fuel processing, in *Hydrogen and Syngas Production and Purification Technologies*, K. Liu, C. Song, V. Subramani (Eds.), AIChE-Wiley, New Jersey, US, Chapter 8, 357–384.

Gallucci, F., Fernandez, E., Corengia, P., van S. Annaland, M., 2013. Recent advances on membranes and membrane reactors for hydrogen production, *Chemical Engineering Science*, 92, 40–66.

García-García, F.R., Torrente-Murciano, L., Chadwick, D., Li, K., 2012. Hollow fibre membrane reactors for high H_2 yields in the WGS reaction, *Journal of Membrane Science*, 405–406, 30–37.

García-García, F.R., León, M., Ordóñez, S., Li, K., 2014. Studies on water–gas-shift enhanced by adsorption and membrane permeation, *Catalysis Today*, 236, 57–63.

Ghenciu, A.F., 2002. Review of fuel processing catalysts for hydrogen production in PEM fuel cell systems, *Current Opinion in Solid State and Materials Science*, 6, 389–399.

Giessler, S., Jordan, L., Diniz da Costa, J.C., Lu, G.Q. (Max), 2003. Performance of hydrophobic and hydrophilic silica membrane reactors for the water gas shift reaction, *Separation and Purification Technology*, 32, 255–264.

Gorte, R.J., Zhao, S., 2005. Studies of the water-gas-shift reaction with ceria-supported precious metals, *Catalysis Today*, 104, 18–24.

Gradisher, L., Dutcher, B., Fan, M., 2015. Catalytic hydrogen production from fossil fuels via the water gas shift reaction, *Applied Energy*, 139, 335–349.

Hsieh, H.P., 1996. *Inorganic Membranes for Separation and Reaction*, Elsevier, Amsterdam, The Netherlands.

Huang, J., El-Azzami, L., Ho, W.S.W., 2005. Modeling of CO_2-selective water gas shift membrane reactor for fuel cell, *Journal of Membrane Science*, 261, 67–75.

Hulteberg, C., 2012. Sulphur-tolerant catalysts in small-scale hydrogen production — A review, *International Journal of Hydrogen Energy*, 37, 3978–3992.

Hwang, H.T., Harale, A., Liu, P.K.T., Sahimi, M., Tsotsis, T.T., 2008. A membrane-based reactive separation system for CO_2 removal in a life support system, *Journal of Membrane Science*, 315, 116–124.

Laan, V.D.G.P., Beenackers, A.A.C.M., 1999. Kinetics and selectivity of the Fischer–Tropsch synthesis: A literature review, *Catalysis Reviews — Science and Engineering*, 41, 255–318.

Leal, A.L., Soria, M.A., Madeira, L.M., 2016. Autothermal reforming of impure glycerol for H_2 production: Thermodynamic study including *in situ* CO_2 and/or H_2 separation, *International Journal of Hydrogen Energy*, 41, 2607–2620.

Lee, D.-W., Lee, M.S., Lee, J.Y., Kim, S., Eom, H.-J., Moon, D.J., Lee, K.-Y., 2013. The review of Cr-free Fe-based catalysts for high-temperature water-gas shift reactions, *Catalysis Today*, 210, 2–9.

Lee, S.H., Kim, J.N., Eom, W.H., Ryi, S.-K., Park, J.-S., Baek, I.H., 2012. Development of pilot WGS/multi-layer membrane for CO_2 capture, *Chemical Engineering Journal*, 207–208, 521–525.

Levalley, T.L., Richard, A.R., Fan, M., 2014. The progress in water gas shift and steam reforming hydrogen production technologies — A review, *International Journal of Hydrogen Energy*, 39, 16983–17000.

Li, H., Dijkstra, J.W., Pieterse, J.A.Z, Boon, J., van den Brink, R.W., Jansen, D., 2011. WGS-mixture separation and WGS reaction test in a bench-scale multi-tubular membrane reactor, *Energy Procedia*, 4, 666–673.

Lu, G.Q., Diniz da Costa, J.C., Duke, M., Giessler, S., Socolow, R., Williams, R.H., Kreutz, T., 2007. Inorganic membranes for hydrogen production and purification: A critical review and perspective, *Journal of Colloid and Interface Science*, 314, 589–603.

Maroño, M., Barreiro, M.M., Torreiro, Y., Sánchez, J.M., 2014. Performance of a hybrid system sorbent–catalyst–membrane for CO_2 capture and H_2 production under pre-combustion operating conditions, *Catalysis Today*, 236, 77–85.

Mejdell, A.L., Jøndahl, M., Peters, T.A., Bredesen, R., Venvik, H.J., 2009. Experimental investigation of a microchannel membrane configuration with a 1.4 μm Pd/Ag23 wt.% membrane — Effects of flow and pressure, *Journal of Membrane Science*, 327, 6–10.

Mendes, D., Mendes, A., Madeira, L.M., Iulianelli, A., Sousa, J.M., Basile, A., 2010a. The water-gas shift reaction: from conventional catalytic systems to Pd-based membrane reactors — A review, *Asia-Pacific Journal of Chemical Engineering*, 5, 111–137.

Mendes, D., Chibante, V., Zheng, J.-M., Tosti, S., Borgognoni, F., Mendes, A., Madeira, L.M., 2010b. Enhancing the production of hydrogen via water-gas shift reaction using Pd-based membrane reactors, *International Journal of Hydrogen Energy*, 35, 12596–12608.

Mendes, D., Sá, S., Tosti, S., Sousa, J.M., Madeira, L.M., Mendes, A., 2011. Experimental and modeling studies on the low-temperature water-gas shift reaction in a dense Pd-Ag packed-bed membrane reactor, *Chemical Engineering Science*, 66, 2356–2367.

Miguel, C.V., Mendes, A., Tosti, S., Madeira, L.M., 2012. Effect of CO and CO_2 on H_2 permeation through finger-like Pd-Ag membranes, *International Journal of Hydrogen Energy*, 37, 12680–12687.

Moe, J.M., 1962. Design of water-gas shift reactors, *Chemical Engineering Progress*, 58, 33–36.

Newsome, D.S., 1980. Water-gas shift reaction, *Catalysis Reviews*, 21, 275–281.

Nguyen-Phan, T.-D., Baber, A.E., Rodriguez, J.A., Senanayake, S.D., 2015. Au and Pt nanoparticle supported catalysts tailored for H_2 production: From models to powder catalysts, *Applied Catalysis A: General*, 518, 18–47.

Ockwig, N.W., Nenoff, T.M., 2007. Membranes for hydrogen separation, *Chemical Reviews*, 107, 4078–4110.

Oyama, S.T., Yamada, M., Sugawara, T., Takagaki, A., Kikuchi, R., 2011. Review on mechanisms of gas permeation through inorganic membranes, *Journal of the Japan Petroleum Institute*, 54, 298–309.

Paglieri, S.N., Way, J.D., 2002. Innovations in palladium membrane research. *Separation & Purification Methods*, 31, 1–169.

Patil, C.S., van Sint Annaland, M., Kuipers, J.A.M., 2007. Fluidised bed membrane reactor for ultrapure hydrogen production via methane steam reforming: Experimental demonstration and model validation, *Chemical Engineering Science*, 62, 2989–3007.

Pérez, P., Cornaglia, C.A., Mendes, A., Madeira, L.M., Tosti, S., 2015. Surface effects and CO/CO_2 influence in the H_2 permeation through a Pd-Ag membrane: A comprehensive model, *International Journal of Hydrogen Energy*, 40, 6566–6572.

Peters, T.A., Stange, M., Klette, H., Bredesen, R., 2008. High pressure performance of thin Pd-23%Ag/stainless steel composite membranes in water gas shift gas mixtures; influence of dilution, mass transfer and surface effects on the hydrogen flux, *Journal of Membrane Science*, 316, 119–127.

Peters, T.A., Stange, M., Sunding, M.F., Bredesen, R., 2015. Stability investigation of micro-configured Pd-Ag membrane modules — Effect of operating temperature and pressure, *International Journal of Hydrogen Energy*, 40, 3497–3505.

Platon, A., Wang, Y., 2010. Water-gas shift technologies, in *Hydrogen and Syngas Production and Purification Technologies*, K. Liu, C. Song, V. Subramani (Eds.), AIChE-Wiley, New Jersey, US, Chapter 6, 311–328.

Rahman, M.A., García-García, F.R., Hatim, M.D.I., Kingsbury, B.F.K., Li, K., 2011. Development of a catalytic hollow fibre membrane micro-reactor for high purity H_2 production, *Journal of Membrane Science*, 368, 116–123.

Ramasubramanian, K., Song, M., Ho, W.S.W., 2013. Spiral-wound water-gas-shift membrane reactor for hydrogen purification, *Industrial and Engineering Chemistry Research*, 52, 8829–8842.

Ratnasamy, C., Wagner, J., 2009. Water gas shift catalysis, *Catalysis Reviews — Science and Engineering*, 51, 325–440.

Rothfleisch, J.E., 1964. Hydrogen from methanol for fuel cells, *Society of Automotive Engineers Technical Papers*, 640377.

Sanz, R., Calles, J.A., Alique, D., Furones, L., Ordóñez, S., Marín, P., 2015. Hydrogen production in a pore-plated Pd-membrane reactor: Experimental analysis and model validation for the water gas shift reaction, *International Journal of Hydrogen Energy*, 40, 3472–3484.

Scholes, C.A., Smith, K.H., Kentish, S.E., Stevens, G.W., 2010. CO_2 capture from pre-combustion processes — Strategies for membrane gas separation, *International Journal of Greenhouse Gas Control*, 4, 739–755.

Shirasaki, Y., Yasuda, I., 2013. Membrane reactor for hydrogen production from natural gas at the Tokyo Gas company: A case study, in *Handbook of Membrane Reactors, Vol. 2 – Reactor Types and Industrial Applications*, A. Basile (Ed.), Woodhead Publishing Series in Energy, Cambridge, UK, 487–507.

Sieverts, A., Krumbhaar, W., 1910. Über die löslichkeit von gasen in metallen und legierungen, *Berichte der Deutschen Chemischen Gesellschaft*, 43, 893.

Silva, J.M., Soria, M.A., Madeira, L.M., 2015. Thermodynamic analysis of glycerol steam reforming for hydrogen production with *in situ* hydrogen and carbon dioxide separation, *Journal of Power Sources*, 273, 423–430.

Smart, S., Lin, C.X.C., Ding, L., Thambimuthu, K., Diniz da Costa, J.C., 2010. Ceramic membranes for gas processing in coal gasification, *Energy & Environmental Science*, 3, 268–278.

Smith, R.J.B., Loganathan, M., Shantha, M.S., 2010. A review of the water gas shift reaction kinetics, *International Journal of Chemical Reactor Engineering*, 8, R4.

Soria, M.A., Tosti, S., Mendes, A., Madeira, L.M., 2015. Enhancing the low temperature water–gas shift reaction through a hybrid sorption-enhanced membrane reactor for high-purity hydrogen production, *Fuel*, 159, 854–863.

Temkin, M.I., 1979. The kinetics of some industrial heterogeneous catalytic reactions, in *Advances in Catalysis*, Vol. 28., D.D. Eley, H. Pines, P.B. Weisz (Eds.), Academic Press, New York, US, 173–281.

Tosti, S., Borgognoni, F., Santucci, A., 2010. Multi-tube Pd-Ag membrane reactor for pure hydrogen production, *International Journal of Hydrogen Energy*, 35, 11470–11477.

Uemiya, S., Sato, N., Ando, H., Kikuchi, E., 1991. The water gas shift reaction assisted by a palladium membrane reactor, *Industrial and Engineering Chemistry Research*, 30, 585–589.

Uemiya, S., Kato, W., Uyama, A., Kajiwara, M., Kojima, T., Kikuchi, E., 2001. Separation of hydrogen from gas mixtures using supported platinum-group metal membranes, *Separation and Purification Technology*, 22–23, 309–317.

Vadrucci, M., Borgognoni, F., Moriani, A., Santucci, A., Tosti, S., 2013. Hydrogen permeation through Pd-Ag membranes: Surface effects and Sievert's law. *International Journal of Hydrogen Energy*, 38, 4144–4152.

van Berkel, F., Hao, C., Bao, C., Jiang, C., Xu, H., Morud, J., Peters, T., Soutif, E., Dijkstra, J.W., Jansen, D., Song, B., 2013. Pd-membranes on their way towards application for CO_2-capture, *Energy Procedia*, 37, 1076–1084.

Wu, X., Wu, C., Wu, S., 2015. Dual-enhanced steam methane reforming by membrane separation of H_2 and reactive sorption of CO_2, *Chemical Engineering Research and Design*, 96, 150–157.

Zhu, M., Wachs, I.E., 2016. Iron-based catalysts for the high-temperature water-gas shift (HT-WGS) reaction: A review, *ACS Catalysis*, 6, 722–732.

Zou, J., Huang, J., Ho, W.S.W., 2007. CO_2-selective water gas shift membrane reactor for fuel cell hydrogen processing, *Industrial and Engineering Chemistry Research*, 46, 2272–2279.

Chapter 4

Liquid Phase Simulated Moving Bed Reactor

4.1. Simulated Moving Bed Technology: The Concept

Chromatography is generally recognized as one of the most relevant purification techniques, finding numerous applications at industrial level for the separation of complex multicomponent mixtures. The specific interactions established between the separation media and the different species present in the fluid phase is the underlying principle governing this separation technique. These interactions can occur through diverse mechanisms as adsorption, ion-exchange or size-exclusion, for instance. The separation media can either be a solid material, a liquid or a bonded-liquid while the fluid phase can be in the gas, liquid or supercritical state. Moreover, throughout the years, several practical realizations of this separation technique have been successfully developed and implemented. The ability to control and optimize such a large number of process variables allows the application of this technique to a virtually unlimited number of systems since it might be able to adapt to all of its specificities. These facts evidence the versatility of chromatography and justify the success it has achieved (Nicoud, 2015).

Single column batch elution chromatography is probably the most representative example of earlier industrial chromatographic processes. It consists in injecting a well-defined amount of the multicomponent mixture to be separated into a column packed with a solid stationary phase to which a fluid mobile phase, generally denominated as eluent or desorbent, is continuously being fed. The different species of the feed mixture will have different affinities towards the stationary phase. Hence, the species with higher affinity will be retained inside the column for longer periods of time than those with weaker interactions which will elute first. This allows collecting purified fractions of the target products at the outlet of the column at different time intervals. The process proceeds through periodic injections of the feed mixture. A schematic representation of the separation of a binary mixture through a single column batch elution chromatographic process is presented in Fig. 4.1.

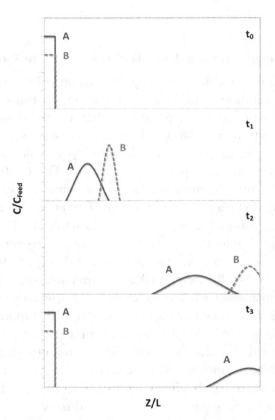

Figure 4.1 Single column batch elution chromatographic process for the separation of a binary mixture containing species "A" and "B" ("A" has a higher affinity towards the solid phase than "B").

The purity of the final product will be directly dependent on the rigorous definition of the fractions collection times as one could easily infer from the description of the single column batch elution operating mode. The main advantages associated with this technology are its simple configuration, operation and scalability but on the other hand, as expected due to its batch nature, its productivity is relatively limited and its operation requires significant amounts of eluent, which can result in high operating costs.

The development of continuous counter-current chromatographic processes represented a crucial stage on the evolution of chromatography as an effective and reliable separation technique at industrial scale (Nicoud, 2014). The innovations introduced by this concept allowed to overcome the limitations of single column batch processes through the maximization of the separation driving force and stationary phase usage, mainly as a consequence of the counter-current operating mode. The True Moving Bed (TMB) is one of the most prominent and simple examples within this class of chromatographic separation processes. In the TMB, the desorbent is fed at one end of the unit while the moving solid phase is fed at the other end so that the liquid and the solid stream flow in opposite directions. If the appropriate operating conditions are met, both streams can be recycled back to the unit. The stream containing the species to be separated is fed in the central area of the TMB while the most and the least retained products are collected through the extract and raffinate streams, respectively, according to the scheme presented in Fig. 4.2.

As can be observed from Fig. 4.2, the TMB inlet and outlet streams divide the unit in four sections operating under different

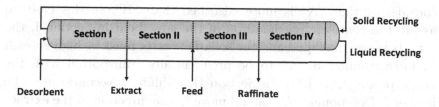

Figure 4.2 Schematic representation of a TMB unit.

Table 4.1 Description of the specific function of each of the four sections of a conventional TMB.

Section	Main function	Separation details
I	Solid phase regeneration	Desorption of the most retained products to allow recycling the solid phase
II	Purification of the most retained products	Adsorption of the most retained products and desorption of the least retained ones to avoid the contamination of the extract stream
III	Purification of the least retained products	Desorption of the least adsorbed products and adsorption of the most retained ones to avoid the contamination of the raffinate stream
IV	Desorbent regeneration	Adsorption of the least retained products to allow recycling the desorbent

internal liquid flow rates. The specific role of each section is described in Table 4.1.

The design of a TMB process consists in ensuring that all of the assumptions presented in Table 4.1 are accomplished through the precise definition of the net fluxes for all the compounds in all sections considering the relative velocities between the liquid and the solid phase and the compounds affinity towards the adsorbent material. To write the design equations for the TMB, the separation between compound "A" and compound "B" will be used as example, considering that "A" is more adsorbed than "B". In this case, to assure the complete regeneration of the adsorbent in Section I, the ratio between the liquid and the solid flow rates must be high enough to force compound "A" to be preferentially transported with the liquid phase (Eq. (4.1)). In Section II, a different scenario must be imposed. Compound "A" should move in the direction of the extract port where it will be collected and compound "B" should move in the opposite direction to avoid the contamination of this stream. Therefore, "A" and "B" must be carried by the solid and the liquid streams, respectively (Eq. (4.2)). A similar behaviour is expected in Section III; however, this time to obtain a purified fraction of "B" in the raffinate stream (Eq. (4.3)). Finally, at the end of Section IV the desorbent shall not contain neither "A" nor "B" so it can be recycled back to Section I. For that purpose, it is necessary to assure that compound "B" is mainly transported by the solid phase (Eq. (4.4)). The set of equations that describe the previous assumptions in terms of net fluxes are presented below.

$$\frac{Q_I C_{A,I}}{Q_s q_{A,I}} > 1 \tag{4.1}$$

$$\frac{Q_{II} C_{A,II}}{Q_s q_{A,II}} < 1 \quad \wedge \quad \frac{Q_{II} C_{B,II}}{Q_s q_{B,II}} > 1 \tag{4.2}$$

$$\frac{Q_{III} C_{A,III}}{Q_s q_{A,III}} < 1 \quad \wedge \quad \frac{Q_{III} C_{B,III}}{Q_s q_{B,III}} > 1 \tag{4.3}$$

$$\frac{Q_{IV} C_{B,IV}}{Q_s q_{B,IV}} < 1 \tag{4.4}$$

In Eqs. (4.1)–(4.4), Q_s represents the solid phase flow rate, Q_J represents the volumetric liquid flow rate in section j (with $j = $ I, II, III, IV) and $C_{i,j}$ and $q_{i,j}$ represent the concentration of compound i (with $i = $ A, B) in section j in the liquid and solid phases, respectively.

The analysis of the separation of the binary mixture composed by "A" and "B" will proceed by assuming that the concentration of these compounds in the solid phase is directly proportional to its concentration in the liquid phase, following the behaviour of a linear adsorption isotherm ($q_i = K_i C_i$), as verified for countless real mixtures, especially under diluted conditions. Moreover, it will also be considered that this adsorption equilibrium can be established instantaneously, neglecting the effect of mass transfer resistances in the interface between the two phases. Defining γ_j as the ratio between the liquid and the solid phases interstitial velocities in section j $(\gamma_j = u_j/u_s)$, Eqs. (4.5)–(4.8) can be rewritten as follows:

$$\gamma_{\mathrm{I}} > \frac{(1-\varepsilon)}{\varepsilon} K_{\mathrm{A}} \tag{4.5}$$

$$\frac{(1-\varepsilon)}{\varepsilon} K_{\mathrm{B}} < \gamma_{\mathrm{II}} < \frac{(1-\varepsilon)}{\varepsilon} K_{\mathrm{A}} \tag{4.6}$$

$$\frac{(1-\varepsilon)}{\varepsilon} K_{\mathrm{B}} < \gamma_{\mathrm{III}} < \frac{(1-\varepsilon)}{\varepsilon} K_{\mathrm{A}} \tag{4.7}$$

$$\gamma_{\mathrm{IV}} < \frac{(1-\varepsilon)}{\varepsilon} K_{\mathrm{B}} \tag{4.8}$$

As previously stated, Sections II and III of the TMB are mainly responsible for the separation of the binary mixture. Hence, by plotting Eqs. (4.6) and (4.7) on the $\gamma_{\mathrm{II}} - \gamma_{\mathrm{III}}$ plane, together with the constraint $\gamma_{\mathrm{III}} > \gamma_{\mathrm{II}}$ (since $Q_{\mathrm{III}} = Q_{\mathrm{II}} + Q_{\mathrm{Feed}}$ as depicted from Fig. 4.2), a triangular region, generally denominated as separation region, is defined as shown in Fig. 4.3. All the combinations of γ_{II} and γ_{III} values that fall within this area will lead to the separation of the binary mixture if the process restrictions associated with the regeneration of the adsorbent and the desorbent, in Sections I and IV, respectively, are satisfied. These restrictions correspond to Eqs. (4.5) and (4.8) and can be plotted in the $\gamma_{\mathrm{I}} - \gamma_{\mathrm{IV}}$ plane, forming the so-called regeneration region. The separation and regeneration regions presented in Fig. 4.3 constitute the graphical representation of the TMB design methodology previously described, based on the "Equilibrium Theory" for linear adsorption isotherms.

The TMB separation and regeneration regions represent the most relevant tools for the optimization of this process since the vertex of these regions indicates the unit's optimum operating conditions in a very direct and intuitive manner. The vertex of the separation region corresponds to the operating point at which the largest difference between γ_{III} and γ_{II} can be attained or, in another perspective, indicates the highest feed flow rate that can be introduced in the TMB without compromising the separation. Consequently, the maximum productivity can be achieved at these conditions. On the other side, the vertex of the regeneration region corresponds to the operating point that minimizes the difference

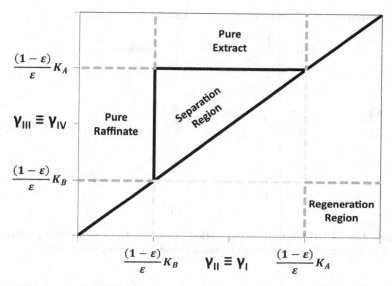

Figure 4.3 TMB separation and regeneration regions (species "A" has higher affinity towards the solid phase than species "B").

between γ_I and γ_{IV}. Therefore, this will allow the regeneration of both the solid and the liquid phases at the highest recycling flow rate possible leading to minimum desorbent consumption. Nevertheless, it is important to notice that, for a real TMB unit, selecting the operating conditions corresponding to the vertex of the separation or the regeneration regions would be highly discouraged since the slightest disturbance on any of the inlet or outlet streams flow rate would lead to the immediate contamination of the final products. Instead, the use of safety factors in all the internal flow rate ratios, typically in the range of 2–5%, is advisable. This will enhance the process robustness, since the operating points will be located in the inner part of the separation and regeneration regions and not on its boundaries, at the cost of a slight loss on the productivity and desorbent consumption.

The design methodology previously described has a direct impact on the distribution of the different species inside the TMB. The concentration bands start developing from the initial moments of the operation and stabilize when the steady state is achieved. Figure 4.4

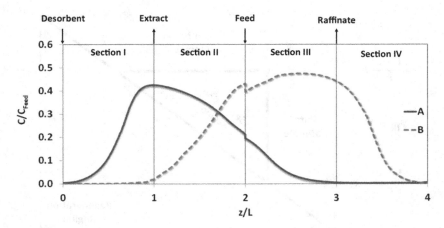

Figure 4.4 TMB internal concentration profiles in the liquid phase (compound "A" has higher affinity towards the solid phase than compound "B").

presents a hypothetical internal concentration profile for a TMB at the steady state.

As the design methodology imposed, compound "A" shall be preferentially present in Section II of the TMB, in the surroundings of the extract port. Compound "B", due to its lower affinity towards the moving solid phase, shall present higher concentration values in Section III, near the raffinate port through which it is collected. It can also be observed that neither "A" nor "B" concentration bands reach the other compound outlet port or the unit's ends as a consequence of the separation and regeneration constraints defined (Eqs. (4.5)–(4.8)). The liquid phase concentration of "A" throughout the unit is generally lower than the observed for "B" since it is the most adsorbed compound. The inverse relation is verified for the solid phase.

Considering everything that has been described previously, if on one hand, the continuous counter-current operating mode of the TMB theoretically represented a considerable improvement for chromatographic separation processes in general, the truth is that the industrial implementation of this technology was never accomplished due to the restrictive technical issues associated with the transport of the solid phase. The fluidization of the bulk phase is extremely dependent on the fluid velocity which may cause the solid movement in the TMB to be non-uniform, discontinuous or even to prevent it at all. Moreover, negative backmixing effects can be generated and mechanical attrition might severely damage the solid material and wear the equipment out. These problems would considerably aggravate with the increase of the unit's size and the amount of solid to be displaced.

However, the advantages associated with the TMB technology were too evident to be neglected. Aware of this fact, Broughton and Gerhold (1961) developed and patented an innovative concept denominated Simulated Moving Bed (SMB), consisting in a set of interconnected chromatographic columns and interchanging inlet and outlet streams. This equipment was able to overcome the problems associated with the movement of the solid phase in the TMB by simulating it through the synchronous periodic shift of the position of all the inlet and outlet streams in the direction of the fluid flow (Fig. 4.5).

Designating the period of time between two consecutive shifts as switching time, t_s, the relative velocity of the solid phase, u_s, can be determined by dividing the length of a column, L_c, by the switching time (Eq. (4.9)).

$$u_s = \frac{L_c}{t_s} \qquad (4.9)$$

An operation cycle is completed when the number of switches equals the total number of columns of the SMB.

The SMB follows the same basic principles of the TMB. The desorbent and the feed mixture are continuously fed to the unit while the most and the least retained products are simultaneously collected in the extract and the raffinate streams, respectively. All the streams keep the same relative positions dividing the unit into four sections that perform the functions previously described in Table 4.1; however, the number of columns is independent for each section.

From the analysis of the SMB operating mode, it can be easily concluded that this unit does not achieve a "standard" steady state.

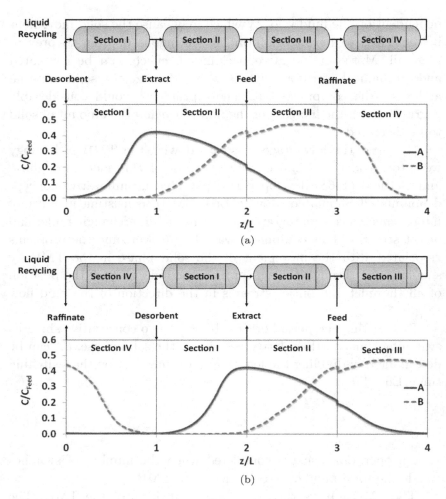

Figure 4.5 Schematic representation of the SMB operating principle considering two consecutive time switches (a) and (b) (compound "A" has higher affinity towards the solid phase than compound "B").

Instead, a cyclic steady state is achieved in which the extract and raffinate concentration histories dynamically change during a switching time being periodically repeated in all the subsequent switches. The internal concentration profiles throughout the unit follow a similar pattern if one considers that the first column is always the column to which the desorbent is being fed. Alternatively, if one considers the static arrangement of the columns, the internal concentration profiles will be exactly the same after a complete operation cycle and during that period all the columns will sequentially execute the tasks associated with each of the four different sections of the SMB.

Hence, conceptually, the SMB can be interpreted as a discrete approach to the TMB, whereby the performance and internal profiles of the two technologies will progressively become more similar with the increase of the number of SMB columns and port switch frequency; however, as a reference, for most systems, an SMB with eight or more columns will provide very similar results to an analogue TMB (Ruthven and Ching, 1989; Pais *et al.*, 1998; Sá Gomes *et al.*, 2009). If the appropriate equivalence between a TMB and an SMB unit can be established, an expeditious determination of the SMB process variables may be accomplished through the design methodology previously presented for the TMB without the necessity of accounting for the complex dynamics of the unit resulting from the periodic discontinuities introduced by the SMB operating mode. In fact, this is one of the reasons why the TMB technology has been so extensively studied. The equivalence between the TMB and the SMB can be attained by keeping the relative interstitial velocity between the solid and the liquid phases. Since the SMB solid phase is not actually moving, the velocity of the liquid mobile phase in section j must be corrected according to Eq. (4.10).

$$u_j^{\text{SMB}} = u_j^{\text{TMB}} + u_s \tag{4.10}$$

In terms of interstitial velocities ratios, Eq. (4.10) can be rewritten as follows, by dividing both members of Eq. (4.11) by u_s.

$$\gamma_j^{\text{SMB}} = \gamma_j^{\text{TMB}} + 1 \tag{4.11}$$

Thus, after obtaining the fundamental adsorption equilibrium data for all the species, it is possible to determine the separation region for the SMB and, with this information, all operating parameters can be defined, taking into account the geometry of the equipment.

Since it was first patented, the SMB technology has definitely caught the attention of the scientific community and the industry (Rodrigues *et al.*, 2015). Historically, the development of this continuous counter-current chromatographic process was associated with the petrochemical industry, more specifically, the separation of aromatic compounds. UOP Inc. (United States of America), which was responsible for creating the SMB concept in 1961, designed and constructed the first large-scale equipment (Carson and Purse, 1962) that was latter applied to the Parex® process for the purification of p-xylene (Broughton *et al.*, 1970; Minceva and Rodrigues, 2005b; Silva *et al.*, 2015), which is still one of the most relevant processes within the petrochemical industry. The SMB influence increased in the subsequent years but it was only after a decade, approximately, that its application was extended to other industries (Broughton, 1983). However, in its initial stage the use of the SMB was limited to the separation of bulk chemicals in large-scale units. The introduction of the SMB in the pharmaceutical industry occurred in the end of the 90s and was mainly driven by the interest generated in highly pure enantiomers in that period. New and highly selective stationary phases had to be developed; however, due to its limited adsorption capacity, the typical productivity values attained by the SMB in the petrochemical industry were reduced by one order of magnitude, approximately. More recently, this technology has been applied to the biopharmaceuticals sector with great success.

Although the development of advanced materials has represented an important strategic factor, the remarkable evolution experienced by the SMB was predominantly related with the innovations introduced in terms of the technical and instrumental aspects of this technology (Faria and Rodrigues, 2015). The development of alternative operating modes, for instance, was of extreme relevance since it considerably increased the efficiency of some of the existing processes and it allowed to accomplish separations that were impossible with the conventional SMB. The most important alternative operating modes, classified according to the major modifications introduced, are the following (the explanatory details for each one of them are provided in the cited literature):

- Dynamic configuration variations: Varicol® (Adam *et al.*, 2000; Ludemann-Hombourger *et al.*, 2000), Pseudo-SMB (Masuda *et al.*, 1993);
- Flow rate modulation: Power Feed (Kearney and Hieb, 1992), Improved-SMB® (Lutin *et al.*, 2002), Partial-Feed, Partial-Discard and Partial-Withdrawal (Zang and Wankat, 2002), Outlet Swing Stream (Sá Gomes and Rodrigues, 2007);
- Concentration modulation: ModiCon (Schramm *et al.*, 2002), Enriched Extract SMB (Paredes *et al.*, 2006);
- Gradient operation: Solvent Gradient (Antos and Seidel-Morgenstern, 2001), Temperature Gradient (Migliorini *et al.*, 2001);
- Alternative SMB configurations: SMB with reduced number of sections (Ruthven and Ching, 1989), SMB with extended number of sections (Nicolaos *et al.*, 2001a, 2001b), SMB cascades (Wankat, 2001).

It is important to notice that some of these unconventional operating modes previously described can be applied in the classical SMB units conceptualized by UOP, while others require adaptations to these units. In this context, the design and construction of the SMB has also evolved considerably. There is a virtually unlimited number of possible configurations for the SMB valves, pumps and transfer lines but most of them can be grouped into two major categories: central valves design and distributed valves design. Conventionally, central valves were large and very complex rotary valves that periodically changed the position of all the inlet and outlet streams in relation to static columns or, alternatively, used a carrousel mechanism to move the columns in relation to static transfer lines. Those two types of valves were the most commonly used in the first SMB units and still represent some of the most reliable solutions in terms of valve designs, finding long-term applications in the petrochemical industry, for instance. However, changing the number of columns of an existing unit or implementing some unconventional operating modes may be extremely difficult or even impossible. The distributed valves design, on the other hand, allow this flexibility. Within this valve design, there are two main possibilities of combining the valves, the columns and the transfer lines: using a two-position valve-to-column approach or a multi-position valve-to-stream approach. One can easily conclude that depending on the strategy chosen, the total number of valves can be quite different. In fact, for some units the number of valves becomes so high that the control system complexity increases considerably. Given its characteristics, the distributed valves design is more frequent in fine chemicals industries.

Finally, one of the topics that has been more intensively investigated is the development and application of hybrid SMB units which result from the combination of the SMB separation with other physical–chemical processes (Rodrigues *et al.*, 2015). This concept has demonstrated the ability to significantly improve the performance of the overall process. Among the most interesting process intensification (PI) strategies reported is the combination of chromatographic reactors with the SMB, generally denominated Simulated Moving Bed Reactor (SMBR). Due to its relevance, this technology will be comprehensively studied in the following sections.

4.2. Combining Reaction and Adsorption: Simulated Moving Bed Reactor

Industrial chemical processes consist of an association of several reaction and separation stages that allow the synthesis and purification of target products from specific raw materials. Traditionally, each unit within the industrial complex was responsible for performing one of the necessary chemical reactions or a separation step, independently. However, recently, an innovative concept, generally designated as PI, has promoted significant changes on the paradigm of process engineering, mainly through the development and implementation of hybrid technologies capable of performing several steps of the conventional industrial processes in a single unit with the intention of improving its efficiency, decreasing plant size and reducing its environmental impact.

In this context, several reactive separation technologies were generated. These new class of reactors, known as multifunctional reactors, combine any separation principle with chemical reaction, particularly reversible ones, in order to promote the continuous removal of one or more products from the media which, according to Le Chatelier's law, will favour the products formation resulting in the displacement of the chemical reaction equilibrium. Consequently, the thermodynamic limitations associated with the reaction and the separation phenomena (for instance, the formation of azeotropes) might be overcome, increasing the reactants conversion, improving the products selectivity and reducing the overall energy consumption. Furthermore, as two or more steps of the conventional process may be performed in a single unit, the reduction of the capital investment required is another interesting advantage associated with reactive separations.

Sorption enhanced reactors (addressed in detail in Chapter 2) are generally recognized by the scientific community as one of the most relevant multifunctional reactors. Chromatography is the separation principle behind the improved performance of these reactors; therefore, as discussed in the previous section, the implementation of a continuous counter-current operation mode will maximize the unit's separation capacity, increasing its ability to displace the thermodynamic equilibrium. Thus, the concept of adding to the SMB the capacity for carrying out chemical reactions while performing the chromatographic separation gave rise to a new and efficient technology, denominated SMBR.

In order to detail the conventional SMBR operating principle in a simple way, it must be mentioned that it shares most of the same basic features of the SMB, starting from the fact that both are implemented in the same type of units, generally constituted by a series of interconnected chromatographic columns, in a closed loop circuit, with interchanging inlet and outlet streams. The reactants are introduced in the SMBR through the feed stream, reacting in its central sections while the most and the least retained products are eluted by the desorbent (which, most commonly, is one of the reactants), being collected through the extract and the raffinate streams, respectively. In most of its applications, the SMBR presents a classical four-section configuration delimited by these two pairs of inlet and outlet streams. As in the SMB, Sections I and IV are responsible for the regeneration of the solid phase and the desorbent, respectively, however Sections II and III, besides performing the products purification, are also responsible for carrying out the reaction steps. Table 4.2 presents a detailed description of the specific functions executed by each of the four sections of the SMBR.

Table 4.2 Description of the specific function of each of the four sections of a conventional SMBR.

Section	Main function	Separation details	Reaction details
I	Solid phase regeneration	Desorption of the most retained products to allow recycling the solid phase	Negligible chemical reaction
II	Purification of the most retained products	Adsorption of the most retained products and desorption of the least retained ones to avoid the contamination of the extract stream	Chemical reaction occurs and the thermodynamic equilibrium might be overcome
III	Purification of the least retained products	Desorption of the least adsorbed products and adsorption of the most retained ones to avoid the contamination of the raffinate stream	Chemical reaction occurs and the thermodynamic equilibrium might be overcome
IV	Desorbent regeneration	Adsorption of the least retained products to allow recycling the desorbent	Negligible chemical reaction

As suggested before, the different concentration profiles observed for reactants and products inside the unit, resulting from the differences in the adsorbent selectivity towards each species, potentiate the thermodynamic equilibrium displacement that allow the SMBR to achieve 100% conversion with 100% recovery of the desired product. This represents one of the most relevant advantages of the SMBR; however, it must be noticed that the products collected in the extract and raffinate streams are normally diluted in the desorbent, which implies that further separation steps must be performed to recycle the desorbent back to the unit and purify the products. In Section 4.5, this subject is addressed in further detail by presenting some examples and effective strategies to optimize the SMBR unit considering the global process.

The most significant difference between the SMB and the SMBR relies on its stationary phases. While the earlier technology only requires an adsorbent material capable of separating the products, the SMBR, in addition to that, requires a solid phase that exhibits a considerable catalytic activity. The combination of these two capabilities can be accomplished through three main approaches:

- using a hybrid solid material with dual behaviour that acts, simultaneously, as adsorbent and catalyst;
- packing the SMBR columns with a mixture of an adsorbent and a catalyst in the appropriate ratio;
- performing the reaction and separation steps in distinct columns packed with the catalyst and the adsorbent materials, respectively.

The use of a hybrid stationary phase with catalytic and adsorptive properties is probably the most common and straightforward approach for implementing an SMBR process. In this scenario, the SMBR can be operated exactly like an SMB without requiring any adaptation to the instrumentation, valves design or transfer lines. Moreover, as all the chromatographic columns are identical, the process design, the prediction of its performance and the description of its dynamic behaviour through mathematical models is simplified when compared with the remaining possible configurations of the SMBR stationary phases. On the other hand, as the catalyst is present in all the columns, reaction may occur in the surroundings of the product collection ports which might lead to the contamination of the target products. Nevertheless, this drawback can be overcome by selecting the appropriate flow rate ratios in each section of the SMBR, usually at a cost of a slight loss in the productivity.

Still considering the use of hybrid materials, particularly interesting results have been reported for the production of organic oxygenated compounds using ion-exchange resins, since these materials present a high catalytic activity for acetalization (Silva and Rodrigues, 2005; Pereira *et al.*, 2008; Graça *et al.*, 2011a, 2012), etherification (Zhang *et al.*, 2001), esterification and transesterification (Lode *et al.*, 2001; Pereira *et al.*, 2009; Geier and Soper, 2010) reactions, for instance, and a substantial adsorption selectivity for some of the synthesized compounds. In fact, the application of the SMBR concept to these systems revealed to be so successful that it led to the development of a few patented processes (Rodrigues and Silva, 2005; Dubois, 2008; Geier and Soper, 2010). Focusing exclusively on the production of acetals, as an example, the use of acid ion exchange resins, as Amberlyst-15, demonstrated to be an extremely effective approach. The selected stationary phase is highly selective to water, which is the side product of these exothermic reactions that occur under acidic conditions. Furthermore, these catalysts generally present a noteworthy activity at near room temperature or even below (down to 283 K). The SMBR technology demonstration was performed for the synthesis of 1,1-dimethoxyehtane (Pereira *et al.*, 2008), 1,1-diethoxyehtane (Silva and Rodrigues, 2005) and 1,1-dibuthoxyehtane (Graça *et al.*, 2011a, 2012), which are produced by the reaction between methanol, ethanol and 1-butanol with acetaldehyde, respectively. The most impressive results were obtained for the production of 1,1-dibuthoxyehtane at 298 K, using 1-butanol as desorbent in an SMBR with 3 columns per section packed with Amberlyst-15 which, under optimized conditions, achieved a productivity value of approximately $55.3 \, \text{kg}_{\text{Prod}} \cdot \text{L}_{\text{Ads}}^{-1} \cdot \text{day}^{-1}$ and a desorbent consumption of $7.0 \, \text{L}_{\text{Des}} \cdot \text{kg}_{\text{Prod}}^{-1}$.

An alternative approach for combining adsorption and chemical reaction in the SMBR consists in packing all its columns with a mixture of an adsorbent and a catalyst, as previously suggested. In this case, the beds can have one of the following arrangements: either the two materials are homogeneously distributed throughout the columns or a multilayer system is implemented, in which the column is packed with a definite number of alternate layers of adsorbent and catalyst. An immediate consequence of using different types of particles with different physical–chemical properties in the SMBR is the problems that arise during the columns packing procedure and after long-term operation periods. For instance, if there are significant differences in the particles size or density, the distribution of the adsorbent and the catalyst might become non-uniform. In this situation, some segregation might be observed for theoretically homogeneous beds or, alternatively, multilayer columns might start presenting an area near the interface of the two solid phases where the catalyst and the adsorbent are mixed. Moreover, these imperfections might also affect the hydrodynamics of the chromatographic columns generating by-passes or dead-volumes that might have a negative impact on the unit's performance.

Similarly to SMBR units packed with hybrid materials, no special adaptations to the SMB apparatus or operating mode are necessary; however, as all the columns within the unit have the same stationary phase with catalytic and adsorptive properties, it is impossible to completely prevent reactants conversion from occurring near the extract and raffinate ports, which can cause the contamination of these streams if the operating parameters are not accurately defined, as already mentioned. Using a mixture of materials also increases, to some extent, the complexity of the mathematical models that allow estimating the unit's steady-state performance and internal concentration profiles since the mass balance equations for the adsorbent and catalyst particles must be solved independently and supplementary boundary conditions must be defined for the interface of the different layers of adsorbent and catalyst. Additionally, two new process variables must be considered in the design and optimization of an SMBR with this type of columns.

The ideal catalyst to adsorbent ratio has to be determined and the most effective bed configuration shall be selected since, even if the same mass of each of the stationary phases is used, different performances can be attained by modifying its arrangement. Despite the slight increase on the overall process complexity, considerable advantages can be withdrawn from operating an SMBR with a mixture of catalyst and adsorbent instead of using hybrid materials. The possibility of defining the relative amounts of adsorbent and catalyst in each column might enhance the process performance since it allows to precisely control the contribution of the adsorption and reaction phenomena independently. Another advantage is related with the capacity of increasing the SMBR selectivity towards a specific product by setting the appropriate layers sequence and contact time between all the species and the different stationary phases. By-products formation and side reactions can be minimized or completely avoided by the optimum definition of these parameters, consequently resulting in the maximization of the unit's productivity.

The use of a multilayer configuration has been proposed for several chromatographic systems as single column fixed bed adsorptive reactors (Oliveira *et al.*, 2011) or pressure swing adsorption units (Bárcia *et al.*, 2010); however, its application for SMBR processes is almost negligible since, for this chromatographic reactor, homogeneously packed beds with an uniform distribution of the adsorbent-catalyst mixture are generally preferred. Nevertheless, remarkable results were published by Gonçalves and Rodrigues (2016), considering the use of an SMBR constituted by multilayer fixed bed columns for the production of p-xylene. The authors suggest that the ideal bed for this process should have an initial zone, corresponding to 15% of the total column volume, containing a mixture of the adsorbent and catalyst particles (in this case, low silica X zeolite exchanged with barium and H-ZSM-5, respectively) in an adsorbent to adsorbent plus catalyst weight ratio of 0.4 and a second zone containing only the adsorbent particles. This innovative chromatographic column configuration that was denominated as dual bed is schematically represented in Fig. 4.6.

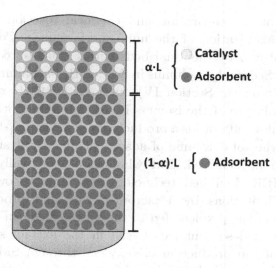

Figure 4.6 Optimized dual bed configuration for the production of p-xylene by SMBR. Adapted from Gonçalves and Rodrigues (2016).

To understand the advantages associated with this unusual bed configuration, the function of each of its subdivisions will be explained in detail. The first part of the bed was designed to maximize the production of p-xylene. By defining the appropriate proportion between the adsorbent and the catalyst, it is possible to reduce the thermodynamic equilibrium constraints of this isomerization reaction by removing p-xylene from the reaction media, as expected for this type of sorption enhanced reactors. However, if the catalyst is not restricted to a minor portion of the unit, the extension of the reverse reaction may become considerable, which might prevent the system from satisfying the purity requirements or significantly limit its performance. In fact, this could be confirmed by the same authors by comparing the results of this study with those obtained in a previous work in which the possibility of implementing a SMBR process considering only homogeneously packed columns was addressed (Gonçalves and Rodrigues, 2014, 2015b). The function of the second part of the bed, which is exclusively constituted by the adsorbent particles, is performing the separation between p-xylene and the remaining isomers. In this way, a balanced contribution of the adsorption and chemical reaction phenomena allows the maximization of the unit's performance. Considering a SMBR constituted by this dual beds with a 2-6-14-2 configuration (2 columns in Section I, 6 columns in Section II, 14 columns in Section III and 2 columns in Section IV) operated at 473 K to which an equilibrium mixture of the isomers is fed, after the optimization of the sectional flow rate ratios a productivity of $5.4\,\mathrm{kg_{Prod}} \cdot \mathrm{L_{Solid}^{-1}} \cdot \mathrm{day}^{-1}$ (considering the total volume of adsorbent plus catalyst within the columns) can be achieved. But what should be really noticed is that the SMBR dual bed technology was able to overcome the equilibrium limitations by a factor of 2.08 and produce in the extract 175% of the p-xylene fed to the unit, which can be collected 70% pure (in a desorbent free basis) in the extract stream. The prominence of the production of p-xylene through SMB or SMB-based technologies at industrial level increases the relevance of these results.

The first application of an SMBR that combined separate adsorption and reaction columns was proposed by Hashimoto *et al.* (1983), aiming the production of higher fructose syrup. Thus, this configuration his commonly named after its inventor. The basic operating mode of this class of SBMR is slightly different from the remaining since the reactors generally keep their relative position towards the unit's inlet and outlet streams during the entire operation while the counter current movement is exclusively simulated for the adsorption columns through consecutive periodical switches (Fig. 4.7), as explained for the SMB. Consequently, it is possible to restrict the reaction to specific sections of the SMBR, typically Sections II and III, since for most chemical systems reaction does not take place in the absence of a catalyst. This partial integration of the adsorption and reaction phenomena represents the most effective SMBR configuration for conducting reactions of the type A ↔ B, for instance, because the extension of the reverse reaction can be controlled, in opposition to what happens for SMBR packed with hybrid particles or mixed adsorbent and catalyst particles in which the catalyst is present in all the columns.

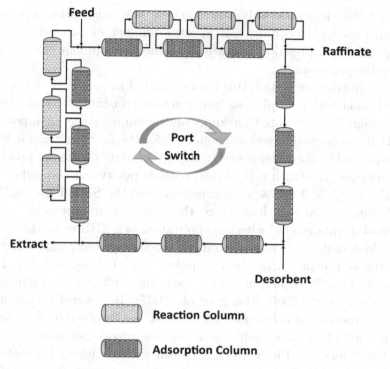

Figure 4.7 SMBR based on the configuration proposed by Hashimoto *et al.* (1983), with independent adsorption and reaction columns.

Due to the particularities associated with its operating mode, the implementation of the Hashimoto's configuration in a standard SMB unit is not possible without performing an adjustment to its transfer lines; however, in most cases, the valves design does not require significant modifications.

The design and optimization of an SMBR with alternate adsorption and reaction columns is a rather challenging task. Besides the standard SMB design variables, a new set of variables must also be determined: the reactors operating temperature (that might be different from the adsorbers temperature); the adsorbent to catalyst ratio; the total number of reactors per section; and the relative arrangement between the reactors and the adsorption columns. Nevertheless, if this drawback can be overcome, the possibility of defining independently the reaction and adsorption operating conditions and adapting such a large number of process variables to a specific application might lead to the development of extremely effective processes.

As previously stated, this kind of SMBR units was developed for the isomerization of glucose into fructose (Hashimoto *et al.*, 1983). Although it represented an important advance, since an innovative SMBR configuration was implemented for the first time and it led to considerable improvements relatively to the state-of-the-art processes at the time, the truth is that the products purity was somewhat low, in the range of 45–65% approximately, and the SMBR productivity was limited. Nonetheless, since then, the open literature reports several studies related with the application of SMBR units that adopt the Hashimoto *et al.* scheme to this particular system, mainly focused on the optimization of the original process (Ching and Lu, 1997; Zhang *et al.*, 2004) and in the pursue of more effective configurations (da Silva *et al.*, 2006; Zhang *et al.*, 2007). In general terms, all of these processes use enzymes as catalyst (as Swetase or Sweetzyme IT), zeolites or ion exchange resins as adsorbents (as Y zeolite in Ca^{2+} form or Finex CG11CS resin in Na^+ form, for instance) and aqueous buffer solutions as desorbent. Among these works, a note must be addressed to the results obtained by Borges da Silva *et al.* (2006), that basically proposed that the SMBR feed solution should be directly introduced in an enzymatic reactor instead of an adsorption column, in opposition to what was established in the prior processes. This simple adjustment allowed replacing the glucose–fructose solution previously used as feed by a pure glucose solution. After the appropriate optimization of the process, it was possible to collect 90% pure product which resulted in a considerable enhancement of the SMBR performance. This study clearly evidences the relevance and complexity of the optimization of this particular kind of SMBR units, due to the considerable increase of the number of process variables imposed by its operating mode, and it shows that studying the combination of all of its effects in simultaneous is essentially unmanageable. Although the production of fructose can be considered as the most representative application of SMBR units constituted by decoupled adsorption and reaction columns, this was not the only reported application. In fact, interesting results have also been obtained by this technology in the production of p-xylene (Minceva *et al.*, 2008; Bergeot *et al.*, 2009a, 2009b).

Finally, one last note must be addressed to the fact that, although the SMBR is generally operated as a conventional SMB (closed loop, four sections, synchronous switching time, constant inlet and outlet flow rates, etc.), any of the alternative operating modes described in Section 4.1 can be implemented in this chromatographic reactor. The number of studies published in the open literature regarding this issue is rather scarce, though. Nevertheless, it is possible to find a few applications of SMBR units with asynchronous switching times, commonly known as the Varicol® operating mode, for the synthesis of methyl tertiary butyl ether (Subramani *et al.*, 2004) or for the inversion of sucrose (Kurup *et al.*, 2005), for instance. SMBR units with reduced number of sections have also been reported, including the process for the production of higher fructose syrup originally proposed by Hashimoto (1983) or the synthesis of other added value chemicals (Kawase *et al.*, 1999). Although for some specific applications the implementation of alternative operating modes can positively affect the SMBR performance, its relevance is still relatively reduced.

4.3. Process Development

The SMBR is an extremely interesting technology, capable of accomplishing outstanding performances and overcoming severe thermodynamic limitations associated with several separations and chemical reactions, through the integration of these two phenomena in a single equipment, as previously exposed. The unit's efficiency also relies in the counter–current operating mode implemented through the periodically shifting of its inlet and outlet streams that simulate the movement of the solid phase in the opposite direction of the liquid. However, this also makes the SMBR a considerably complex technology due to its intrinsic dynamic nature and the high number of physical–chemical processes involved in its operation.

Hence, for developing an SMBR process, all of these factors must be studied in detail, and, preferably, through independent experiments, so one can be able to predict and control their impact in the overall performance.

Before starting the actual process development, a preliminary screening stage is, most of the times, necessary for determining the most suitable SMBR desorbent, adsorbent and catalyst, considering the studied reactive system. It is virtually impossible to implement a feasible selection procedure that allows an expeditious determination of the best conjugation of these three components of the SMBR; therefore, commonly, independent screening studies are performed for all of them or, alternatively, some of them are predefined based on the literature review or previous know-how regarding that particular system while the remaining are independently submitted to screening studies. Nevertheless, the general idea is to perform simple experiments, using a set of conditions as representative of the final process as possible, which allow an easy determination of the most suitable solid materials or desorbent.

For instance, considering the adsorbent, it can be easily concluded that the performance of the SMBR will benefit from the maximization of the selectivity between the target products, which might increase the productivity, and from limiting the adsorption of the most retained species, which might decrease the desorbent consumption. Simple elution chromatography tests can be performed by packing fixed bed columns with different solid phases and injecting the target products to assess these parameters. Focusing on the reaction, the objective of the screening will be determining the most active catalyst for that particular system that might avoid limitations in the products purity and in the unit's productivity due to the presence of high amounts of the reactants near the extract and raffinate ports. This can be achieved by testing different materials in a batch reactor or a fixed bed reactor and assessing which one provides the highest limiting reactant conversion and which is less prone to the formation of side-products. Some guidelines and reference values have been reported in the literature for the effect of adsorption selectivity/capacity and reaction rate (Fricke *et al.*, 1999).

Most of the reported SMBR applications use one of the reactants as desorbent, which not only eliminates the desorbent screening step and avoids additional investments in raw materials as it also favours the reaction rate within the unit and, consequently, improves its performance. However, there are some reactive systems for which an external compound must be used as desorbent. Several reasons can motivate this decision like the formation of by-products when high amounts of a reactant contact with the catalyst, or the high viscosity of those compounds, or even the high costs associated with the separation of the selected desorbent and the target products in the complementary units. The selection a desorbent for an SMBR can be a critical issue since the addition of new species to the reactive system can negatively affect the adsorption, the reaction rate and equilibrium and lead to miscibility problems. Although some systematic desorbent screening methodologies have been published for generic chromatographic processes, for the particular case of the SMBR, this information is rather scarce. One of the few studies on this subject (Faria *et al.*, 2013a), proposes a two stage methodology consisting in an initial theoretical pre-selection of a reduced number of desorbents from a large list of common solvents used in chemical processes, followed by a detailed analysis of the influence of that short-list of compounds in the adsorption of the products, in the reactants conversion and in the miscibility of the reactive mixture through simple experimental tests.

After selecting the most suitable desorbent, adsorbent and catalyst for the SMBR process, fundamental data related with the hydrodynamics, adsorption, reaction and mass transfer must be determined. For that purpose, an independent study of all of these phenomena must be carried out.

The main goal of the study of the hydrodynamics of the adsorptive and reactive beds of the SMBR is the identification of the appropriate flow model and the determination of its specific parameters. Generally, this is accomplished by tracer experiments, consisting in the injection of an inert compound in a column to which the SMBR desorbent is being fed at a constant flow rate and monitoring its outlet concentration histories. For most cases, a plug flow model considering axial dispersion can accurately describe the hydrodynamics of the columns. Less frequently, the dispersion of the compounds according to the radial dimension must also be assumed. Either way, the parameters to be determined in these situations are the packing porosity and the dispersion coefficients (Guiochon *et al.*, 2006).

Regarding the adsorption, the aim of the study should be the determination of the competitive adsorption equilibrium isotherms for all the compounds in the selected adsorbent. The most frequently adopted techniques are elution chromatography and frontal analysis (Ruthven, 1984; Seidel-Morgenstern, 2004). The first consists in injecting different mixtures or amounts of all the pure species in a column to which the SMBR desorbent is continuously being fed while the second consists in performing stepwise changes in the concentration of the feed stream and waiting for the saturation chromatographic column and repeating the procedure for new compositions. The history of concentrations at the outlet of the chromatographic column allows the determination of the distribution of the multicomponent mixture between the liquid and the solid phases (Ruthven, 1984; Guiochon *et al.*, 2006). These data are then fitted by the appropriate adsorption equilibrium model and the specific parameters of the adsorption isotherm for each compound are estimated (Foo and Hameed, 2010). It must be mentioned that, for cases in which the adsorbent also presents catalytic activity, the determination of the competitive adsorption isotherms for reactive pairs cannot be accomplished by the proposed techniques.

The study of chemical reactions should be focused in the determination of the reaction kinetics and, for thermodynamically limited reactions, the chemical equilibrium. The investigation can be accomplished by performing a set of experiments in a batch reactor, for instance, operated at different conditions and monitoring the evolution of the concentration of all the compounds with time. A suitable reaction rate law must be identified and its parameters determined based on the results obtained in the initial instants of the batch experiments (Froment and Bischoff, 1979). For reversible reactions, the equilibrium constant can be estimated through the composition of the media for very long reaction times, after which the concentration of all the species remains unchanged. The set of experiments performed shall include reactions carried out at different temperatures, initial reaction media compositions and catalyst loadings so the effect of these variables can be determined and the extrapolation of the results for the SMBR operating conditions can be more reliable. It should be noticed that these experiments should be performed in the absence of internal and external mass and heat transfer resistances or, alternatively, the determination of the reaction kinetics parameters must account for their influence (Froment and Bischoff, 1979). Moreover, if the catalyst can adsorb a considerable amount of any of the species present in the reaction media, these tests shall be carried out with the lowest catalyst amount possible, to assure that the adsorption will not influence and generate misinterpretations of the reactions results.

The rigorous experimental determination of mass transfer coefficients is generally a rather difficult task with significant uncertainties associated. Furthermore, as the open literature reports numerous correlations for the estimation of these values with a satisfactory accuracy this is usually the preferable route for computing these variables.

All the fundamental data gathered in the previous steps will allow the development of an accurate mathematical model for the description of the SMBR internal concentration profiles and outlet streams composition evolution. The model described hereinafter takes into consideration the most common assumptions for the fixed bed adsorptive reactors that represent the elementary units of the SMBR, including an axially dispersed flow accounting for velocity variations, resulting from the continuous modification of the liquid phase composition, and a linear driving force (LDF) mechanism for the estimation of the mass transfer resistances. Moreover, all the beds should be packed with hybrid particles that act simultaneously as adsorbent and catalyst, its temperature, length and packing were assumed constant and the chemical reaction was considered to occur

exclusively at the active sites inside the catalyst pores. As imposed
by the previously described assumptions, separate bulk and particle
mass balances must be written (Eqs. (4.12) and (4.13)) together with
the respective initial and boundary conditions (Eqs. (4.14)–(4.16)),

$$\frac{\partial C_{b,i,j}}{\partial t} + \frac{\partial (u_j C_{b,i,j})}{\partial z} + \frac{1 - \varepsilon_b}{\varepsilon_b} \frac{6}{d_p} k_{\text{LDF},i,j}(C_{b,i,j} - \overline{C_{p,i,j}})$$

$$= D_{\text{ax},j} \frac{\partial}{\partial z} \left(C_{T,j} \frac{\partial x_{i,j}}{\partial z} \right) \tag{4.12}$$

$$\varepsilon_p \frac{\partial \overline{C_{p,i,j}}}{\partial t} + (1 - \varepsilon_p) \frac{\partial \overline{q_{i,j}}}{\partial t} - \nu_i \frac{\rho_b}{1 - \varepsilon_b} r(\overline{C_{p,i,j}})$$

$$= \frac{6}{d_p} k_{\text{LDF},i,j}(C_{b,i,j} - \overline{C_{p,i,j}}) \tag{4.13}$$

$$t = 0, \forall z \quad \rightarrow \quad C_{b,i,j} = \overline{C_{p,i,j}} = C_{b,i,j_0} \tag{4.14}$$

$$z = 0, \forall t \quad \rightarrow \quad u_{\text{in},j} C_{\text{in},i,j} = u_j \left. C_{b,i,j} \right|_{z=0} - D_{\text{ax},j} \left. \frac{\partial C_{b,i,j}}{\partial z} \right|_{z=0} \tag{4.15}$$

$$z = L, \forall t \quad \rightarrow \quad \left. \frac{\partial C_{b,i,j}}{\partial z} \right|_{z=L} = 0, \tag{4.16}$$

where $C_{b,i,j}$, $\overline{C_{p,i,j}}$ and $\overline{q_{i,j}}$ represent the bulk, the average pore
and adsorbed concentrations of compound i and $C_{T,j}$ is the total
concentration of the liquid phase, $x_{i,j}$ is the molar fraction of
compound i in the liquid phase, t is time, z is the axial coordinate, u_j
is the interstitial fluid velocity, $D_{\text{ax},j}$ is the axial dispersion coefficient,
$k_{\text{LDF},i,j}$ is the global mass transfer coefficient of compound i, ε_b and
ε_p are the bed and particle porosity, ρ_b is the bulk density, d_p is the
particle diameter, ν_i is the stoichiometric coefficient of compound i
and r is the reaction rate described in terms of concentrations. The
liquid phase velocity variation and the respective boundary condition
can be expressed by Eqs. (4.17) and (4.18),

$$\frac{\partial u_j}{\partial z} = -\frac{1 - \varepsilon_b}{\varepsilon_b} \frac{6}{d_p} \sum_{i=1}^{\text{NC}} k_{\text{LDF},i,j} V_{M,i} \left(C_{b,i,j} - \overline{C_{p,i,j}} \right) \tag{4.17}$$

$$z = 0, \quad \forall t \quad \rightarrow \quad u_j = u_{\text{in},j}, \tag{4.18}$$

where $V_{M,i}$ represents the molar volume of compound i. The determination of the global mass transfer coefficient, $k_{\text{LDF},i,j}$, as a function of the internal and external mass transfer coefficients, $k_{\text{int},i,j}$ and $k_{\text{ext},i,j}$, respectively, can be carried out by establishing an analogy to the association of resistors in series in electrical systems, according to Eq. (4.19).

$$\frac{1}{k_{\text{LDF},i,j}} = \frac{1}{k_{\text{ext},i,j}} + \frac{1}{\varepsilon_p k_{\text{int},i,j}} \qquad (4.19)$$

In all previous equations, j represents the index of the chromatographic column.

For the complete description of the SMBR unit, the mass balance to its desorbent, extract, feed and raffinate nodes must be considered, as described in Eqs. (4.20)–(4.27).

$$Q_I C_{b,i,n_D}|_{z=0} = Q_{\text{Rec}} C_{b,i,n_D-1}|_{z=L} + Q_D C_{D,i} \qquad (4.20)$$

$$Q_I = Q_{\text{Rec}} + Q_D \qquad (4.21)$$

$$C_{b,i,n_X}|_{z=0} = C_{b,i,n_X-1}|_{z=L} \qquad (4.22)$$

$$Q_{II} = Q_{\text{Rec}} + Q_D - Q_X \qquad (4.23)$$

$$Q_{III} C_{b,i,n_F}|_{z=0} = Q_{II} C_{b,i,n_F-1}|_{z=L} + Q_F C_{F,i} \qquad (4.24)$$

$$Q_{III} = Q_{\text{Rec}} + Q_D - Q_X + Q_F \qquad (4.25)$$

$$C_{b,i,n_R}|_{z=0} = C_{b,i,n_R-1}|_{z=L} \qquad (4.26)$$

$$Q_{IV} = Q_{\text{Rec}} + Q_D - Q_X + Q_F - Q_R \qquad (4.27)$$

The variable Q, represents a flowrate, the indexes I–IV refer to the corresponding SMBR section (cf. for instance Fig. 4.2) while the indexes Rec, D, X, F and R refer to the recycling, desorbent, extract, feed and raffinate streams, respectively. The number of the column immediately after the desorbent, extract, feed and raffinate node is represented by n_D, n_X, n_F and n_R. The discontinuities associated with the periodic shift of the SMBR inlet and outlet streams position are integrated in the mathematical model by updating the previous variables values (n_D, n_X, n_F and n_R) at the end of each switching time, according to the new position of all the streams.

Finally, taking as example a reaction of the type A $\leftrightarrow$ B + C (where B is more retained than C, the compound of interest) using E as desorbent, the performance of an SMBR with n_c columns can be assessed taking into account quantitative parameters including the raffinate and extract streams purities (Pur_R and Pur_X), limiting reactant conversion (X_{Lim}), raffinate productivity (PR) and desorbent consumption (DC), which can be determined by Eqs. (4.28)–(4.32), correspondingly.

$$\text{Pur}_R = \frac{\int_t^{t+n_c \cdot t_s} C_{C,R}(t)dt}{\int_t^{t+n_c \cdot t_s} \sum_{i=A,B,C} C_{i,R}(t)dt} \cdot 100\% \tag{4.28}$$

$$\text{Pur}_X = \frac{\int_t^{t+n_c \cdot t_s} C_{B,X}(t)dt}{\int_t^{t+n_c \cdot t_s} \sum_{i=A,B,C} C_{i,X}(t)dt} \cdot 100\% \tag{4.29}$$

$$X_{\text{Lim}} = \left[1 - \frac{\begin{array}{c} Q_X \cdot \int_t^{t+n_c \cdot t_s} C_{\text{Lim},X}(t)dt \\ + Q_R \cdot \int_t^{t+n_c \cdot t_s} C_{\text{Lim},R}(t)dt \end{array}}{Q_F \cdot C_{\text{Lim},F} \cdot n_c \cdot t_s} \right] \cdot 100\% \tag{4.30}$$

$$\text{PR}(\text{kg}_C \cdot L_{\text{Ads}}^{-1} \cdot \text{day}^{-1}) = \frac{Q_R \cdot M_C \cdot \int_t^{t+n_c \cdot t_s} C_{C,R}(t)dt}{(1 - \varepsilon_b) \cdot V_{\text{unit}} \cdot n_c \cdot t_s} \tag{4.31}$$

$$\text{DC}(L_E \cdot \text{kg}_C^{-1}) = \frac{(Q_D \cdot C_{D,E} + Q_F \cdot C_{F,E}) \cdot V_{M,E} \cdot n_c \cdot t_s}{M_C \cdot Q_R \cdot \int_t^{t+n_c \cdot t_s} C_{C,R}(t)dt} \tag{4.32}$$

where M_C represents the molar mass of compound C which was assumed to be the desired product.

To achieve the intended reactants conversion and products separation in the SMBR, several design variables must be assigned, among which the switching time and the sectional flow rates are, probably, the most relevant. In Section 4.1, an expeditious design methodology for TMB separations was presented, assuming as valid the hypothesis formulated by the "Equilibrium Theory", which neglects the effects of axial and radial dispersion and mass transfer resistances. Extrapolating this methodology for the SMBR process and considering the equivalence between TMB and SMB systems (Eq. (4.11)), the following constraints (Eqs. (4.33)–(4.35)) can be imposed to the relative velocity ratio between the liquid and the solid phase for each of the SMBR sections.

$$\gamma_{\text{I}}^{\text{TMB}} > \frac{1 - \varepsilon_b}{\varepsilon_b} \left[\varepsilon_p + (1 - \varepsilon_p) \frac{q_B}{C_B} \right] \qquad (4.33)$$

$$\frac{1 - \varepsilon_b}{\varepsilon_b} \left[\varepsilon_p + (1 - \varepsilon_p) \frac{q_C}{C_C} \right] < \gamma_{\text{II}}^{\text{TMB}} < \gamma_{\text{III}}^{\text{TMB}}$$

$$< \frac{1 - \varepsilon_b}{\varepsilon_b} \left[\varepsilon_p + (1 - \varepsilon_p) \frac{q_B}{C_B} \right] \qquad (4.34)$$

$$\gamma_{\text{IV}}^{\text{TMB}} < \frac{1 - \varepsilon_b}{\varepsilon_b} \left[\varepsilon_p + (1 - \varepsilon_p) \frac{q_C}{C_C} \right] \qquad (4.35)$$

However, in a SMBR, besides the chromatographic separation, chemical reaction and mass transfer also play an important role. Therefore, to account for the effect of these phenomena, a safety factor shall be added to the sectional flow rate ratios. Commonly the safety factor represents 5–20% of the value of these variables depending on their expected impact, which may be predicted from the analysis of the results obtained in the previously performed adsorption and reaction experiments.

These constraints can be considered as the starting point for the SMBR design methodology. At this point, an extensive simulation study using the mathematical model previously described shall be performed, to determine the most efficient conjugation of γ_{II} and γ_{III} values that assures the desired limiting reactant conversion and products purity, considering predefined values for γ_{I} and γ_{IV} that lead to the complete regeneration of the adsorbent and the desorbent, respectively. Plotting exclusively in the $\gamma_{\text{II}}-\gamma_{\text{III}}$ plane the combinations of flow rate ratios that meet the process specifications defines the denominated SMBR reactive-separation region (Minceva and Rodrigues, 2005a) (in an analogy to the TMB and SMB separation regions). The vertex of this region indicates the operating conditions that maximize the productivity and minimize the desorbent consumption of the process.

Despite the efforts of several research groups, the optimization of the SMBR is still an extremely challenging problem due to the complexity of the physical–chemical processes involved and the difficulties in formulating realistic objective functions based on economic parameters.

Although the previously reported design methodology is sufficiently reliable to provide practical results that outcome the performance of most alternative technologies as demonstrated for several systems (Silva and Rodrigues, 2005; Pereira *et al.*, 2008, 2009; Graça *et al.*, 2011a), basically, it only allows to determine the optimum flow rate ratios in each sections and the switching time. Some of the remaining process variables are defined based on qualitative criteria or theoretical assumptions. For instance, considering the SMBR configuration, it is easy to predict that the process will benefit from increasing the number of columns in the reactive sections (II and III) if reaction is the limiting step due to its slow kinetics. The usual procedure to optimize this parameter consists on setting an initial SMBR configuration and then performing a sensitivity analysis to the number of columns in each section by repeating the design methodology for different configurations. Instead, for other variables, simplified mathematical models or assumptions can be defined for the determination of its optimum value. An example of that is the optimization of the dual bed configuration proposed by Gonçalves and Rodrigues (2016).

More comprehensive optimizations of the SMBR process based on the determination of its reactive-separation regions have also been reported in the open literature (Azevedo and Rodrigues, 2001; Minceva and Rodrigues, 2005a) but usually require extensive computational resources. Due to the large number of optimization variables that can be assumed, multiobjective optimization approaches using genetic algorithms have been proposed as well (Ziyang *et al.*, 2002; Yu *et al.*, 2003). Nevertheless, even using these methodologies, doubts might still persist regarding the selection of the most favourable operating conditions, since many times the optimization leads to conflicting solutions and there is a generalized lack of information about the raw materials and purified products market value and of the global operation costs.

Finally, one of the most relevant aspects regarding the optimization of the SMBR that has been neglected is the integration of this unit with its complementary reaction and separation units in more complex production plants. Only a few recent studies addressed this issue (Silva *et al.*, 2009; Constantino *et al.*, 2015b; Gonçalves and Rodrigues, 2015a). A more detailed analysis of these processes is provided in Section 4.5.

4.4. SMBR for the Production of Organic Oxygenated Compounds: Acetals and Esters

As stated in Section 4.2, several research groups have identified the SMBR as one of the most efficient technologies for the production of organic oxygenated compounds (Rodrigues *et al.*, 2012), among which, acetals and esters, in particular. The main reasons for this will be addressed through the development of SMBR processes for two case studies: the synthesis of glycerol ethyl acetal, a cyclic acetal, and butyl acrylate, an ester.

4.4.1. Glycerol ethyl acetal

Cyclic acetals are generally recognized as effective fuel additives. Glycerol ethyl acetal (GEA), in particular, demonstrated the ability to reduce the particles emissions associated with the exhaust gases, simultaneously allowing to control the fuel fluid properties.

The most common route for producing GEA is the acetalization of glycerol with acetaldehyde under acidic conditions. As the equilibrium conversion for acetalization reactions is thermodynamically limited, the SMBR might represent an efficient and sustainable technology for the production of this compound. Moreover, the open literature reports successful application of this sorption enhanced reactor in the synthesis of several acetals (Silva and Rodrigues, 2005; Pereira *et al.*, 2008; Graça *et al.*, 2011a).

According to the design methodology described in the previous section, the first step on the implementation of the SMBR process should be the selection of the most effective desorbent, adsorbent and catalyst. Analysing the previous SMBR processes for the production of acetals it is possible to conclude that in all of them one of the reactants was used as desorbent, namely, the alcohol, and Amberlyst-15, an acid ion-exchange resin, was used as adsorbent/catalyst. Considering this, Amberlyst-15 shall also be the selected stationary phase for this process since it presents a high adsorption selectivity towards water, the only by-product formed, and a significant catalytic activity. However, glycerol, due to its high viscosity cannot be used as desorbent in this system because it would significantly increase the pressure drop along the unit and the mass transfer resistances. Its high boiling point would also result in an increase of the costs associated with the purification of the species collected in the outlet streams of the SMBR. Moreover, acetaldehyde does not represent a valid alternative as desorbent because it is very reactive and might oligomerize if large amounts of the compound get in contact with the catalyst. Hence, after applying a desorbent screening procedure specifically developed for sorption enhanced reactors (Faria *et al.*, 2013a), it was possible to identify dimethyl sulfoxide (DMSO) as the most suitable desorbent.

As already addressed, the overall performance of the SMBR is significantly affected by the chemical reaction; therefore, its thermodynamic equilibrium and kinetics in the presence of the desorbent must be carefully determined. A large set of independent batch reactor experiments were performed analysing the effects of the initial reaction media composition, temperature and catalyst loading (Faria *et al.*, 2013a, 2013b). The values of the reaction standard molar enthalpy and reaction Gibbs free energy estimated through these experiments were $-9.6 \, \text{kJ} \cdot \text{mol}^{-1}$ and $-13.0 \, \text{kJ} \cdot \text{mol}^{-1}$. The reaction kinetics, in terms of concentration, was described by a Langmuir–Hinshelwood–Hougen–Watson model, considering the surface reaction between the adsorbed reactants as the rate determining step and that water and DMSO would be the preferentially adsorbed species

(Eq. (4.36)).

$$r = 5.97 \times 10^8 \cdot \exp\left(-\frac{E_a}{RT}\right)$$

$$\times \frac{\left(C_{\text{Acetaldehyde}} \cdot C_{\text{Glycerol}} - \frac{C_{\text{GEA}} \cdot C_{\text{Water}}}{K_{\text{eq}}}\right)}{\left(1 + K_{s,\text{Water}} C_{\text{Water}} + K_{s,\text{DMSO}} C_{\text{DMSO}}\right)^2} \quad (4.36)$$

The value estimated for the reaction activation energy, E_a, was $77.7\,\text{kJ} \cdot \text{mol}^{-1}$, while the adsorption equilibrium constants for water and DMSO, $K_{S,i}$, were $0.345\,\text{L} \cdot \text{mol}^{-1}$ and $0.106\,\text{L} \cdot \text{mol}^{-1}$, respectively.

To determine the adsorption equilibrium established between the solid and the liquid phases, several non-reactive binary breakthrough experiments were performed at 303 K in a chromatographic column packed with Amberlyst-15 through the frontal analysis technique. The multicomponent equilibrium isotherms obtained were fitted by a competitive Langmuir model and the estimated parameters are presented in Table 4.3 (Faria *et al.*, 2014).

After determining all the required adsorption and reaction data the production of GEA by SMBR at 303 K can be evaluated for a 12-column lab-scale unit with the average bed characteristics presented in Table 4.4.

For that purpose, a 2-4-4-2 configuration was assumed, providing larger reaction sections to compensate the reduction on the reaction

Table 4.3 Competitive multicomponent Langmuir parameters for the adsorption of glycerol, water, GEA and DMSO in Amberlyst-15 at 303 K (Faria *et al.*, 2014).

Compound	$Q_{\max}$ (mol $\cdot$ L$_{\text{Ads}}^{-1}$)	K (L $\cdot$ mol^{-1})
Glycerol	5.8	19.8
Water	23.9	8.8
GEA	2.7	5.3
DMSO	6.3	29.5

Table 4.4 Lab-scale SMBR columns properties.

Property	Value
Column length	23 cm
Column diameter	2.6 cm
Bulk density	0.39 g · cm^{-3}
Bed porosity	0.4
Péclet number	300

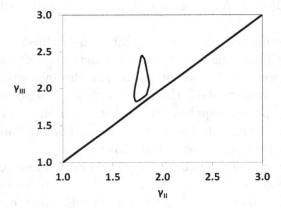

Figure 4.8 Reactive-separation region for the production of GEA in a SMBR (2-4-4-2 configuration, $\gamma_I = 2.25$, $\gamma_{IV} = 1.46$, $T = 303$ K and 90% pure products).

rate promoted by the dilution of the reactants in the DMSO fed at a high flow rate through the desorbent port. The feed consisted in an equimolar mixture of glycerol and acetaldehyde (without desorbent). To assure complete regeneration of the adsorbent and the desorbent in Sections I and IV of the SMBR, respectively, Eqs. (4.33) and (4.35) were used to determine the flow rate ratios between the solid and the liquid phase; the values obtained, considering a 10% security factor, were 2.25 and 1.46, correspondingly. With this information, and defining a minimum raffinate and extract purities of 90%, it was possible to determine the reactive-separation region for this system (Fig. 4.8).

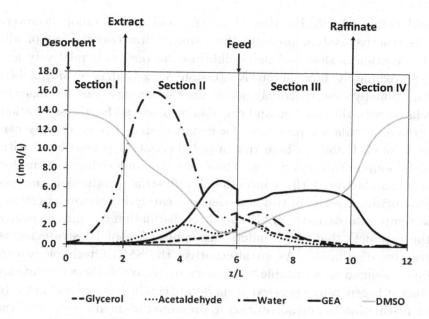

Figure 4.9 SMBR steady-state internal concentration profiles for the production of GEA (2-4-4-2 configuration, $\gamma_I = 2.25$, $\gamma_{IV} = 1.46$, $T = 303$ K).

At the optimum operating conditions (reactive-separation vertex), the productivity of the SMBR reached $15.4 \, \mathrm{kg_{Prod}} \cdot \mathrm{L_{Ads}^{-1}} \cdot \mathrm{day}^{-1}$ *day* with a desorbent consumption of $1.5 \, \mathrm{L_{Des}} \cdot \mathrm{kg_{Prod}^{-1}}$. The simulated internal concentration profiles within the unit at these conditions are presented in Fig. 4.9.

Although only marginally, the SMBR was able to overcome the equilibrium conversion for this reaction (by approximately 2%). Hence, the synthesis of GEA can be accomplished in the SMBR achieving an interesting performance.

4.4.2. Butyl acrylate

Butyl acrylate is an acrylate monomer which has been extensively used as comonomer in polyethene production and as precursor of several products as textiles, plastics, coatings and adhesives, among other applications. This compound can be synthesised through the acid catalysed esterification of acrylic acid and butanol; however, this reactive system presents three major drawbacks. First of all, the reaction is slow and its equilibrium conversion is relatively low (approximately 65% at 363 K) (Ostaniewicz-Cydzik *et al.*, 2014). Additionally, its thermodynamic limitations are not exclusively related with the reaction and can also be observed for the separation through distillation processes, for instance, since five azeotropes can be formed. Finally, a high risk of polymerization is associated with high temperature operation. These are the underlying reasons for the complexity of the conventional industrial production process (comprising a series of two homogeneous catalytic reactors operating at temperatures above 373 K and three distillation columns to purify the product) that also hinders the application of PI strategies as reactive distillation. As an alternative, the SMBR technology can be considered as a suitable technology to overcome these limitations since it is generally operated at moderate conditions and will not rely on distillation but on adsorption to promote the chemical equilibrium displacement.

A fixed bed adsorptive reactor unit has already been used to accomplish this esterification reaction with success (Constantino *et al.*, 2015a); hence, the same solid phase, Amberlyst-15, is proposed for the SMBR process. Moreover, unlike glycerol for the GEA system, the alcohol used in this reaction, butanol, can also be used as desorbent. In this scenario, the desorbent, adsorbent and catalyst screening steps can be neglected.

The required information regarding the esterification reaction equilibrium and kinetics and the multicomponent adsorption of all the species of this reactive system has been previously attained through an independent study of these phenomena. After performing several experiments in a batch reactor under different operating conditions, the reaction standard enthalpy and entropy estimated values were $12.4 \, \text{kJ} \cdot \text{mol}^{-1}$ and $60.0 \, \text{kJ} \cdot \text{mol}^{-1}$ and a simplified Langmuir–Hinshelwood–Hougen–Watson model was selected as the best model to describe the experimental reaction results, in terms of activities, a_i (Eq. (4.37)).

$$r = 1.52 \times 10^7 \cdot \exp\left(-\frac{E_a}{RT}\right)$$

$$\times \frac{\left(a_{\text{Acrylic Acid}} \cdot a_{\text{Butanol}} - \frac{a_{\text{Butyl Acrylate}} \cdot a_{\text{Water}}}{K_{\text{eq}}}\right)}{(1 + K_{s,\text{Water}} a_{\text{Water}})^2} \quad (4.37)$$

An activation energy of $67.0\,\text{kJ}\cdot\text{mol}^{-1}$ was determined for this reaction and the value of $K_{S,\text{Water}}$ was 1.589 (Ostaniewicz-Cydzik *et al.*, 2014). A frontal analysis methodology was implemented to determine the adsorption equilibrium isotherms at 363 K for all the compounds present in the reaction media, fitting the results with a multicomponent competitive Langmuir adsorption model. The parameters estimated for each species are presented in Table 4.5 (Constantino *et al.*, 2015a).

The adsorption and reaction data previously presented can be included in the model described in Section 4.3 in order to proceed with the SMBR design methodology considering a 12-column industrial scale unit with a 2-4-4-2 configuration (for similar reasons as those presented for the GEA system) operated at 363 K. The properties of the SMBR columns are the same as the ones described in Table 4.4, except for its length and diameter that were changed to 0.62 m and 2.23 m, respectively.

Table 4.5 Competitive multicomponent Langmuir parameters for the adsorption of butanol, acrylic acid, butyl acrylate and water in Amberlyst-15 at 363 K (Constantino *et al.*, 2015a).

Compound	Q_{max} $(\text{mol}\cdot\text{L}_{\text{Ads}}^{-1})$	K $(\text{L}\cdot\text{mol}^{-1})$
Butanol	4.55	7.1
Acrylic acid	6.09	1.9
Butyl acrylate	2.91	1.9
Water	24.18	22.7

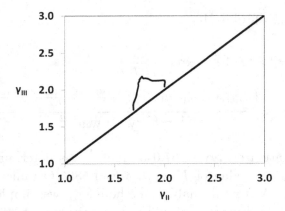

Figure 4.10   Reactive-separation region for the production of butyl acrylate in a SMBR (2-4-4-2 configuration, $\gamma_I = 9.10$, $\gamma_{IV} = 1.52$, $T = 363\,\mathrm{K}$, 99.5% pure raffinate and 95% pure extract). Adapted from Constantino *et al.* (2015b).

Defining the limit flow rate ratio for Sections I and IV as previously described and introducing a safety factor of 5% ($\gamma_I = 9.10$, $\gamma_{IV} = 1.52$), it is possible to obtain the reactive separation region presented in Fig. 4.10.

The maximum productivity that can be obtained for this system is 7.2 $\mathrm{kg_{Prod} \cdot L_{Ads}^{-1} \cdot day^{-1}}$ with a desorbent consumption of 27.6 $\mathrm{L_{Des} \cdot kg_{Prod}^{-1}}$. Although the desorbent consumption is considerably high, the SMBR process had a very interesting performance in what concerns its ability to overcome the thermodynamic limitations associated with the chemical reaction. At the reported operating conditions, a limiting reactant conversion of approximately 99.5% was reached, while the equilibrium conversion is only slightly higher than 60%. Moreover, the raffinate and the extract could be collected with a high purity degree (99.5% and 95%, respectively, in a solvent-free basis). The simulated distribution of the compounds along the unit is presented in Fig. 4.11, which complements the interpretation of the results previously described.

Comparing the SMBR performance for the two presented examples it can be concluded that this is in fact a very versatile technology that bases its success on the ability to overcome diverse limitations associated with different reactive systems by adopting an effective

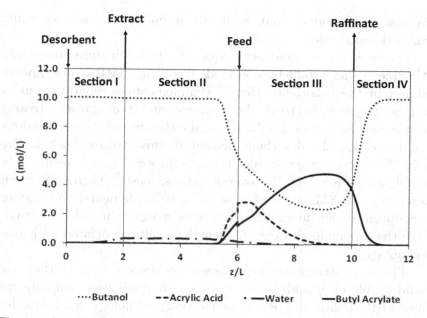

Figure 4.11 SMBR steady-state internal concentration profiles for the production of butyl acrylate (2-4-4-2 configuration, $\gamma_{I} = 9.10$, $\gamma_{IV} = 1.52$, $T = 363\,\mathrm{K}$). Adapted from Constantino *et al.* (2015b).

PI strategy through the combination of the reaction and adsorption. In further detail, it is possible to conclude that for these particular systems, the SMBR was able to achieve higher productivities and lower desorbent consumptions for the acetalization reaction. However, the deviation from equilibrium was more prominent for the esterification reaction, partially due to its much lower equilibrium conversion, and higher product purities could be obtained.

4.5. An Integrated Process

The previous sections were dedicated to the study and optimization of the SMBR as an independent unit. The advantages of this multifunctional reactor were evidenced, with particular emphasis on its ability to overcome the equilibrium limitations existing in several reversible chemical reactions, which leads to an increase on the conversion values that would be impossible to achieve using conventional reactors.

However, as a direct consequence of its intrinsic operating principles, the products collected from an SMBR are always diluted in the desorbent. Hence, complementary separation units will be required to treat the raffinate and the extract streams to recover the target products within the intended specifications and to recycle the desorbent present in this streams back to the unit. This strict need for downstream processing can result in a significant increase of the overall process costs. Therefore, when developing an SMBR-based process, it is fundamental to consider and optimize the integrated process as a whole in order to maximize the economic and environmental benefits associated with this technology.

The most straightforward process integration strategy that one could think of would be to couple two distillation columns to the SMBR (since this is the separation technology with broader acceptance at industrial level) and, after separating the desorbent from the products, recycle it through the desorbent stream. However, this is not necessarily the most efficient way to proceed to the process integration: on one hand, distillation might not be the most suitable technology for separating the compounds present in the extract or the raffinate streams and, on the other hand, one must assure that the contaminants that might be recycled through the desorbent stream will not compromise the SMBR performance.

Notwithstanding the relatively large number of publications and scientific works dedicated to SMBR optimization, the fact is that the information concerning the global process integration is rather scarce. Among the few works published is the development of an industrial scale SMBR process for the synthesis of butyl acrylate (Fig. 4.12) (Constantino *et al.*, 2015b). To produce this ester (as it was already addressed in Section 4.4.2), acrylic acid and butanol must react in the presence of an acid catalyst in stoichiometric amounts to form the desired product and water. Hence, a mixture of acrylic acid and butanol in a molar ratio of 1:1 was fed to a 12-column SMBR packed with Amberlyst-15 operating at 363 K. Butanol was simultaneously

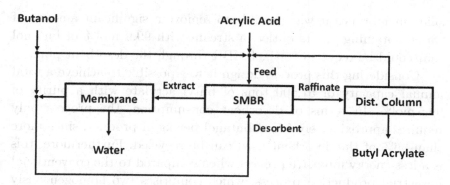

Figure 4.12 Process integration for the production of butyl acrylate: conventional SMBR process.

used as reactant and as desorbent. At optimum operating conditions it was possible to obtain a raffinate stream with 99.9 mol% purity of butyl acrylate (in a solvent free basis) and an extract stream composed only by water and butanol (3.3 mol% and 96.7 mol%, respectively).

The purification of the target product from the raffinate could be accomplished by distillation. The bottom stream of this unit is composed by butyl acrylate with a purity of 99.5 mol%, as required for most of its commercial applications, while the top stream of the distillation column is composed by butanol and almost 4 mol% of butyl acrylate that cannot be directly recovered. If this stream was recycled through the desorbent port, the performance of the SMBR would be affected since the extract would be contaminated with butyl acrylate, the less retained component in this system. Hence, the top stream of the distillation column was recycled through the feed port. This design makes it possible to recover 100% of the butyl acrylate produced in the SMBR without having any negative impact on the extract purity.

To recover butanol from the extract stream, the authors suggest the application of pervaporation since it has been demonstrated that this is the most cost-effective technology for the dehydration of alcohols when the water content does not exceed 10 wt.%, as in this process. By using a module of Pervap® SMS tubular amorphous silica membranes it was possible to remove a significant amount of water obtaining, at its outlet, a stream with 99.9 mol% of butanol that could be recycled to the SMBR through the desorbent port.

Considering this process design it was possible to achieve a total annual capacity of 51,500 tons of butyl acrylate with a purity of 99.5 mol%. In terms of desorbent consumption, the process only requires approximately 1 L of butanol per kg of product, since more than 96% of the desorbent used can be recycled. Furthermore, this is a less energy intensive process when compared to the conventional industrial production process, which comprises two homogeneously catalysed reactors and three distillation columns (Niesbach *et al.*, 2013).

An alternative design, including a fixed bed adsorptive reactor before the SMBR (Fig. 4.13) and keeping the remaining complementary units, can lead to a significant improvement on the overall process performance (Constantino *et al.*, 2016). This configuration is extremely effective, since the equilibrium conversion can be achieved in the first reactor and the unreacted acrylic acid can then be completely converted in the SMBR. In this way, it is possible to overcome the major drawback in this process which is related with its slow reaction kinetics. The impact of this new process

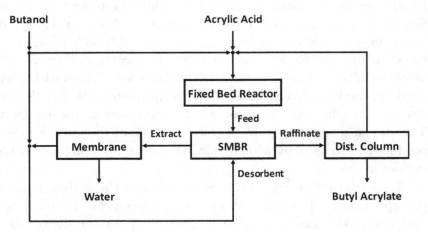

Figure 4.13 Process integration for the production of butyl acrylate: coupled fixed bed adsorptive reactor — SMBR process.

Table 4.6 Main performance parameters of the conventional SMBR and the coupled fixed bed reactor — SMBR processes for the production of butyl acrylate (BAc).

Performance parameter	Conventional SMBR process	Coupled fixed bed reactor — SMBR process
BAc purity (%)	99.5	99.5
Production capacity ($\text{ton}_{\text{Prod}} \cdot \text{year}^{-1}$)	55,000	67,000
Desorbent consumption ($\text{L}_{\text{Des}} \cdot \text{kg}_{\text{Prod}}^{-1}$)	1.0	0.75
Energy consumption ($\text{MJ} \cdot \text{kg}_{\text{Prod}}^{-1}$)	1.7	1.3

integration approach is substantial as the productivity increases 30%, to 67,000 tons of butyl acrylate per year, the total energy consumption decreases approximately 30%, and the dimensions of the distillation column and the pervaporation unit decrease 40% and 10%, respectively. Table 4.6 presents a direct comparison of the most relevant variables and performance indicators for the two processes.

Implementing an effective process integration strategy is particularly relevant for systems that encompass exceptionally difficult separations. For instance, 1,1-diethoxyethane can be synthesized by reacting acetaldehyde and ethanol in an acidic medium, forming water as by-product, but this system presents a very complex thermodynamic behaviour since the acetal, water and ethanol form one ternary and three binary azeotropes. Azeotropic distillation is a very expensive technology that would compromise the process competitiveness. Moreover, the difference between the boiling points of 1,1-diethoxyethane and water is not sufficiently large to ensure a feasible separation or to implement a reactive distillation process. In this context, the synthesis of 1,1-diethoxyethane by SMBR was proposed as an alternative route and was experimentally demonstrated at pilot scale (Silva and Rodrigues, 2005). At the optimum operating conditions (Silva *et al.*, 2009), by feeding and equimolar mixture of acetaldehyde and ethanol to an SMBR packed with Amberlyst-15 and using ethanol as desorbent, it was possible to collect the acetal at the raffinate port with a purity of 99 mol%

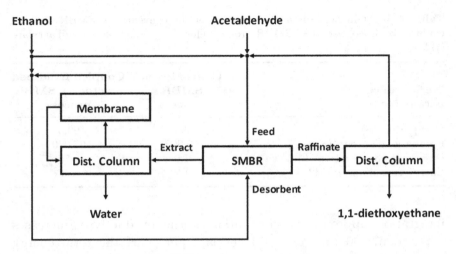

Figure 4.14   Process integration for the production of 1,1-diethoxyethane.

(solvent-free) and a productivity of $23.5 \, \mathrm{kg_{Prod}} \cdot \mathrm{L_{Ads}^{-1}} \cdot \mathrm{day^{-1}}$ while water was withdrawn from the unit through the extract port. However, as both streams were mainly composed by ethanol, the vapour–liquid equilibrium limitations for this system could only be partially eliminated. Nevertheless, the designed process integration strategy, presented in Fig. 4.14, represented a major improvement for the global process. To recover the desorbent from the extract stream a distillation column coupled with a pervaporation unit was used to separate water from ethanol, overcoming the problems with this binary azeotrope. To purify 1,1-diethoxyethane, the raffinate stream was fed to a distillation column and the target product was collected in the bottom stream with a purity of 99 mol%. The ethanol–diethoxyethane azeotropic mixture obtained in the top stream of the distillation column was mixed with acetaldehyde and recycled through the feed port.

Finally, one of the most noteworthy investigations regarding SMBR process integration was reported by Gonçalves and Rodrigues (2015a), in which the authors propose a revamping of an existing aromatics complex for the production of xylenes and benzene. The relevance of this work is related with two main issues: first, p-xylene and benzene are currently the two most important building blocks in the aromatics industry, representing a huge market that for p-xylene, for instance, is expected to grow at a rate larger than 7% in the next few years; and second, because it proposes the conversion of one of the most important SMB processes, the UOPs Parex®, into an SMBR based process.

The new aromatics complex configuration results from performing minor adaptations to the existing units and introducing new ones, namely, a set of distillation columns for the separation of benzene, toluene and the xylenes isomeric mixture, a selective toluene disproportionation unit and a single-stage crystallizer, as shown in Fig. 4.15. Only two modifications were implemented on the existing SMB unit. Toluene was used as eluent instead of the traditional p-diethylbenzene and the original homogeneous adsorbent beds were replaced by dual beds, in which the initial 15% of the bed is composed

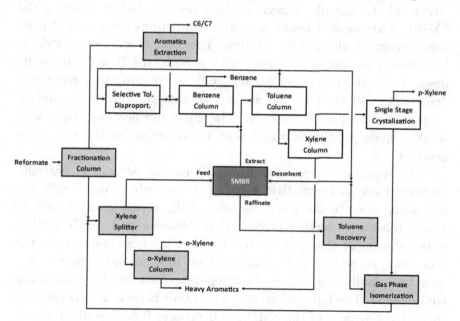

Figure 4.15 Simplified flowsheet of the enhanced aromatics complex comprising an SMBR (white shapes correspond to new units, light grey shapes correspond to the existing units and the dark grey shape represents the SMBR based on the existing SMB unit). Adapted from Gonçalves and Rodrigues (2015a).

by a homogeneous mixture of adsorbent and catalyst (in an adsorbent
to adsorbent plus catalyst weight ratio of 0.4), followed by a second
section exclusively composed by the adsorbent, according to the
optimal bed configuration proposed by Gonçalves and Rodrigues
(2016).

In opposition to the conventional SMB processes, strictly used
for the separation of p-xylene from the remaining compounds of the
isomeric mixture, the main function of the SMBR unit integrated in
this complex is to maximize the conversion of m-xylene and o-xylene
into p-xylene, purifying it at the same time. The SMBR feed stream,
containing only 25 wt.% of p-xylene and around 50 wt.% of m-xylene
(the remaining constituents are o-xylene and ethylbenzene), results
from the combination of the bottom stream of the reformate
fractionation column, in which benzene, toluene, and non-aromatics
are separated from xylenes and heavier aromatics, with the outlet
stream of the gas phase isomerization reactor. Before entering the
SMBR, o-xylene and heavy aromatic compounds are removed from
this stream in the xylene splitter. The target product, p-xylene,
is collected at the extract port with a purity of 70 wt.% (solvent-
free). This stream is sent to the toluene column, to remove the
desorbent, and then to the xylenes column in which the amount
of p-xylene surpasses 75 wt.%. The final purification stage is a
single-stage crystallization unit that allows obtaining 100 wt.% pure
product.

The desorbent separated from the extract stream is partially
recycled back to the SMBR through the desorbent port while the
remaining goes to the selective toluene disproportionation unit in
which toluene is further converted to benzene and p-xylene. The
outlet stream of this unit is mixed with a stream coming from the
aromatics extraction unit, composed by toluene and benzene, and
goes to a distillation column that separates benzene from the other
components. The bottom stream of this unit is then incorporated in
the extract stream of the SMBR and processed as described above.
The selection of toluene as desorbent in this process had a remarkable
impact in the overall process performance and competitiveness. Not
only was it possible to increase p-xylene and benzene production by

adopting an appropriate process integration strategy and extending toluene conversion in the selective disproportionation unit, as it was also possible to mitigate the risk of xylenes dismutation that occurs when p-diethylbenzene is used as desorbent.

To complete this comprehensive process integration, the raffinate stream of the SMBR, mainly composed by m-xylene, o-xylene and ethylbenzene (14 wt.%, 50 wt.% and 33 wt.%, in a solvent-free basis, respectively), is separated from the desorbent in the toluene recovery unit. The desorbent is recycled to the SMBR and the stream containing the xylenes is sent to the gas phase isomerization reactor in which the thermodynamic equilibrium is re-established by producing p-xylene and, as previously mentioned, this stream is then used as a part of the SMBR feed.

In terms of the global performance of this new aromatics complex, keeping the current feed of reformate from the existing plant would result in an increase of 72% of the p-xylene production capacity up to a value of 158,000 tons per year. It is important to underline that 15% of this improvement was a direct consequence of the transition from an SMB to a sorption enhanced reactive unit, the SMBR, which presents a productivity of $3.6 \, \mathrm{kg_{Prod}} \cdot \mathrm{L_{Solid}^{-1}} \cdot \mathrm{day^{-1}}$. The remaining increment was accomplished by the introduction of the selective toluene disproportionation unit together with the effective process integration adopted (including the choice of toluene as desorbent). Moreover, the benzene production reached 117,500 tons per year corresponding to an increase of 170% of the installed capacity while the o-xylene production remained unchanged.

However, the improvements on the aromatics complex performance where so noticeable that the existing isomerization units became oversized. Therefore, a new study was performed assuming that the reformate flow rate was twice as high (Gonçalves and Rodrigues, 2015a). The complementary upstream units were scaled accordingly and the SMBR dual-bed configuration was slightly modified, accounting now for only 10% of the initial section containing the homogeneous mixture of adsorbent and catalyst. The results obtained were even more impressive, since it was possible to increase the production of p-xylene by a factor of approximately

2.4 (314,000 tons per year) while the production of benzene was almost 4.5 times higher (238,000 tons per year). At the same time, the production of o-xylene doubled. Analysing the performance of the SMBR, independently, the adjustment of the feed flow rate resulted in an increase of the unit productivity, which reached $7.0 \, \mathrm{kg_{Prod}} \cdot \mathrm{L_{Solid}^{-1}} \cdot \mathrm{day^{-1}}$, without having a significant impact on the desorbent consumption. Nevertheless, it is interesting to notice that, despite the evident increase on the overall productivity achieved with a higher reformate flow rate, if the current aromatics complex feed conditions where maintained, the SMBR would be more efficient from the PI stand point, since it would promote a higher deviation from the equilibrium of the isomerization reaction (1.9% against 2.3%).

Summarizing, the process integration strategy adopted will depend on a large number of aspects that must be analysed in detail for each particular process. In this section, three distinct examples were presented that emphasise the versatility and effectiveness of the SMBR as a multifunctional reactor and demonstrate how one can take advantage of these features to use the SMBR as the core technology in the design of competitive and environmental-friendly industrial processes.

4.6. Combining Reaction, Adsorption and Membrane Permeation: The PermSMBR

Reactive distillation columns, membrane reactors and chromatographic reactors, particularly the SMBR, are currently the most representative reactive separation technologies. All of this multifunctional reactors adopt the most efficient concepts of PI and combine reaction with a specific separation technique aiming to overcome the thermodynamic limitations of the system. However, there is no restriction to combine multiple separation techniques together with reaction. In fact, this might bring significant advantages to the final process since the separation of target products can be further extended or, alternatively, several species can be separated simultaneously through the different techniques applied. In this context, a new technology was recently developed, the Simulated Moving Bed Membrane Reactor (PermSMBR) (Silva *et al.*, 2012), which combines reaction, adsorption and membrane permeation in a single unit.

The PermSMBR equipment and operating principles are based on the SMB and SMBR technologies. Hence, typically, the unit shall have two inlet streams, to introduce the feed and the desorbent, and two outlet streams, to withdraw the extract and the raffinate. The counter current movement of the solid phase is simulated by the synchronous shift off all the streams at regular time intervals in the direction of the liquid flow and the inlet and outlet streams divide the unit in four sections as explained for the SMBR earlier on this chapter (Table 4.2 in Section 4.2). The most relevant difference between the SMBR and the PermSMBR is related with its elementary units: while the SMBR consists in a set of fixed bed chromatographic reactors, the PermSMBR adds to each of this elements a membrane module selective to one or more reaction products which can be based in any membrane separation principle as pervaporation, vapour permeation or ultrafiltration, for instance. Therefore, the elementary units of a PermSMBR can present one of the following configurations:

1. Integrated PermSMBR — each chromatographic reactor of the SMBR is replaced by a membrane module and the stationary phase (a mixture of catalyst and adsorbent or a material with dual behaviour) is packed inside or outside the membranes, according to the location of its active layer (Fig. 4.16(a));
2. Coupled PermSMBR — each chromatographic reactor of the SMBR is followed and/or preceded by an independent membrane module (Fig. 4.16(b)).

Comparing the two configurations one can say that the Integrated PermSMBR might represent the most interesting configuration under the scope of PI, since it reduces the unit's size and its capital costs, promoting at the same time a more efficient combination of the separation and the reaction phenomena. On the other hand, the Coupled PermSMBR presents crucial advantages from an industrial

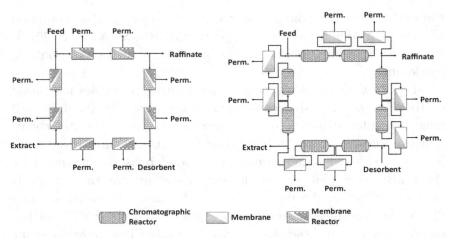

Figure 4.16 Schematic representation of the Integrated and Coupled PermSMBR units.

implementation point of view. As the membrane modules are independent from the fixed bed columns, the installation, replacement and maintenance of any of these elements will be considerably simplified and the unit becomes more flexible, allowing a wider range of operating conditions. Moreover, as the direct contact between the membranes and the solid stationary phase is avoided, the stability of the membranes can be preserved and its lifetime will not be compromised.

The introduction of the membrane modules generates a new stream, the permeate stream, that results from the combination of the fluxes withdrawn from each module and is mainly composed by the products to which the membranes are selective to. To obtain this stream, it is necessary to supply vacuum or sweep gas flow. Nevertheless, it should be noted that the flux through a membrane can be deactivated by cutting off these supplies.

Similarly to the SMBR, due to the presence of a solid phase with catalytic and adsorptive properties, in Sections II and III of the PermSMBR the reactants are converted and the products are simultaneously separated based on their adsorption selectivity. The products that establish stronger interaction with the solid phase are collected at the extract port while the less retained products are carried with the liquid and collected at the raffinate port. However, the presence of membrane modules selective to a specific product (or set of products) in the PermSMBR enhances the unit's separation capacity. The products separation does not only rely on the adsorbent capacity and selectivity but is also accomplished by membrane permeation. Consequently, due to the combination of these two separation principles, the thermodynamic equilibrium displacement can be extended, increasing the reactants conversion even further. Moreover, the SMBR desorbent requirements can be significantly decreased. If the membranes used in the PermSMBR are selective to the most adsorbed products, the adsorbent regeneration in Section I will become easier and will demand less desorbent since these products are also being removed from the unit through the membranes. If, on the other hand, the membranes are selective to the least adsorbed products of the reaction, its permeation will simplify the desorbent regeneration in Section IV, allowing the unit to operate at a higher internal recycle flow rate, which will also result in a decrease of the desorbent consumption. Actually, for some systems the flux through the membranes is so high that it might justify the elimination of one of the PermSMBR sections. Instead of operating with the standard four sections, depending if the membranes are permeable to the most or the least adsorbed products, Section I or Section IV can be eliminated, respectively. The reduction of the number of sections is accomplished by eliminating the extract or the raffinate stream. The three sections configuration is particularly effective and has been denominated as PermSMBR-3s. Nevertheless, as PermSMBR is an SMB-based technology, it can operate with a virtually unlimited number of sections, by increasing or decreasing the number of its inlet or outlet streams, and it can also make use of alternative operating modes as Varicol®, Power Feed or ModiCon, for instance.

Considering its characteristics, one can conclude that the PermSMBR shall be a particularly efficient technology for systems comprising reversible reactions, parallel reactions, sequential reactions or (pseudo-)ternary separations within reactive systems, if the appropriate conjugation of membranes, adsorbent and catalyst can be found. The case of thermodynamic limited reactions was already approached and, as it was stated, the membranes and the adsorbent must be selective to one of the reaction products (either the same product or different ones) and the combination of the two separation techniques will increase the rate of removal of the products from the reaction media increasing the PermSMBR productivity. In another scenario, if one of the reactants or the desorbent reacts with the target product as it is formed, this compound must be immediately removed from the reaction media in order to avoid its conversion into undesirable by-products. Using a PermSMBR constituted by membranes modules that are highly selective to the target product can prevent these sequential or parallel reactions from occurring overcoming, in this way, the loss of productivity that it would imply, while the remaining components would still be separated by adsorption. Finally, in the case of ternary or pseudo-ternary separations, the more adsorbed products are separated from the least adsorbed ones by the PermSMBR stationary phase based on the SMB separation principles. In this situation, the PermSMBR membranes shall be responsible for the removal of the intermediately adsorbed products in order to collect the purified target product (or products mixture) in the permeate stream or, alternatively, avoid the contamination of the extract and raffinate streams with these species.

Although, as far as our knowledge goes, the PermSMBR has not been applied to any industrial process up to the moment, its potential has been demonstrated in several studies mainly focused on acetalization and esterification reactions, particularly, for the synthesis of 1,1-diethoxyethane, 1,1-dibutoxyethane (Pereira and Rodrigues, 2013) (acetals) and ethyl lactate (Silva *et al.*, 2011) (ester). The successful development of PermSMBR processes for these type of reactions is related with the fact that both produce water as by-product and there is a large number of adsorbent materials and membranes that are highly selective to water and allow an effective displacement of the thermodynamic equilibrium.

The mentioned studies used comprehensive mathematical models to describe the cyclic operating mode of the PermSMBR, the flow through its fixed beds and membranes, the mass transfer between the different phases and the reaction, adsorption and permeation phenomena. As expected, the mathematical model is very similar to the SMBR model presented in Section 4.3 (Eqs. (4.12)–(4.32)). However, it must account for the flux of the species that permeate through each membrane, J_i, which, considering, for instance, a pervaporation membrane, can be described according to the following equation:

$$J_i = k_{g,i}(x_i \gamma_i^* P_i^{\text{sat}} - y_i P_{\text{Perm}}) \tag{4.38}$$

where $k_{g,i}$ is the global membrane mass transfer coefficient for compound i, x_i, γ_i^* and P_i^{sat} are its liquid molar fraction, activity coefficient and saturation pressure, respectively, y_i is the molar fraction of compound i in the vapour phase in the permeate and P_{Perm} is the permeate pressure. The global membrane mass transfer coefficient results from the contributions of the diffusion of each compound through the boundary layer and its permeance through the membrane, $Q_{m,i}$ (Wijmans *et al.*, 1996)

$$\frac{1}{k_{g,i}} = \frac{1}{Q_{m,i}} + \frac{\gamma_i^* P_i^{\text{sat}} V_{M,i}}{k_{\text{bl},i}} \tag{4.39}$$

where $k_{\text{bl},i}$ is the membrane boundary layer mass transfer coefficient that can be determined through the Lévêque correlation (Lévêque, 1928).

The mathematical model will also present slight differences depending if an Integrated or Coupled PermSMBR configuration is considered. For the Integrated PermSMBR, only the mass balance to the liquid phase and the interstitial velocity variation equations must be changed to account for the compounds permeation. Hence, Eqs. (4.12) and (4.17) of the SMBR model must be replaced by

Eqs. (4.40) and (4.41), respectively.

$$\frac{\partial C_{b,i,j}}{\partial t} + \frac{\partial (u_j C_{b,i,j})}{\partial z} + \frac{1 - \varepsilon_b}{\varepsilon_b} \frac{6}{d_p} k_{\text{LDF},i,j} (C_{b,i,j} - \overline{C_{p,i,j}})$$

$$= D_{ax} \frac{\partial}{\partial z} \left(C_{T,j} \frac{\partial x_{i,j}}{\partial z} \right) - \frac{A_m}{\varepsilon} J_{i,j} \tag{4.40}$$

$$\frac{\partial u_j}{\partial z} = -\frac{1 - \varepsilon_b}{\varepsilon_b} \frac{6}{d_p} \sum_{i=1}^{\text{NC}} k_{\text{LDF},i,j} V_{M,i} \left(C_{b,i,j} - \overline{C_{p,i,j}} \right)$$

$$- \frac{A_m}{\varepsilon} \sum_{i=1}^{N.C.} J_{i,j} V_{M,i} \tag{4.41}$$

In these equations, A_m, represents the membrane specific area.

The Coupled PermSMBR configuration introduces other modifications to the SMBR model since, in this case, the chromatographic reactors and the membrane modules are placed sequentially. Therefore, the mass balance equations for each of these elements are independent. Equations (4.12) and (4.17) of the SMBR mathematical model remain unchanged and the membrane mass balance (Eq. (4.42)) and velocity variation (Eq. (4.43)) equations must be added to the model, together with the corresponding initial and boundary conditions (Eqs. (4.44)–(4.46), respectively).

$$\frac{\partial C_{i,m}}{\partial t} + \frac{\partial (u_m C_{i,m})}{\partial z} = D_{ax,m} \frac{\partial}{\partial z} \left(C_{T,m} \frac{\partial x_{i,m}}{\partial z} \right) - A_m J_{i,m} \tag{4.42}$$

$$\frac{\partial u_m}{\partial z} = -A_m \sum_{i=1}^{N.C.} J_{i,m} V_{M,i} \tag{4.43}$$

$$t = 0, \ \forall z \rightarrow C_{i,m} = C_{i,m_0} \tag{4.44}$$

$$z = 0, \ \forall t \rightarrow u_{in,m} C_{in,i,m} = u_m C_{i,m}|_{z=0} - D_{ax,m} \left. \frac{\partial C_{i,m}}{\partial z} \right|_{z=0} \tag{4.45}$$

$$z = L_m, \ \forall t \rightarrow \left. \frac{\partial C_{i,m}}{\partial z} \right|_{z=L_m} = 0 \tag{4.46}$$

For this configuration, the mass balance to the nodes can also be described by Eqs. (4.20)–(4.27) but within each section the following equations must be considered (assuming that the membranes modules are positioned after the fixed bed columns):

$$C_{b,i,j}|_{z=L} = C_{i,m}|_{z=0} \qquad (4.47)$$

$$Q_j|_{z=L} = Q_m|_{z=0} \qquad (4.48)$$

$$C_{i,m}|_{z=L_m} = C_{b,i,j+1}|_{z=0} \qquad (4.49)$$

$$Q_m|_{z=L_m} = Q_{j+1}|_{z=0} \qquad (4.50)$$

The production of 1,1-dibutoxyethane by PermSMBR has been thoroughly studied in the open literature; therefore, this example will be used to illustrate the procedure for the development and optimization of a PermSMBR process, to demonstrate the influence of its major operating parameters and to compare the performance of the different PermSMBR configurations. This acetal is formed through the reaction between acetaldehyde and butanol (in a stoichiometry of 1:2) producing water as by-product. All the necessary chemical reaction (Graça *et al.*, 2010), adsorption (Graça *et al.*, 2011b) and pervaporation (Pereira *et al.*, 2011) data can be found in the literature. Moreover, its production by SMBR was demonstrated to be feasible and the mathematical model implemented (similar to the model presented in Section 4.3) accurately described the experimental results (Graça *et al.*, 2011a). The SMBR process developed based on this results (Silva *et al.*, 2011) consisted in a 12-column unit packed with Amberlyst-15 (used as catalyst and adsorbent) operating at 323 K. The properties of the chromatographic beds are presented in Table 4.4. Butanol was selected as desorbent, the feed consisted in an equimolar mixture of acetaldehyde and butanol and a 3-3-3-3 configuration was used. Setting a minimum raffinate and extract purity of 95% as well as an acetaldehyde conversion above 95%, and determining the appropriate flow rate ratios between the liquid and solid phases through the equilibrium theory that assure complete regeneration of the adsorbent and the solid ($\gamma_I = 8.00$, $\gamma_{IV} = 1.33$), it was possible to determine the SMBR reactive-separation region (Fig. 4.17). The results indicated that a maximum productivity of $53.2\,\text{kg}_{\text{Prod}} \cdot \text{L}_{\text{Ads}}^{-1} \cdot \text{day}^{-1}$ could be achieved with a desorbent consumption of $2.7\,\text{L}_{\text{Des}} \cdot \text{kg}_{\text{Prod}}^{-1}$.

Given these data, the development of a four zone Integrated PermSMBR becomes a straightforward procedure. If the SMBR configuration, operating temperature and flow rate ratios in Sections I and IV are kept unchanged, the procedure is limited to the determination of the vertex of the PermSMBR reactive-separation region accounting for the effect of the permeation through the membranes. For the specific case of the synthesis of 1,1-dibutoxyethane, the designed Integrated PermSMBR unit was composed by 12 modules, each containing 13 commercial silica pervaporation membranes from Pervatech, as described in Table 4.7.

Each membrane was packed with Amberlyst-15 and vacuum was provided to obtain the permeate stream (5 mbar). The number and size of the membranes within each module was defined in order to impose a similar residence time in the SMBR chromatographic reactors and in the PermSMBR modules. In this way, a baseline of comparison could be established between the two technologies. However, as the dimensions of the SMBR and PermSMBR elementary units were not exactly the same, since commercial membranes with standard dimensions were used, the switching time was adjusted (from 3.1 min for the SMBR to 3.4 min for PermSMBR) keeping the recycling and the desorbent flow rates unchanged, to assure the required flow rate ratios in the regeneration sections. Figure 4.17 presents a direct comparison between the SMBR and the PermSMBR performances through its reactive-separation regions.

Table 4.7   PermSMBR membrane properties.

Property	Value
Membrane length	25.45 cm
Membrane diameter	0.7 cm
Bulk density	$0.374\ \text{g} \cdot \text{cm}^{-3}$
Bed porosity for membranes packed w/A-15	0.424
Number of membranes per module	13

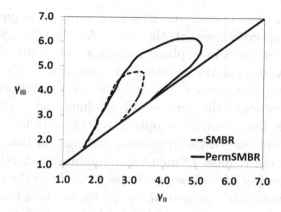

Figure 4.17 SMBR and PermSMBR reactive-separation regions. Adapted from Pereira *et al.* (2011).

As it can be observed from Fig. 4.17, the PermSMBR allows a much wider range of operating conditions that satisfy the purity and conversion requirements imposed. This enlarged reactive-separation region was already expected since more water is being removed from the reaction media by the introduction of the pervaporation membranes. Consequently, the productivity of the SMBR can be increased by 21%, to $64.2\,\mathrm{kg_{Prod}} \cdot \mathrm{L_{Ads}^{-1}} \cdot \mathrm{day}^{-1}$, using the PermSMBR, while the desorbent consumption is simultaneously decreased by 20%. The performance of the PermSMBR can be further improved by adjusting the switching time to take full advantage of the combination of the adsorption and permeation phenomena. Under the optimum operating conditions ($t_s = 2.5\,\mathrm{min}$) the Integrated PermSMBR achieved a maximum productivity of $69.7\,\mathrm{kg_{Prod}} \cdot \mathrm{L_{Ads}^{-1}} \cdot \mathrm{day}^{-1}$ and a desorbent consumption of $2.0\,\mathrm{L_{Des}} \cdot \mathrm{kg_{Prod}^{-1}}$, representing an improvement of 31% and 26%, respectively, comparing with the SMBR technology.

The PermSMBR using only three sections has been identified earlier as one of the most promising non-conventional PermSMBR configurations and the results reported for the production of 1,1-dibutoxyethane through the Integrated PermSMBR-3s confirmed its potential (Pereira *et al.*, 2014). The optimum operating conditions of the four sections unit previously addressed were used as starting point for the development of this new process, maintaining the geometrical specifications of the unit. As the PremSMBR used Amberlyst-15 as stationary phase and silica membranes, both highly selective to water, the three section configuration could be achieved by eliminating the extract stream. Due to the reduction of the number of sections, the number of modules per section had to be adjusted. The strategy adopted by the authors consisted in keeping the position of the remaining streams unchanged in order to maximize the number of membrane modules in Section I, which will be responsible for removing all the water from the unit. Hence, a 6-3-3 configuration was used. However, under these conditions, the flux through the membranes would have to be so high that it became impossible to satisfy the purity and conversion requirements keeping, at the same time, a reasonable productivity. As reported for similar systems (Silva *et al.*, 2011), the solution for this problem consisted in increasing the operating temperature from 323 K to 343 K. At this temperature, the process performance was significantly improved. Alternatively, the permeance could also be increased by reducing the operating pressure (Silva *et al.*, 2011).

As there is no direct equivalence between the four sections and the three sections PermSMBR units in terms of its internal flow rate ratios, the optimization of the Integrated PermSMBR-3s was based on the determination of the recycling flow rate value that maximizes the unit performance keeping the remaining conditions. For that purpose, an entire reactive-separation region was determined for each recycling flow rate value. The results obtained are presented in Fig. 4.18.

These data revealed that the Integrated PermSMBR-3s under the optimum operating conditions for the production of 1,1-dibutoxyethane is able to achieve a productivity of $70.8 \, \mathrm{kg_{Prod}} \cdot \mathrm{L_{Ads}^{-1}} \cdot \mathrm{day^{-1}}$ demanding a desorbent amount as low as $0.3 \, \mathrm{L_{Des}} \cdot \mathrm{kg_{Prod}^{-1}}$, considering a minimum raffinate purity of 95% and a acetaldehyde conversion of 95%. In terms of productivity, the three and the four section Integrated PermSMBR units present similar performances. However, the desorbent consumption for the Integrated PermSMBR-3s is 85% lower than the required for the

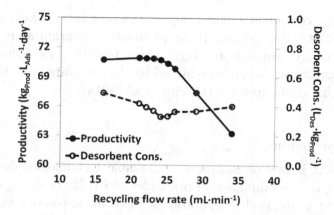

Figure 4.18 Effect of the recycling flow rate in the Integrated PermSMBR-3s productivity and desorbent consumption. Adapted from Pereira *et al.* (2011).

four section PermSMBR and almost 90% lower than the SMBRs. In terms of the global process performance, although the Integrated PermSMBR-3s must operate at higher temperatures, its lower desorbent consumption and the fact that only its raffinate stream requires further processing might lead to a significant reduction of the downstream separation costs. Therefore, only a detailed study considering appropriate process integration would allow to determine which technology is able to produce 1,1-dibutoxyethane more effectively.

The development of a Coupled PermSMBR process for the synthesis of 1,1-dibutoxyethane has also been approached (Pereira *et al.*, 2014). However, the process optimization is much more complex than for the Integrated PermSMBR. An equivalence can be assumed by setting the same global membrane area for both configurations but a large number of process variables still needs to be defined for the Coupled configuration as, for instance, the number of membrane modules and its arrangement relative to each other and to the fixed bed columns. Moreover, while the transition from an optimized SMBR process to an Integrated PermSMBR process can be accomplished by keeping the internal flow rate ratios, for the Coupled configuration this is not possible since the membrane modules will have an uneven influence on the residence time of the compounds depending on the flow rate of each section. Nevertheless, after setting these parameters appropriately, based on an extensive simulation study, it was possible to conclude that similar performances can be achieved by the Integrated and Coupled PermSMBR configurations (Pereira *et al.*, 2014).

4.7. Conclusions

The SMBR is one of the most remarkable technologies developed according to the fundamental principles of PI. By coupling counter current chromatographic separation with chemical reaction this sorption enhanced reactor is capable of overcoming the thermodynamic barriers that limit the performance of several reactive systems. For that reason, the SMBR can extend the conversion of the limiting reactant and effectively separate the products, presenting, at the same time, high productivities and satisfactory desorbent consumption values.

An expeditious design methodology for the determination of the flow rate ratios and the switching time of the SMBR was described; however, the simultaneous optimization of all of its design variables is a rather complex task that usually requires a very high computation capacity level. Nevertheless, it must be noticed that, beyond the relevance of optimizing the SMBR as an independent unit, one should focus in the definition of the most suitable process integration strategy, to allow the maximization of the advantages associated with the use of this technology in more complex production plants.

Finally, it was demonstrated that SMBR process re-intensification can generate new and even more efficient technologies, as the PermSMBR. The PermSMBR combines chemical reaction with two simultaneous separation principles, SMB chromatographic separation and pervaporation, and reveals the potential to significantly exceed the performance of the SMBR. The application of this technology to the production of 1,1-dibuthoxyehtane was studied in detail but, given the satisfactory results obtained, it shall be possible to apply it in the production of several organic oxygenated compounds, among other added value chemicals.

Although it has been intensively studied by the scientific community throughout the years, the industrial application of the SMBR concept is extremely limited or non-existent, in opposition to what is currently verified for the SMB technology, which has already attained a particularly significant relevance in the pharmaceutical and petrochemical industries. This probably represents the most important challenge to the evolution of the SMBR technology in the near future.

Nomenclature

a_i	Activity of compound i	
A_m	Membrane specific area	$m^2 \cdot dm^{-3}$
$C_{b,i,j}$	Concentration of compound i in section j of the SMB/SMBR in the liquid phase	$mol \cdot dm^{-3}$
$\overline{C_{p,i,j}}$	Average pore concentrations of compound i in column j of the SMB/SMBR	$mol \cdot dm^{-3}$
$C_{T,j}$	Total concentration of the liquid phase in column j of the SMB/SMBR	$mol \cdot dm^{-3}$
$D_{ax,j}$	Axial dispersion coefficient in column j of the SMB/SMBR	$m^2 \cdot s^{-1}$
DC	Desorbent consumption	$L_{Des} \cdot kg_{Prod}^{-1}$
d_p	Particle diameter	m^2
E_a	Reaction activation energy	$kJ \cdot mol^{-1}$
J_i	Permeate flux of compound i	$mol \cdot m^{-2} \cdot s^{-1}$
$k_{bl,i}$	Membrane boundary layer mass transfer coefficient for compound i	$m \cdot s^{-1}$
$k_{ext,i,j}$	External mass transfer coefficients of compound i in column j of the SMB/SMBR	$m \cdot s^{-1}$
$k_{g,i}$	Global membrane mass transfer coefficient for compound i	$mol \cdot m^{-2} \cdot s^{-1} \cdot bar^{-1}$
$k_{int,i,j}$	Internal mass transfer coefficients of compound i in column j of the SMB/SMBR	$m \cdot s^{-1}$
K_i	Equilibrium constant for the Langmuir adsorption isotherm of compound i	$L \cdot mol^{-1}$
$k_{LDF,i,j}$	Global mass transfer coefficient of compound i in column j of the SMB/SMBR	$m \cdot s^{-1}$
$K_{S,i}$	Equilibrium constant for the adsorption isotherm of compound i (reaction rate law)	$L \cdot mol^{-1}$
L_c	Length of an SMB/SMBR column	m
M_i	Molar mass of compound i	$kg \cdot mol^{-1}$
P_i^{sat}	Saturation pressure compound i	bar

P_{Perm}	Permeate pressure	bar
PR	Productivity	$\text{kg}_{\text{Prod}} \cdot \text{L}_{\text{Ads}}^{-1} \cdot \text{day}^{-1}$
Pur_R	Raffinate stream purity	%
Pur_X	Extract stream purity	%
$q_{i,j}$	Concentration of compound i in section j of the SMB/SMBR in the liquid phase	$\text{mol} \cdot \text{L}_{\text{sol}}^{-1}$
$\overline{q_{i,j}}$	Average adsorbed concentrations of compound i in column j of the SMB/SMBR	$\text{mol} \cdot \text{L}_{\text{sol}}^{-1}$
Q_j	Volumetric liquid flow rate in section j	$\text{L} \cdot \text{min}^{-1}$
$Q_{\max}$	Maximum monolayer capacity for the Langmuir adsorption isotherm of compound i	$\text{mol} \cdot \text{L}_{\text{Ads}}^{-1}$
Q_s	Solid phase flow rate	$\text{L} \cdot \text{min}^{-1}$
r	Reaction rate	$\text{mol} \cdot \text{g}_{\text{cat}}^{-1} \cdot \text{min}^{-1}$
t	Time	min
T	Temperature	K
t_s	Switching time	min
u_j	Interstitial fluid velocity in column j of the SMB/SMBR	$\text{m} \cdot \text{min}^{-1}$
u_s	Velocity of the solid phase	$\text{m} \cdot \text{min}^{-1}$
$V_{M,i}$	Molar volume of compound i	$\text{L} \cdot \text{mol}^{-1}$
x_i	Molar fraction of compound i in the liquid phase	
X_{Lim}	Limiting reactant conversion	%
y_i	Molar fraction of compound i in the vapour phase	
z	Axial coordinate	m

Greek symbols

γ_i^*	Activity coefficient of compound i	
γ_j	Ratio between the liquid and the solid phases interstitial velocities in section j of the SMB/SMBR	
ε_b	Bed porosity	
ε_p	Particle porosity	
ν_i	Stoichiometric coefficient of compound	
ρ_b	Bulk density	$\text{kg} \cdot \text{L}^{-1}$

Subscripts

b	Bulk/bed
bl	Boundary layer
c	Column
D	Desorbent
ext	External
F	Feed

g	Global
i	Compound i
int	Internal
j	Column j of the SMB/SMBR
LDF	Linear driving force
Lim	Limiting reactant
m	Membrane
M	Molar
p	Particle
Perm	Permeate
Prod	Product
R	Raffinate
Rec	Recycling
s	Solid
T	Total
X	Extract

Superscripts

sat	Saturation

Acronyms

BAc	Butyl acrylate
DMSO	Dimethyl sulfoxide
GEA	Glycerol ethyl acetal
LDF	Linear driving force
PermSMBR	Simulated moving bed pervaporation membrane reactor
SMB	Simulated moving bed
SMBR	Simulated moving bed reactor
TMB	True moving bed

References

Adam, P., Nicoud, R.N., Bailly, M., Ludemann-Hombourger, O., 2000. Process and device for separation with variable-length, *US 6136198 A*.

Antos, D., Seidel-Morgenstern, A., 2001. Application of gradients in the simulated moving bed process, *Chemical Engineering Science*, 56, 23, 6667–6682.

Azevedo, D.C., Rodrigues, A.E., 2001. Design methodology and operation of a simulated moving bed reactor for the inversion of sucrose and glucose–fructose separation, *Chemical Engineering Journal*, 82, 1, 95–107.

Barcia, P.S., Silva, J.A., Rodrigues, A.E., 2010. Octane upgrading of C5/C6 light naphtha by layered pressure swing adsorption, *Energy & Fuels*, 24, 9, 5116–5130.

Bergeot, G., Laroche, C., Leflaive, P., Leinekugel, L.C.D., Wolff, L., 2009a. Reactive simulated mobile bed for producing paraxylene, *WO 2009130402 A1*.

Bergeot, G., Leinekugel-Le-Cocq, D., Leflaive, P., Laroche, C., Muhr, L., Bailly, M., 2009b. Simulated moving bed reactor for paraxylene production, *Chemical Engineering Transactions*, 17, 87–92.

Broughton, D.B., 1983. Sucrose extraction from aqueous solutions featuring simulated moving bed, *US 4404037 A*.

Broughton, D.B., Gerhold, C.G., 1961. Continuous sorption process employing fixed bed of sorbent and moving inlets and outlets, *US 2985589*.

Broughton, D.B., Neuzil, R.W., Pharis, J.M., Brearley, C.S., 1970. Parex process for recovering paraxylene, *Chemical Engineering Progress*, 66, 9, 70.

Carson, D.B., Purse, F., 1962. Rotary valve, *US 3040777 A*.

Ching, C., Lu, Z., 1997. Simulated moving-bed reactor: Application in bioreaction and separation, *Industrial & Engineering Chemistry Research*, 36, 1, 152–159.

Constantino, D.S., Faria, R.P., Pereira, C.S., Loureiro, J.M., Rodrigues, A.E., 2016. Enhanced simulated moving bed reactor process for butyl acrylate synthesis: Process analysis and optimization, *Industrial & Engineering Chemistry Research*, 55, 40, 10735–10743.

Constantino, D.S.M., Pereira, C.S.M., Faria, R.P.V., Ferreira, A.F.P., Loureiro, J.M., Rodrigues, A.E., 2015a. Synthesis of butyl acrylate in a fixed-bed adsorptive reactor over Amberlyst 15, *AIChE Journal*, 61, 4, 1263–1274.

Constantino, D.S.M., Pereira, C.S.M., Faria, R.P.V., Loureiro, J.M., Rodrigues, A.E., 2015b. Simulated moving bed reactor for butyl acrylate synthesis: From pilot to industrial scale, *Chemical Engineering and Processing: Process Intensification*, 97, 153–168.

da Silva, E.A.B., de Souza, A.A.U., de Souza, S.G.U., Rodrigues, A.E., 2006. Analysis of the high-fructose syrup production using reactive SMB technology, *Chemical Engineering Journal*, 118, 3, 167–181.

Dubois, J.-L., 2008. Procede de synthese d'acetals par transacetalisation dans un reacteur a lit mobile simulé, *FR2909669*.

Faria, R.P.V., Pereira, C.S.M., Silva, V.M.T.M., Loureiro, J.M., Rodrigues, A.E., 2013a. Glycerol valorisation as biofuels: Selection of a suitable solvent for an innovative process for the synthesis of GEA, *Chemical Engineering Journal*, 233, 159–167.

Faria, R.P.V., Pereira, C.S.M., Silva, V.M.T.M., Loureiro, J.M., Rodrigues, A.E., 2013b. Glycerol valorization as biofuel: Thermodynamic and kinetic study of the acetalization of glycerol with acetaldehyde, *Industrial & Engineering Chemistry Research*, 52, 4, 1538–1547.

Faria, R.P.V., Pereira, C.S.M., Silva, V.M.T.M., Loureiro, J.M., Rodrigues, A.E., 2014. Sorption enhanced reactive process for the synthesis of glycerol ethyl acetal, *Chemical Engineering Journal*, 258, 229–239.

Faria, R.P.V., Rodrigues, A.E., 2015. Instrumental aspects of simulated moving bed chromatography, *Journal of Chromatography A*, 1421, 82–102.

Foo, K., Hameed, B., 2010. Insights into the modeling of adsorption isotherm systems, *Chemical Engineering Journal*, 156, 1, 2–10.

Fricke, J., Meurer, M., Dreisörner, J., Schmidt-Traub, H., 1999. Effect of process parameters on the performance of a simulated moving bed chromatographic reactor, *Chemical Engineering Science*, 54, 10, 1487–1492.

Froment, G., Bischoff, K., 1979. *Chemical Reactor Analysis and Design*, John Willey & Sons Inc, New York, USA.

Geier, D., Soper, J.G., 2010. Simultaneous synthesis and purification of a fatty acid monoester biodiesel fuel, *US 7828978*.

Gonçalves, J.C., Rodrigues, A.E., 2014. Simulated moving bed reactor for p-xylene production: Adsorbent and catalyst homogeneous mixture, *Chemical Engineering Journal*, 258, 194–202.

Gonçalves, J.C., Rodrigues, A.E., 2015a. Revamping an existing aromatics complex with simulated moving bed reactor for p-xylene production, *Chemical Engineering & Technology*, 38, 12, 2340–2344.

Gonçalves, J.C., Rodrigues, A.E., 2015b. Simulated moving bed reactor for p-xylene production: Optimal particle size, *The Canadian Journal of Chemical Engineering*, 93, 12, 2205–2213.

Gonçalves, J.C., Rodrigues, A.E., 2016. Simulated moving bed reactor for p-xylene production: Dual-bed column, *Chemical Engineering and Processing: Process Intensification*, 104, 75–83.

Graça, N.S., Pais, L.S., Silva, V.M.T.M., Rodrigues, A.E., 2010. Oxygenated biofuels from butanol for diesel blends: Synthesis of the acetal 1,1-dibutoxyethane catalyzed by Amberlyst-15 ion-exchange resin, *Industrial and Engineering Chemistry Research*, 49, 15, 6763–6771.

Graça, N.S., Pais, L.S., Silva, V.M.T.M., Rodrigues, A.E., 2011a. Analysis of the synthesis of 1,1-dibutoxyethane in a simulated moving bed adsorptive reactor, *Chemical Engineering and Processing: Process Intensification*, 50, 11–12, 1214–1225.

Graça, N.S., Pais, L.S., Silva, V.M.T.M., Rodrigues, A.E., 2011b. Dynamic study of the synthesis of 1,1-dibutoxyethane in a fixed-bed adsorptive reactor, *Separation Science and Technology*, 46, 4, 631–640.

Graça, N.S., Pais, L.S., Silva, V.M.T.M., Rodrigues, A.E., 2012. Thermal effects on the synthesis of 1,1-dibutoxyethane in a fixed-bed adsorptive reactor, *Chemical Engineering & Technology*, 35, 11, 1989–1997.

Guiochon, G., Felinger, A., Shirazi, D.G., Katti, A.M., 2006. *Fundamentals of Preparative and Nonlinear Chromatography*, Academic Press, Amsterdam, The Netherlands.

Hashimoto, K., Adachi, S., Noujima, H., Ueda, Y., 1983. A new process combining adsorption and enzyme reaction for producing higher-fructose syrup, *Biotechnology and Bioengineering*, 25, 10, 2371–2393.

Kawase, M., Inoue, Y., Araki, T., Hashimoto, K., 1999. The simulated moving-bed reactor for production of bisphenol A, *Catalysis Today*, 48, 1–4, 199–209.

Kearney, M.M., Hieb, K.L., 1992. Time variable simulated moving bed process, *US 5102553 A*.

Kurup, A.S., Subramani, H.J., Hidajat, K., Ray, A.K., 2005. Optimal design and operation of SMB bioreactor for sucrose inversion, *Chemical Engineering Journal*, 108, 1, 19–33.

Lévêque, A., 1928. Les Lois de la transmission de chaleur par convection, *Annales des Mines*, 13, 201.

Lode, F., Houmard, M., Migliorini, C., Mazzotti, M., Morbidelli, M., 2001. Continuous reactive chromatography, *Chemical Engineering Science*, 56, 2, 269–291.

Ludemann-Hombourger, O., Nicoud, R.M., Bailly, M., 2000. The "VARICOL" process: A new multicolumn continuous chromatographic process, *Separation Science and Technology*, 35, 12, 1829–1862.

Lutin, F., Bailly, M., Bar, D., 2002. Process improvements with innovative technologies in the starch and sugar industries, *Desalination*, 148, 1, 121–124.

Masuda, T., Sonobe, T., Matsuda, F., Horie, M., 1993. Process for fractional separation of multi-component fluid mixture, *US 5198120 A*.

Migliorini, C., Wendlinger, M., Mazzotti, M., Morbidelli, M., 2001. Temperature gradient operation of a simulated moving bed unit, *Industrial & Engineering Chemistry Research*, 40, 12, 2606–2617.

Minceva, M., Gomes, P.S., Meshko, V., Rodrigues, A.E., 2008. Simulated moving bed reactor for isomerization and separation of p-xylene, *Chemical Engineering Journal*, 140, 1, 305–323.

Minceva, M., Rodrigues, A.E., 2005a. Simulated moving bed reactor: Reactive–separation regions, *AIChE Journal*, 51, 10, 2737–2751.

Minceva, M., Rodrigues, A.E., 2005b. UOP's Parex: Modeling, simulation and optimization. ENPROMER, Brasil.

Nicolaos, A., Muhr, L., Gotteland, P., Nicoud, R.-M., Bailly, M., 2001a. Application of equilibrium theory to ternary moving bed configurations (four+ four, five+ four, eight and nine zones): I. Linear case, *Journal of Chromatography A*, 908, 1, 71–86.

Nicolaos, A., Muhr, L., Gotteland, P., Nicoud, R.-M., Bailly, M., 2001b. Application of the equilibrium theory to ternary moving bed configurations (4 + 4, 5 + 4, 8 and 9 zones): II. Langmuir case, *Journal of Chromatography A*, 908, 1, 87–109.

Nicoud, R.M., 2014. The amazing ability of continuous chromatography to adapt to a moving environment, *Industrial & Engineering Chemistry Research*, 53, 10, 3755–3765.

Nicoud, R.M., 2015. *Chromatographic Processes*, Cambridge University Press, Cambridge.

Niesbach, A., Kuhlmann, H., Keller, T., Lutze, P., Górak, A., 2013. Optimisation of industrial-scale n-butyl acrylate production using reactive distillation, *Chemical Engineering Science*, 100, 360–372.

Oliveira, E.L.G., Grande, C.A., Rodrigues, A.E., 2011. Effect of catalyst activity in SMR-SERP for hydrogen production: Commercial vs. large-pore catalyst, *Chemical Engineering Science*, 66, 3, 342–354.

Ostaniewicz-Cydzik, A.M., Pereira, C.S., Molga, E., Rodrigues, A.E., 2014. Reaction kinetics and thermodynamic equilibrium for butyl acrylate synthesis from n-butanol and acrylic acid, *Industrial & Engineering Chemistry Research*, 53, 16, 6647–6654.

Pais, L.S., Loureiro, J.M., Rodrigues, A.E., 1998. Modeling strategies for enantiomers separation by SMB chromatography, *AIChE Journal*, 44, 3, 561–569.

Paredes, G., Rhee, H.-K., Mazzotti, M., 2006. Design of simulated-moving-bed chromatography with enriched extract operation (EE-SMB): Langmuir isotherms, *Industrial & Engineering Chemistry Research*, 45, 18, 6289–6301.

Pereira, C.S., Silva, V.M., Rodrigues, A.E., 2014. Coupled PermSMBR–Process design and development for 1,1-dibutoxyethane production, *Chemical Engineering Research and Design*, 92, 11, 2017–2026.

Pereira, C.S.M., Gomes, P.S., Gandi, G.K., Silva, V.M.T.M., Rodrigues, A.E., 2008. Multifunctional reactor for the synthesis of dimethylacetal, *Industrial and Engineering Chemistry Research*, 47, 10, 3515–3524.

Pereira, C.S.M., Rodrigues, A.E., 2013. Process intensification: New technologies (SMBR and PermSMBR) for the synthesis of acetals, *Catalysis Today*, 218–219, 148–152.

Pereira, C.S.M., Silva, V.M.T.M., Rodrigues, A.E., 2011. Green fuel production using the PermSMBR technology, *Industrial & Engineering Chemistry Research*, 51, 26, 8928–8938.

Pereira, C.S.M., Zabka, M., Silva, V.M.T.M., Rodrigues, A.E., 2009. A novel process for the ethyl lactate synthesis in a simulated moving bed reactor (SMBR), *Chemical Engineering Science*, 64, 14, 3301–3310.

Rodrigues, A.E., Pereira, C.S.M., Minceva, M., Pais, L.S., Ribeiro, A.M., Ribeiro, A., Silva, M., Graça, N., Santos, J.C., 2015. Principles of Simulated Moving Bed, in *Simulated Moving Bed Technology: Principles, Design and Process Applications*, Butterworth-Heineman, Elsevier, Oxford.

Rodrigues, A.E., Pereira, C.S.M., Santos, J.C., 2012. Chromatographic Reactors, *Chemical Engineering & Technology*, 35, 7, 1171–1183.

Rodrigues, A.E., Silva, V.M.T.M., 2005. Industrial process for acetals production in a simulated moving bed reactor, *WO 2005/11347 A1*.

Ruthven, D.M., 1984. *Principles of Adsorption and Adsorption Processes*, (Ed.), Wiley, New York.

Ruthven, D.M., Ching, C., 1989. Counter-current and simulated counter-current adsorption separation processes, *Chemical Engineering Science*, 44, 5, 1011–1038.

Sá Gomes, P., Lamia, N., Rodrigues, A.E., 2009. Design of a gas phase simulated moving bed for propane/propylene separation, *Chemical Engineering Science*, 64, 6, 1336–1357.

Sá Gomes, P., Rodrigues, A.E., 2007. Outlet streams swing (OSS) and Multifeed operation of simulated moving beds, *Separation Science and Technology*, 42, 2, 223–252.

Schramm, H., Kaspereit, M., Kienle, A., Seidel-Morgenstern, A., 2002. Improving simulated moving bed processes by cyclic modulation of the feed concentration, *Chemical Engineering & Technology*, 25, 12, 1151–1155.

Seidel-Morgenstern, A., 2004. Experimental determination of single solute and competitive adsorption isotherms, *Journal of Chromatography A*, 1037, 1, 255–272.

Silva, M.S.P., Rodrigues, A.E., Mota, J.P.B., 2015. Modeling and simulation of an industrial-scale parex process, *AIChE Journal*, 61, 4, 1345–1363.

Silva, V.M., Pereira, C.S., Rodrigues, A.E., 2011. PermSMBR — A new hybrid technology: Application on green solvent and biofuel production, *AIChE Journal*, 57, 7, 1840–1851.

Silva, V.M.T., Silva, R., Rodrigues, A.E., 2009. Green diesel additive synthesis: Elimination of azeotropic distillation by coupling simulated moving bed reactor with solvent recovery units. AIChE Annual Meeting, Nashville, TN.

Silva, V.M.T.M., Marques, P.C.S., Rodrigues, A.E., 2012. Simulated moving bed membrane reactor, new hybrid separation process and uses thereof, *EP 2418009 A1*.

Silva, V.M.T.M., Rodrigues, A.E., 2005. Novel process for diethylacetal synthesis, *AIChE Journal*, 51, 10, 2752–2768.

Subramani, H.J., Zhang, Z., Hidajat, K., Ray, A.K., 2004. Multiobjective optimization of simulated moving bed reactor and its modification — Varicol process, *Canadian Journal of Chemical Engineering*, 82, 3, 590–598.

Wankat, P.C., 2001. Simulated moving bed cascades for ternary separations, *Industrial & Engineering Chemistry Research*, 40, 26, 6185–6193.

Wijmans, J., Athayde, A., Daniels, R., Ly, J., Kamaruddin, H., Pinnau, I., 1996. The role of boundary layers in the removal of volatile organic compounds from water by pervaporation, *Journal of Membrane Science*, 109, 1, 135–146.

Yu, W., Hidajat, K., Ray, A.K., 2003. Modeling, simulation and experimental study of a Simulated Moving Bed Reactor for the synthesis of methyl acetate ester, *Industrial and Engineering Chemistry Research*, 42, 26, 6743–6754.

Zang, Y., Wankat, P.C., 2002. Three-zone simulated moving bed with partial feed and selective withdrawal, *Industrial & Engineering Chemistry Research*, 41, 21, 5283–5289.

Zhang, Y., Hidajat, K., Ray, A.K., 2004. Optimal design and operation of SMB bioreactor: Production of high fructose syrup by isomerization of glucose, *Biochemical Engineering Journal*, 21, 2, 111–121.

Zhang, Y., Hidajat, K., Ray, A.K., 2007. Modified reactive SMB for production of high concentrated fructose syrup by isomerization of glucose to fructose, *Biochemical Engineering Journal*, 35, 3, 341–351.

Zhang, Z., Hidajat, K., Ray, A.K., 2001. Application of simulated counter-current moving bed chromatographic reactor for MTBE synthesis, *Industrial & Engineering Chemistry Research*, 40, 23, 5305–5316.

Ziyang, Z., Hidajat, K., Ray, A.K., 2002. Multiobjective optimization of simulated countercurrent moving bed chromatographic reactor (SCMCR) for MTBE synthesis, *Industrial and Engineering Chemistry Research*, 41, 13, 3213–3232.

Chapter 5

Conclusions and Perspectives

5.1. Looking Back...

More than 45 years ago, one of the authors was involved on what could be called Reaction Enhanced Ion Exchange process during his doctoral work; it was in fact the anionic step of water demineralization in which a resin in OH^- form exchanges with Cl^- ion from HCl solution, accompanied by the neutralization reaction between hydroxyl ions liberated from the resin and H^+ from the acid solution, producing demineralized water. The overall result is that the system behaves as having a rectangular ion exchange equilibrium isotherm instead of the favourable one in absence of neutralization.

Much of the past work on sorption enhanced reaction processes (SERP) was focused on hydrogen production and started by Sircar's group at Air Products (Hufton *et al.*, 2000); such effort is summarized in the book edited by Sircar and Lee (2010) and several examples, scientific and technological developments and obstacles are illustrated herein in Chapter 2.

To overcome the thermodynamic limitations of reversible reactions, membrane reactors (MRs) have been proposed as an alternative technological solution, by removing one (or more) reaction product(s) from the reaction medium through permselective membranes. The concept has been proposed as long as in the 1960 s (Rothfleisch, 1964), but many publications refer instead to the work of Uemiya *et al.* (1991) in the 1990s, and others to previous pioneering works, from Gryaznov in the 1960s for Pd-based membranes, or even before. This is however still a challenging topic, addressed in Chapter 3 for a specific industrially relevant gas-phase reaction: the water-gas shift (WGS).

Reactive chromatography has been studied in the area of biodiesel production by Hilaly (2006) from Archer Daniels Midland Company (ADM), and renewed interest on simulated moving bed reactor (SMBR) and SERP technologies for this application was shown by the studies of Kapil *et al.* (2010) at the University of Manchester and Ray (2015) at the University of Western Ontario on the esterification of fatty acids to biodiesel.

SMBR has been studied in the bioengineering area by Pilgrim *et al.* (2006) for the enzyme-catalyzed production of lactosucrose and at Laboratory of Separation and Reaction Engineering (LSRE) in many acetalization and esterification reactions as described in Chapter 4 of this book.

In any SERP, one has to make compatible the operating conditions for reaction and sorption or membrane processes. This is sometimes difficult to achieve, for instance when adsorbent capacity becomes low at high temperatures or the broth composition in bioprocesses deactivates ion exchange resins used for separation.

A key objective in running SERP-related processes is therefore to match reaction rate of product formation with removal rate of that product either by using a sorbent or a membrane.

5.2. What are the Bottlenecks?

The question is then to understand what are the difficulties preventing the adoption of the technologies addressed along this book by industry. Below is given some reasoning for that with a few examples.

Risk of changing existing technology

One of the most studied topics among the technologies addressed herein was Sorption Enhanced Steam Methane Reforming (SE-SMR) for hydrogen production; this is certainly an area where the new technology is going to compete with the existing train of steam reforming, WGS, CO_2 absorption and hydrogen purification by pressure swing adsorption (PSA). The idea of working at lower temperature is certainly appealing because of lifetime increase of reformer tubes; however, the capacity of adsorbents such as modified hydrotalcites at say 450°C is not that high. On the other hand, it is a challenge to introduce a new technology in an area where the existing one is well established; the new technology needs to be robust and much cheaper to be successful and convincing.

Robustness of SERP

As general rule, the robustness of the combined reaction–separation process is lower than that of a process with independent reaction and separation units. Or in other words: a robust integrated process requires a more detailed understanding of reaction kinetics, adsorption/permeation kinetics and this includes mastering catalyst and adsorbent/membrane deactivation. Solving this bottleneck is another obstacle for the scientific and industrial communities to bring such technology into the market.

Lack of tests with real feedstock/close to industrial conditions and scale

In a recent Comment in Nature, Sholl and Lively (2016) listed seven important chemical separations and in the recommendations they start by saying that "researchers and engineers must consider realistic chemical mixtures". This is a critical step to fully consider catalyst/adsorbent/membrane deactivation. Consider, for instance, diethylacetal, a blending agent for diesel. It is made from ethanol and acetaldehyde. For a sustainable process, the reactants can be obtained from biomass but one needs to understand the impact of impurities on the life of the catalyst and adsorbent, which is an acid ion exchange resin that can be deactivated/poisoned by some contaminants potentially present in this type of feedstock. So, the SMBR technology (or PermSMBR technology) needs to be tested at pilot scale with real feed to address these issues.

Long time ago, Levenspiel said that a chemical reactor transforms reactants in products, hopefully with price increase (added value along the chain). So the technology has to match these considerations too.

As a positive example regarding this subject, one can mention a case study at pilot scale (or close to industrial) comprising the use of a membrane reactor by the Tokyo Gas Company for hydrogen production from natural gas in a plant capacity of $40\,\mathrm{Nm}^3/\mathrm{h}$ of H_2 with 99.999% purity and CO_2 as off-gas with purity 90%, which can easily be liquefied and captured. The CO_2 emissions decreased 50% with small loss in energy (Shirasaki and Yasuda, 2013).

5.3. Looking Ahead

Below are highlighted a few areas where the technologies addressed in this book are finding application, or new topics that might emerge.

Bioengineering applications

The bioengineering area is still a field of possible applications of SERP. A recent study of Fuderer *et al.* (2015) illustrates this point. The racemic mixture of Methionine (Met) is separated by chiral simulated moving bed (SMB) giving in the extract the desired product D-Met; the secondary product, L-Met, from the raffinate is concentrated by nanofiltration and racemized in an enzymatic membrane reactor which will feed the SMB.

SMBR and PermSMBR

Applications of SMBR are being studied by several groups. In Mahalani's group, the synthesis of 2-ethylhexyl acetate ester has been addressed (Gyani *et al.*, 2014; Reddy *et al.*, 2014). At Dow Chemical and Georgia Tech, several recent works deal with SMBR for production of propylene glycol methyl ether acetate (Agrawal *et al.*, 2014; Tie *et al.*, 2016; Kawajiri *et al.*, 2015). At LSRE, research effort is concentrated on butyl acrylate and diethoxybutane (DEB) synthesis using SMBR and PermSMBR technologies and on the use of supercritical SMBR for the synthesis of solketal.

SERP and hybrid processes

Sorption enhanced methanation has been studied recently by So and Lee (2016) for the production of synthetic natural gas (SNG) from coal at 350°C with removal of CO_2 by adsorption in K-promoted hydrotalcite allowing a CH_4 product purity >95%.

Other hybrid processes have been proposed recently like adsorptive absorption (Lei *et al.*, 2015) for CO_2 capture using ionic liquids and ZIF-8 adsorbent, or reactive distillation coupled with pervaporation for the synthesis of ethyl acetate (Lee *et al.*, 2016).

One further example of clear process integration resulting in hybrid units are the sorption enhanced membrane reactors. This has been experimentally proved for the WGS reaction (Soria *et al.*, 2015), with CO_2 being adsorbed in a K_2CO_3-promoted hydrotalcite (which capacity is increased in the presence of water vapour, a reactant in this case), while H_2 is selectively removed through a Pd-based membrane. It has been shown that hydrogen production is enhanced compared to either a traditional or a sorption enhanced reactor, allowing overcoming equilibrium limitations while obtaining a pure H_2 stream. The concept revealed also to be promising, from the thermodynamic point of view, for glycerol steam reforming (with an enhancement >200%, in terms of hydrogen yield, comparatively to a traditional reactor) (Silva *et al.*, 2015), but still needs to be tested for this and other reactions.

Another interesting idea to explore is the coupling of SMBR and PSA in gas-phase systems. The idea has been described for SMB–PSA separations by Rao *et al.* (2005); for reaction–separation it was tested by Pakseresht *et al.* (2002) for CO hydrogenation in view of the production of ethylene and ethane (Fischer Tropsch synthesis). These authors used a Fe-Cu-K/ZSM5 catalyst, operating at 18 bar, 280°C with a ratio $CO/H_2 = 0.775$. The adsorbent used was 5A zeolite. The SMBR–PSA configuration is shown in Fig. 5.1.

This can open a window of opportunity in processes leading to olefins/paraffins separation by gas-phase SMB as recently demonstrated by Martins *et al.* (2016).

Paraxylene

"Benzene derivatives from each other" are included in the article in Nature by Sholl and Lively (2016) already mentioned. In particular, p-xylene is the raw material for PET production for which a strong

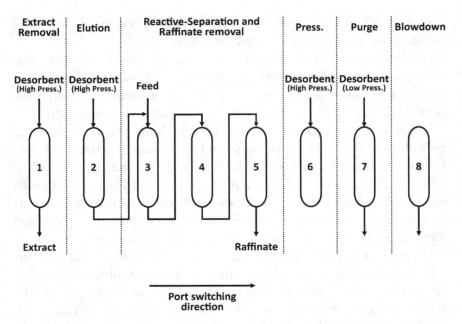

Figure 5.1 SMBR–PSA configuration for FTS. Adapted from Pakseresht *et al.* (2002).

growth is predicted for the coming years. Here is an opportunity to study SMBR for p-xylene production (Gonçalves, 2015). This is an area where adsorption technology using SMB technology (PAREX process from UOP, ELUXYL from AXENS, Toray) competes with crystallization (BP technology commercialized by CB&I Lummus, GTC Corporation, Chevron-Phillips) and hybrid processes of crystallization and adsorption (GT-Hybrid PX from GTC in partnership with Clariant for the adsorbent Zeosorb PX-200). Crystalization processes are driven by the fact that the freezing point of p-xylene is 14°C, much higher than those of other C_8 isomers. One alternative process could be SMBR followed by crystallization.

Purification of diluted streams

A general problem faced by chemical industry concerns diluted streams, either in liquid or in gas phase. One example is the recovery of process water in industry where water appears contaminated by products, as is the case of aniline industry, or in air pollution due to VOCs. In the case of air polluted by VOCs, adsorption can be used as a concentration step before catalytic reaction to destroy VOCs at much higher rate; here there is sorption enhancement of the reaction rate although not driven by the thermodynamic equilibrium displacement.

Towards a decentralized world

A very recent study by Malmali *et al.* (2016) deals with ammonia synthesis at high temperature and relatively low pressure by coupling reaction with absorptive separation of ammonia by calcium chloride; they showed the process is viable at 25 bar with 80% conversion if ammonia is efficiently removed. This can promote the idea of "distributed ammonia synthesis" as announced in a AIChE Webinar by E.L. Cussler on 21st December 2016 where "wind is turned into ammonia used in local farms". It may open "a different chemical industry based on scaling down processes to harvest local energy". One should remember that ammonia synthesis is connected with two Nobel Prizes: one for catalysts and the other for high pressure technology required for the process. The proposal by Cussler's group is a departure in a new direction of a decentralized world.

Nomenclature

FTS	Fischer Tropsch synthesis
MRs	Membrane reactors
PET	Polyethylene terephthalate
PSA	Pressure swing adsorption
SERP	Sorption enhanced reaction processes
SE-SMR	Sorption enhanced steam methane reforming
SMB	Simulated moving bed
SMBR	Simulated moving bed reactor
VOCs	Volatile organic compounds
WGS	Water-gas shift

References

Agrawal, G., Oh, J., Sreedhar, B., Tie, S., Donaldson, M.E., Frank, T.C., Schultz, A.K., Bommarius, A.S., Kawajiri, Y., 2014. Optimization of reactive simulated moving bed systems with modulation of feed concentration for production of glycol ether ester, *Journal of Chromatography A*, 1360, 196–208.

Fuderer, M., Femmer, C., Storti, G., Panke, S., Bechtold, M., 2015. Integration of simulated moving bed chromatography and enzymatic racemization for the production of single enantiomers, *Chemical Engineering Science*, 152, 649–662.

Gonçalves, J.C., 2015. Hybrid separations and adsorption/reaction processes: The case of isomerization/separation of xylenes, PhD Thesis, University of Porto.

Gyani, V.C., Reddy, B., Bhat, R., Mahajani, S., 2014. Simulated moving bed reactor for the synthesis of 2-ethylhexyl acetate ester Part I — Experiments and simulations, *Industrial & Engineering Chemistry Research*, 53, 15811–15823.

Hilaly, A., 2006. Reactive chromatography for biodiesel production, Archer Daniels Midland Company (ADM), Decatur IL https://crelonweb.eec.wustl .edu/files/CRELMEETINGS/2006/Hilaly.pdf.

Hufton, J., Waldrom, W., Weigel, S., Rao, M., Nataraj, S., Sircar, S., 2000. Sorption enhanced reaction (SERP) for the production of hydrogen, Report NREL/CP-570-28890, *Proc. of 2000 DOE Hydrogen Program Review*, 98–109.

Kapil, A., Bhat, S.A., Sadhukhan, J., 2010. Dynamic simulation of sorption enhanced reaction processes for biodiesel production, *Industrial & Engineering Chemistry Research*, 49, 2326–2335.

Kawajiri, Y., Bommarius, A.S., Oh, J., Agrawal, G., Sreedhar, B., Huebsch, J., 2015. Process for operating a simulated moving bed reactor, WO 2015 187931.

Lee, H.Y., Li, S.Y., Chen, C.-L., 2016. Evolutional design and control of the equilibrium-limited ethyl acetate process via reactive distillation-pervaporation hybrid configuration, *Industrial & Engineering Chemistry Research*, 55, 8802–8817.

Lei, Z., Dai, C., Lei, Z., Song, W., 2015. Adsorptive absorption: A preliminary experimental and modeling study on CO_2 solubility, *Chemical Engineering Science*, 127, 260–268.

Malmali, M., Wei, Y., Cormick, A.Mc., Cussler, E.L., 2016. Ammonia synthesis at reduced pressure via reactive separation, *Industrial & Engineering Chemistry Research*, 55, 8922–8932.

Martins, V.F.D., Ribeiro, A.M., Santos, J.C., Loureiro, J.M., Gleichmann, K., Ferreira, A., Rodrigues, A.E., 2016. Development of gas-phase SMB technology for light olefin/paraffin separations, *AIChE Journal*, 62, 2490–2500.

Pakseresht, S., Kazemeini, M., Akbarnejad, M.M., 2002. An experimentally determined configuration for simulated moving beds as a separative-reactor in gas phase, *Scientia Iranica*, 9, 74–85.

Pilgrim, A., Kawase, M., Matsuda, F., Miura, K., 2006. Modeling of the simulated moving bed reactor for the enzyme-catalysed production of lactosucrose, *Chemical Engineering Science*, 61, 353–362.

Rao, D.P., Sivakumar, S.V., Mandal, S., Kota, S., Ramaprasad, B.S.G., 2005. Novel simulated moving-bed adsorber for the fractionation of gas mixtures, *Journal of Chromatography A*, 1069, 141–151.

Ray, M.N., 2015. A comprehensive study of esterification of free fatty acid to biodiesel in a simulated moving bed system, PhD Thesis, University of Western Ontario.

Reddy, B., Gyani, V.C., Mahajani, S., 2014. Simulated moving bed reactor for the synthesis of 2-ethylhexyl acetate ester Part II: Simulation based design, *Industrial & Engineering Chemistry Research*, 53, 15824–15835.

Rothfleisch, J., 1964. Hydrogen from Methanol for Fuel Cells, SAE Technical Paper 640377, doi:10.4271/640377.

Shirasaki, Y., Yasuda, I., 2013. Membrane Reactor for Hydrogen Production from Natural Gas at the Tokyo Gas Company: A Case Study, in *Handbook of Membrane Reactors*, A. Basile (Ed.), Woodhead Publishing Series in Energy, Cambridge, UK, pp. 487–507.

Sholl, D.S., Lively, R.P., 2016. Seven chemical separations to change the world, *Nature*, 532, 435–437.

Silva, J.M., Soria, M.A., Madeira, L.M., 2015. Thermodynamic analysis of glycerol steam reforming for hydrogen production with in situ hydrogen and carbon dioxide separation, *Journal of Power Sources*, 273, 423–430.

Sircar, S., Lee, K.B., 2000. *Sorption Enhanced Reaction Concepts for Hydrogen Production*: Materials & Processes, Research Signpost, Kerala.

So, I.I., Lee, K.B., 2016. New sorption-enhanced methanation with simultaneous CO_2 removal for the production of synthetic natural gas, *Industrial & Engineering Chemistry Research*, 55, 9244–9255.

Soria, M.A., Tosti, S., Mendes, A., Madeira, L.M., 2015. Enhancing the low temperature water-gas shift reaction through a hybrid sorption-enhanced membrane reactor for high-purity hydrogen production, *Fuel*, 159, 854–863.

Tie, S., Sreedhar, B., Donaldson, M.E., Frank, T.C., Bommarius, A.S., Kawajiri, Y., 2016. Process integration for simulated moving bed reactor for the production of glycol ether acetate, Paper 457969, AIChE meeting, San Francisco.

Uemiya, S., Sato, N., Ando, H., Kikuchi, E., 1991. The water gas shift reaction assisted by a palladium membrane reactor, *Industrial and Engineering Chemistry Research*, 30, 585–589.

Index

A

acetaldehyde decomposition, 122
acetalization, 276
acetals, 276, 293
acid ion exchange resins, 276
activation energies, 194, 299
active phases used for WGS, SR of
  methane, methanol and ethanol, 63
adsorbent, 6, 103
  CO_2 adsorbent, 35, 148
  adsorbent regeneration, 18
  calcium oxide-based adsorbents,
    86
  calcium-based sorbents, 59
  high temperature CO_2
    adsorbents, 82
  high temperature carbon dioxide
    sorbents, 59
  hydrotalcite CO_2 adsorbent, 20
  K_2CO_3-promoted hydrotalcite,
    211
  lithium-based adsorbents, 87
  regeneration, 129
adsorption, 13, 17, 38
  adsorption capacity, 15
  adsorption rate, 15
  adsorption selectivity, 15
  chemical adsorption, 83
  physical adsorption, 83
adsorption equilibrium, 295
adsorption equilibrium isotherms, 299

adsorption reactors
  applications, 18
adsorption/regeneration, 86
adsorptive reactors, 6–7, 17, 45
air reactor, 116
Al_2O_3 hollow fibres, 202
Amberlyst-15, 276, 294, 298, 302,
  318, 320
Arrhenius-type equation, 193
asymmetric Al_2O_3 hollow fibres,
  207
ATR process, 52
auto-thermal reforming
  (ATR)/oxidative steam reforming
  (OSR) of methane, 11, 51
auto-thermal state, 116

B

balance, 120
basic alumina, 98
bed chromatographic reactors, 311
benzene, 307, 309
binary azeotrope, 306
binary mixture, 263
biodiesel production, 332
bioengineering applications, 334
bottlenecks, 332
breakthrough, 106
Broughton and Gerhold, 25, 267
butyl acrylate, 297, 300, 302–303

Printed in the United States
By Bookmasters